Max Schaldach

Electrotherapy of the Heart

Technical Aspects in Cardiac Pacing

Springer-Verlag
Berlin Heidelberg NewYork London Paris Tokyo
HongKong Barcelona Budapest

Prof. Dr.-Ing. Max Schaldach

Zentralinstitut für Biomedizinische Technik
der Friedrich-Alexander-Universität Erlangen-Nürnberg
Turnstraße 5
8520 Erlangen
Germany

ISBN 978-3-642-50211-8 ISBN 978-3-642-50209-5 (eBook)
DOI 10.1007/978-3-642-50209-5

Springer-Verlag Berlin Heidelberg 1992
Softcover reprint of the hardcover 1st edition 1992

Coverdesign: Jan Vogel; Typesetting: Camera ready by author

60/3020 - 5 4 3 2 1 0 - Printed on acid-free paper

FOREWORD

Biomedical engineering has significantly contributed to the success of medicine in diagnosis and therapy during the last thirty years. One of the most successful contributions has been the development of pacemaker technology. Since the first implant in 1958, the concept has evolved from a simple pulse generator to a powerful system capable of not only stimulating, but sensing, programmability and computer-based, decision-making technology. Today's implantable pacemakers demonstrate one of the most successful examples of the interdisciplinary task of applying modern technology in the fields of materials and electronics to medicine. The clinical treatment of all types of arrhythmias has reached a standard that renders possible not only emergency treatment but also increases the quality of life. Despite these impressive advances, the artifical pacemaker is still inferior to the physiological pacing system, especially regarding the capability to adjust the stimulation rate to the hemodynamic requirements.

Electrotherapy of the Heart is a unique representation of this fascinating development. The author, Max Schaldach, is one of the founders of pacemaker technology and is internationally known as a researcher and professor in the fields of applied physics and biomedical engineering. He continues to contribute extensively to the field of cardiac pacing in development, design and manufacture, and has strongly influenced the technology to its present advanced state.

There is no doubt that this book will further promote research and development in pacemaker technology with the aim of a more physiological approach for the greatest benefit to the patient.

Graz, January 1992 *Helmut Hutten*

PREFACE

In 1958, the first implantable cardiac pacing system was developed by Elmquist and implanted by Senning. Since then, the exemplary collaboration between medicine and engineering has developed into an extremely successful therapy.

The multidisciplinary nature of the technical task of pacemaker development is based upon the diverse components of physiology, electronics, physics, electrochemistry and the material sciences. Progress in basic technologies such as microelectronics has made it possible to design pacemakers which, in practice, has led to broad clinical applications of treating a wide spectrum of arrhythmias. The information presented here is not generally found in conventional journals and textbooks and, therefore, may contribute to the interdisciplinary collaboration between biomedical engineering and the medical profession.

This book, as its title suggests, highlights many of the recent and most important technological advances and concepts in cardiac electrotherapy to implement different approaches of adapting the stimulation rate from a systems engineering standpoint to meet the cardiovascular requirements. Innovations to solve the rate-adaptation problem include *physiological* rate adaptation which operates as a closed-loop control system. The artificial pacemaker and the Autonomic Nervous system, in concert, reestablish chronotropy by utilizing its physiological control elements. This approach is an important milestone toward development of an "intelligent" pacemaker capable of implementing a much broader concept of electrotherapy for the heart.

As progress in neurocardiology suggests, monitoring of the autonomic balance should result in a preventive pacing technique which can be superior to pharmacological approaches and which will finally replace the defibrillator.

May the multidisciplinary task of pacemaker development become evident to the reader.

Erlangen, January 1992 *M. Schaldach*

Dedicated to

my colleagues and my friends
who have accompanied me
during the past thirty years.

Max Schaldach

CONTENTS

BASIC ANATOMY AND PHYSIOLOGY OF THE HEART

To understand pacemaker technology, the knowledge of certain anatomical, physiological and pathophysiological facts is required. The aim of the following chapters is to briefly explain the basic functions of the heart and its control mechanisms such as the autonomic nervous system as a part of the central nervous system. However, the content should not be considered as a complete and detailed description of all anatomical and physiological details. It is recommended that other textbooks on physiology, electrophysiology, pathophysiology as well as cardiology be studied [1-4].

Position and Structure of the Heart

The heart is situated toward the front of the chest cavity, slightly off center to the left at about the level of the armpit. It is shaped like an inverted cone, with its base toward the upper right and the apex pointing down and to the left. Its anatomical axis (an imaginary line from the middle of the base to the apex) runs from the posterior upper part of the right thorax to the anterior lower part on the left. Its exact position varies among individuals due to differences in shape of the diaphragm and size of the heart [5].

A membrane of connective tissue, the pericardium, completely encloses the heart. The pericardium, which consists of two layers separated by a very narrow fluid-filled space, protects the heart from forces exerted by the adjacent lung. The space between the two pericardial layers allows the heart to move freely as it beats. Around the bases of the large vessels that emerge from the heart, the inner layer fuses with the outer layer.

Figure 1 shows the simplified internal structure of the heart. At its base are the orifices of the pulmonary artery, which carries blood to the lungs, and the aorta supplying the rest of the body. The large veins also open into this region—the four pulmonary veins from the lungs and the superior and inferior venae cavae.

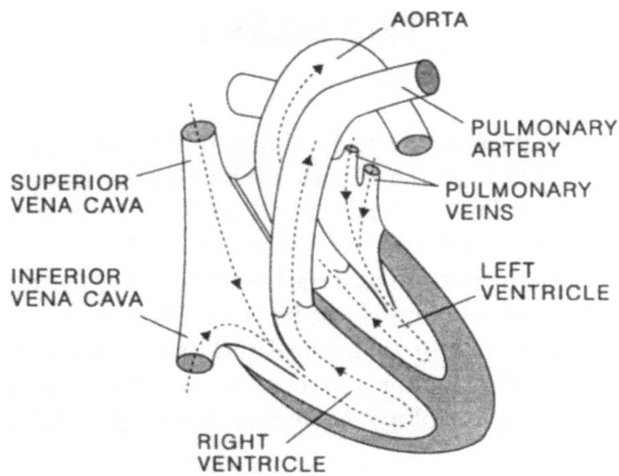

Figure 1. *Schematic drawing of the heart to demonstrate the basic principles of blood flow through the atria and ventricles. The arrows show the direction of blood flow.*

The heart wall is a hollow muscle, the myocardium, in which some fibers spiral around the cone and others run straight from base to apex. A muscular septum along the anatomical axis divides the heart into two cavities so that the heart has a distinct right half and left half. The myocardium of the left half is considerably more massive than that on the right. Each half is further subdivided into two chambers—the atrium and the ventricle—by valves formed of separate connective tissue flaps, the cusps (Figure 2). The valve between the right atrium and ventricle has three cusps and is known as the tricuspid valve; the mitral valve on the left side of the heart has only two cusps. The free ends of the cusps are anchored by tendon filaments to the papillary muscles, which are the small muscles at the tip of the ventricles. It is this arrangement that allows a set of cusps to function as a valve; the filaments prevent the cusps from bending back into the atria when the ventricles contract so the blood can only flow from atrium to ventricle. There are two additional valves in the heart that provide for unidirectional flow out of the ventricles—the pulmonary valve at the junction of the right ventricle and pulmonary artery, and the aortic valve in the left ventricle at the base of the aorta. Each of these is composed of three semilunar segments that fold back against the vessel wall when the blood flows out of the ventricle but fills with blood whenever the flow tends to reverse, so that the passage back into the ventricle is blocked.

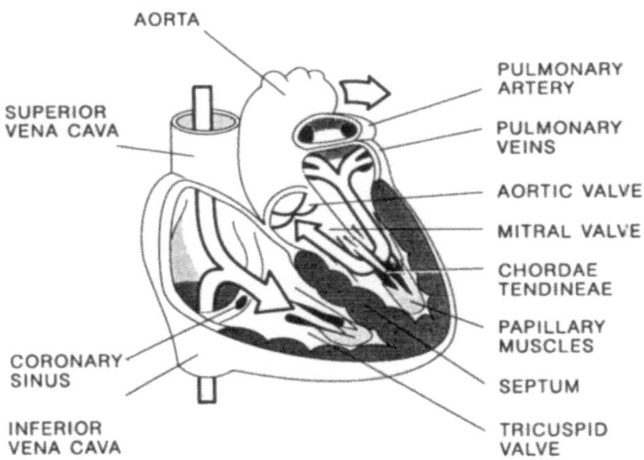

Figure 2. *Functional positions of the valves, their mechanical arrangement and the papillary muscles as located in the right and left heart.*

As shown in Figure 3, the heart is a pump integrated into the vascular system. Its function is to ensure that blood flows continuously through the organs of the body, supplying them with sufficient oxygen and nutrients and effectively removing metabolic products. In functional terms, the right and left halves of the heart can be considered independent of one another. In the right heart, oxygen-depleted blood pours into the atrium from the superior and inferior vena cava and then passes through the tricuspid valve into the ventricle. When the heart contracts, the blood in the right ventricle is expelled into the pulmonary artery. During the subsequent passage through the lungs, the blood releases carbon dioxide into the respired air and takes up oxygen from the air. Now charged with oxygen, the blood flows through the pulmonary veins to the left atrium, passes through the mitral valve into the ventricle, and is pumped into the aorta. It is distributed through large and small arteries, finally reaching the very smallest vessels—the capillaries in the organs—where the blood-borne substances are exchanged with

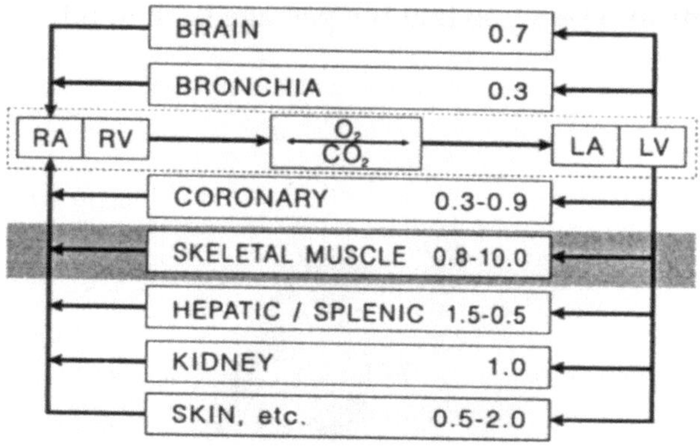

Figure 3. *Diagram of the distribution of blood flow to the various organs of the body (L/min for 70 Kg male) showing the distribution throughout the cardiovascular system. The range of cardiac output (CO = l/min) refers to rest and exercise conditions.*

the tissues. Finally, the blood collects in small and large veins that ultimately empty into the venae cavae. The part of this circuit that lies between the pulmonary artery and pulmonary veins is called the pulmonary (or lesser) circulation; the part that passes through the rest of the body is called the systemic (or greater) circulation.

The heart pumps blood by rhythmic contractions of the entire myocardium at a rate of about 70-80 beats per minute at rest. The muscle cells contract by electrical activity originating in specialized myocardial cells that serve as the heart's intrinsic pacemaker; these spontaneously generated signals are transmitted to the rest of the myocardium in a coordinated manner through special fiber bundles. Figure 4 illustrates the arrangement of this specific pacemaker and conducting system. Normally, excitation originates in the *sinoatrial* (SA) node, a group of muscle fibers in the right atrium near the opening of the superior vena cava. Three fiber bundles (internodal bundles) pass along the atrial wall to the *atrioventricular* (AV) node (also known as the Aschoff-Tawara node) located in the lower septal region of the right atrium. It is not yet clear how the internodal bundles contribute to the conduction of excitation from the SA to the AV node. The AV node gives rise to the His bundle, which passes

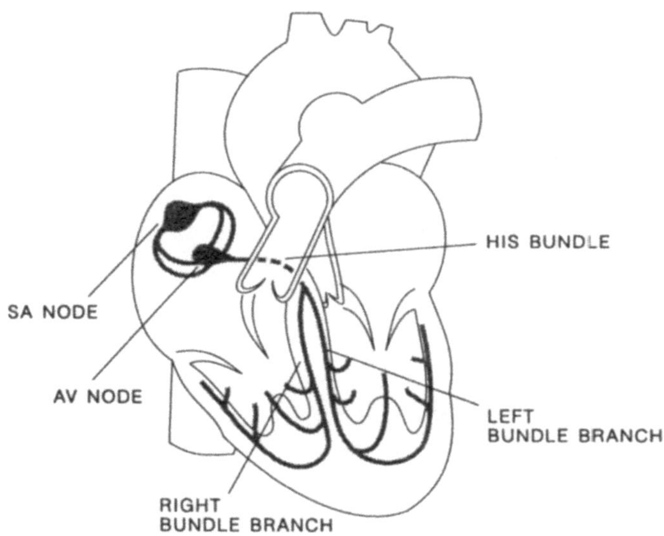

Figure 4. Anatomical location of the natural pacemaker and the conduction system.

through the septum to the left side of the heart, crossing the boundary between the atrium and the ventricle. Shortly thereafter, the His bundle divides into a right and a left branch. A little further on, the left bundle branch bifurcates, producing an anterior and a posterior division. All three of these strands run along the surface of the septum to the tip of the ventricle where they split into small fiber bundles—the Purkinje fibers.

Generation and Time Course of Excitation

All myocardial cells, those of the pacemaker and conducting system as well as the working myocardium, are electrically excitable. In the resting state, the intracellular space is at a lower potential than the extracellular space. This voltage gradient across the cell membrane, a "resting potential" of about , -80 mV, is produced by the diffusion

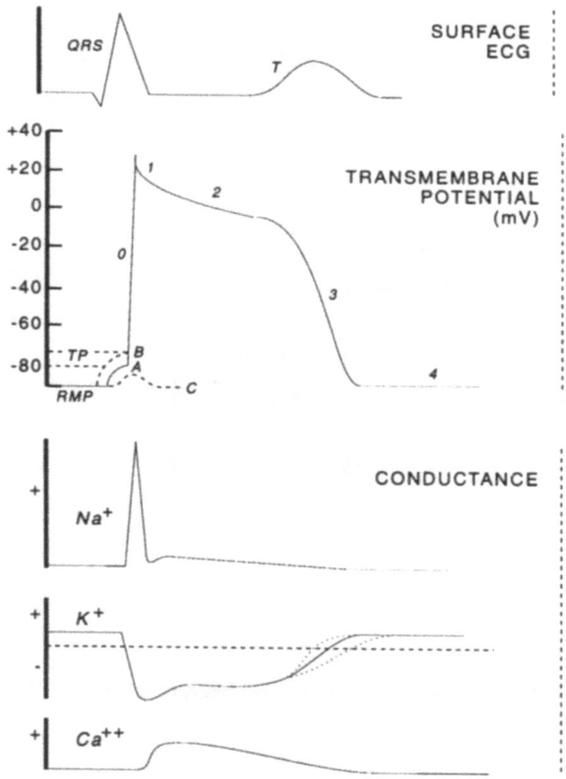

Figure 5. *Relationship between the surface ECG, the transmembrane potential, and the relative changes in conductance responsible for the Action Potential (AP). A fast AP is indicated by phases 0-3, phase 4 is the resting membrane potential (RMP). TP = threshold potential; A = threshold stimulus; B = lower threshold stimulus; C = subthreshold stimulus. Phase 0 depicts the large and rapid increase in Na+ conductance as compared with K+ and Ca++ conductance.*

of potassium ions out of the intracellular space, along their concentration gradient (see Figure 5). This is a passive process; that is, no energy is required to maintain a diffusion potential.

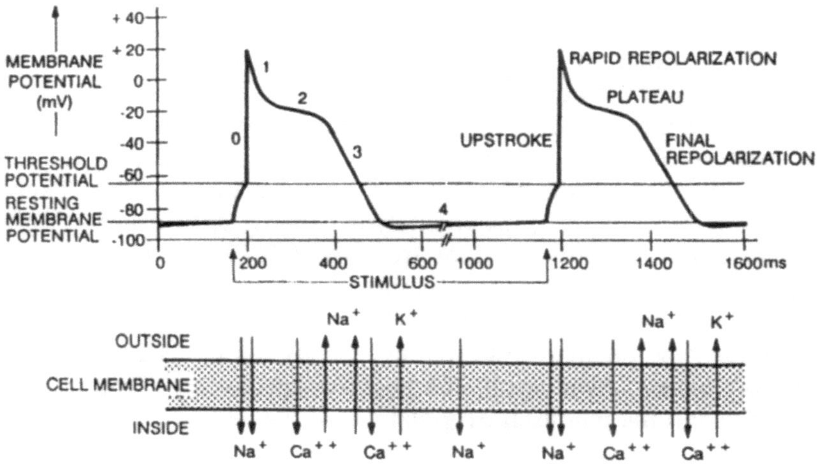

Figure 6. Relationship between phases of the AP and ion flux across the cardiac cell membrane.

All cells in the myocardium are capable of altering the transmembrane potential to develop transient excitation, which is known as *action potential* (AP). The charge distribution across the membrane is briefly reversed, making the intracellular space positive with respect to the extracellular space. The cells of the working myocardium generate action potentials only under the influence of an external stimulus, *i.e.*, the AP of a neighboring cell or a voltage pulse from an artificial pacemaker. This stimulus depolarizes the membrane and if, in the process, the membrane potential reaches a certain threshold value, an AP is generated. Normal APs are always identical in shape and amplitude (all-or-none law). As shown in Figure 6, the AP first rises sharply to +30 mV (the upstroke), then falls slightly to about +10 mV and remains near this level for ca. 200-250 ms. This plateau phase, which is characteristic of the AP of the cardiac muscle, is followed by a repolarization phase lasting about 50 ms, during which the myocardial cell returns to its resting potential. The overall duration of an AP can vary from 180 to 380 ms depending on the heart rate.

The potential changes during the AP are brought about by the transmembrane flow of various ions—sodium, potassium, and calcium. The upstroke results from a very rapid, inward sodium diffusion current. When the voltage across the membrane reverses, making the inside of the cell positive to the outside of the cell, influx of sodium comes to an abrupt halt. At the same time, the potassium outflow increases and an inward diffusion of calcium begins. These two ionic currents are approximately in balance, so that there is only a very slight net afflux of ions from the cell (Figure 6). It is this equilibrium that keeps the membrane potential nearly constant during the long plateau phase. The calcium current, however, is progressively inactivated so that eventually the potassium current predominates. As a result, the negative intracellular potential is restored and the membrane repolarizes to the resting potential.]

The time-dependent changes in the various ion permeabilities are brought about by ion-specific channels with protein molecules embedded in the cell membrane to form a tunnel through it. The conformation of each channel protein depends on the electrical field strength in its vicinity, so that changes in the field can "open" the tunnel and allow specific ions to pass through it.

Throughout the AP, the cell is unresponsive (or less responsive) to a second external stimulus—that is, the myocardial cell is refractory. The refractory phase can be divided into two sections, as shown in Figure 7. The first section is the absolute refractory period in which it is impossible to trigger a new AP with a stimulus of any strength. This period lasts from the beginning of the upstroke until repolarization has returned the membrane potential to ca. -50 mV, which includes the entire plateau phase. It is followed by the relative refractory period, in which a strong stimulus can trigger a new AP, but the amplitude and duration of this potential are distinctly reduced. The cell does not regain its normal excitability until the membrane potential has returned to the resting level.

The long refractory phase of the myocardial cell serves two purposes. First, it ensures that the mechanical events associated with the heartbeat have been completed before a new myocardial contraction can be initiated by an AP. Second, if an AP spreading through the myocardium encounters a region that has already been excited, and these cells are capable of responding again, the activity could continue around the myocardium indefinitely. A long refractory period prevents this reentry from occurring.

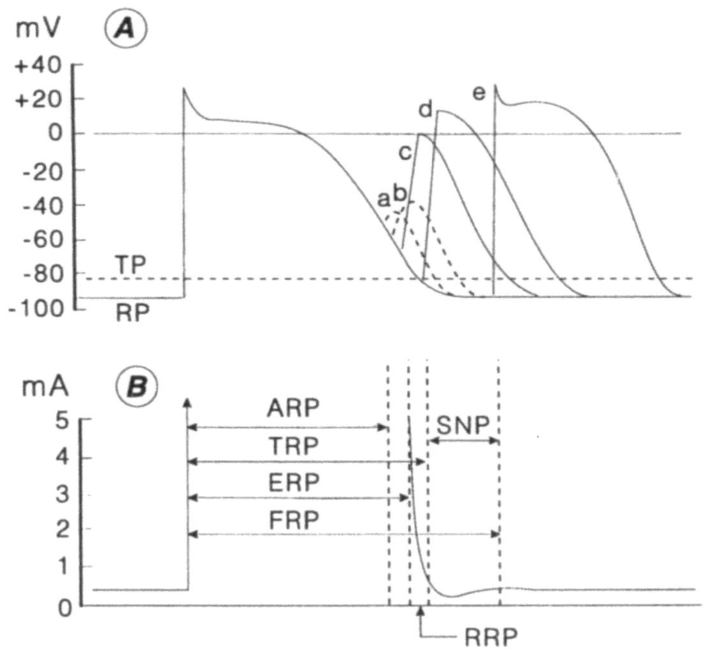

Figure 7. Comparison of the two sections of the refractory period: **A** *= normal AP and responses to stimuli applied during and after repolarization (TP = threshold potential); RP = refractory period).* **B** *= duration of refractory period (RP) (ARP = absolute; TRP = total; ERP = effective; FRP = full; RRP = relative; SNP = supernormal period).*

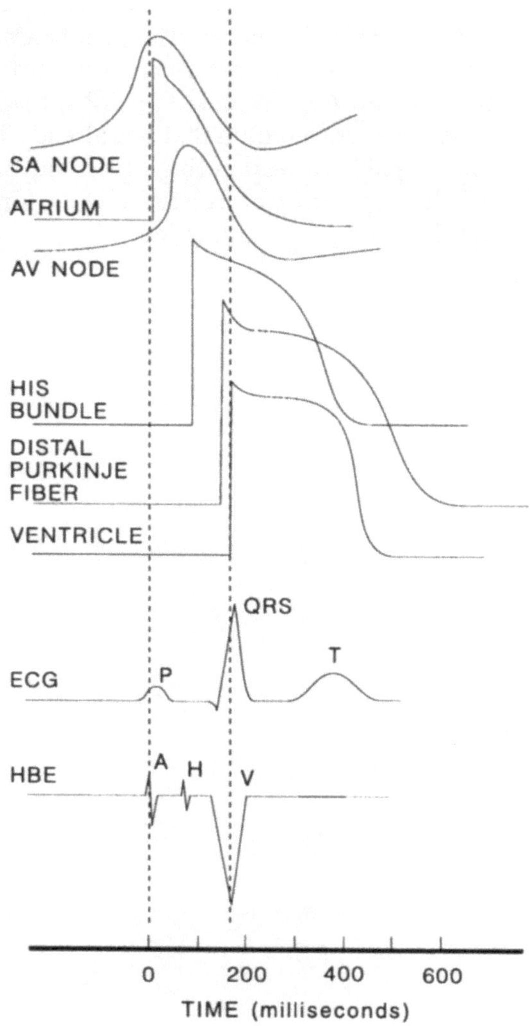

Figure 8. *Surface ECG recording and His bundle electrogram (HBE) in reference to simultaneous activities and corresponding AP at various points in the pacemaking and conducting system and the myocardium.*

The cells of the pacemaking and conducting system are capable of spontaneously developing AP. Figure 8 shows the potential time course for various parts of the pacemaking and conducting system and for the working myocardium. The cells in the autorhythmic tissue do not have a stable resting potential like that of the working myocardial cells. Instead, as soon as the repolarization at the end of an AP is complete, a slow spontaneous reduction of the membrane potential begins. This "diastolic depolarization" is caused by a slight inward sodium current. When it has brought the membrane potential to the threshold level, an AP is elicited. The rate of spontaneous diastolic depolarization varies in different parts of the pacemaking and conducting system; it is most rapid in the SA node, so that the threshold is reached and an AP is generated sooner than in the AV node, His bundle, or bundle branches.

Excitation-Contraction Coupling and the Contraction Process

The electrical excitation causes contraction of the myocardial cell by use of a coupling mechanism in which calcium ions play the major role. As an AP progresses over the cell membrane, calcium diffuses into the cell (Figure 9). At the same time, large amounts of calcium are released into the intracellular matrix (the cytoplasm) from reservoirs within the cell (sarcoplasmic reticulum). The intracellular calcium concentration rises by about fifty-fold. This increase induces the formation of bridgelike connections between filaments lying parallel within the cell, created by the proteins actin and myosin. Then, in an energy-consuming process, these cross-bridges undergo structural changes that shift the actin and myosin filaments past one another, producing a telescopic shortening of the whole muscle fiber. The contraction ceases when ion pumps have transported the calcium out of the cytoplasm, returning it to the sarcoplasmic reticulum and to the extracellular space and the muscle fiber relaxes.

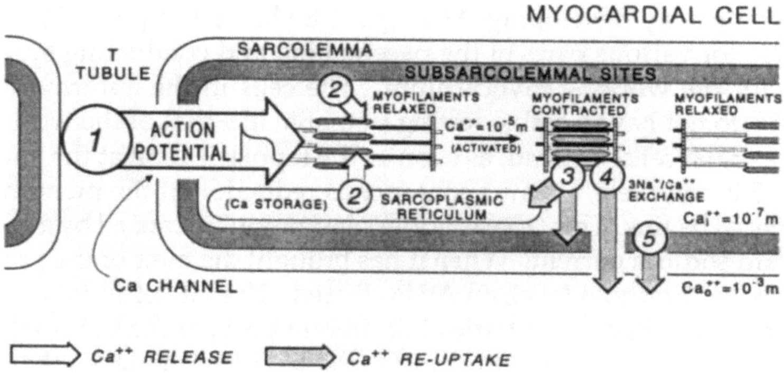

Figure 9. Diagram of the myocardial cell showing Ca^{2+} movement during the cardiac cycle. (1) Inward calcium movement across the sarcolemma during phase 2 of the AP. (2) These events trigger Ca^{2+} release from the sarcoplasmic reticulum and probably from the subsarcolemmal sites. (3) Ca^{2+} reuptake is predominantly into the sarcoplasmic reticulum by an energy-dependent pump, but some Ca^{2+} is also rebound at subsarcolemmal binding sites. (4) Extrusion of Ca^{2+} occurs by Na^+/Ca^{2+} exchange across the sarcolemma and (5) also by a sarcolemmal Ca^{2+} pump.

PHYSIOLOGY OF THE HEARTBEAT

The Pacemaker Potential and the Spread of Excitation

The spontaneous, rhythmic generation of APs in the intrinsic pacemaker centers of the heart is the basis of the autonomous heartbeat rhythm [6]. Normally, the heartbeat is initiated in the SA node. The AP generated propagates from one muscle cell to the next until it has spread over the entire atrial wall and reached the AV node at the boundary between the atrium and the ventricle. Because of

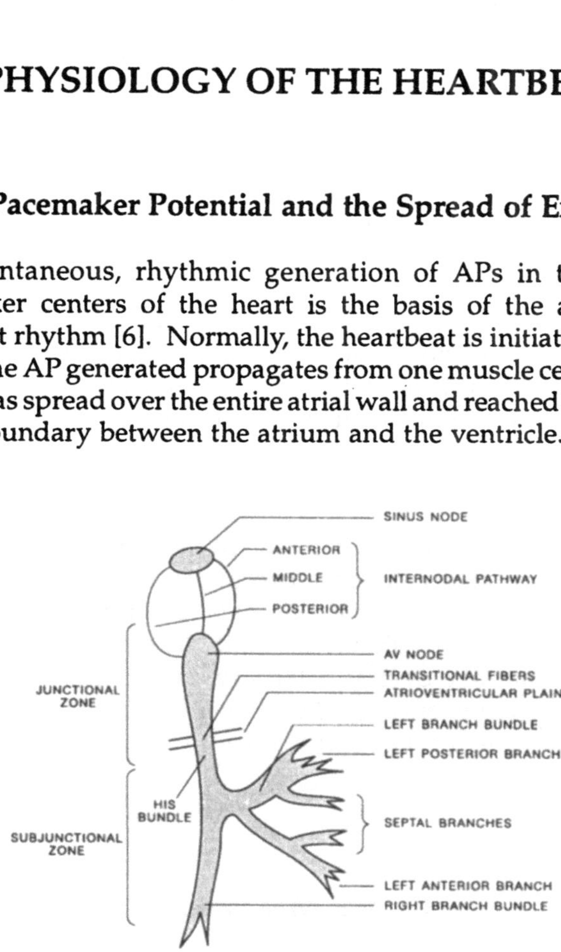

Figure 10. *Diagram of sinus and AV nodes to describe the structure of the conduction system.*

the low conduction velocity in the AV node, there is a delay in transmission of the wave of excitation to the ventricles that allows the atrial contraction to be completed before contraction of the ventricles begins (Figure 10). From the AV node, the excitation passes along the His bundle through the atrioventricular boundary.

Normally, this is the only route for electrical conduction between the two chambers. In the ventricle, the AP rapidly spreads along conducting pathways—the His bundle, the bundle branches, and the Purkinje fibers—to the myocardium at the apex of the heart. Within the ventricular myocardium, the AP wave then propagates from cell to cell toward the base of the heart until all the muscle cells are excited and contract.

Like the SA node, the AV node and the other parts of the conducting system are capable of pacemaking activity, but the spontaneous depolarization here is slower; therefore, the frequency of the AP is lower. Under physiological conditions, the SA node, with a frequency of ca. 70 beats per minute at rest, serves as the primary pacemaker. AP from the SA node reach the other parts of the conducting system before their own depolarization has progressed far enough to generate an AP so that spontaneous rhythm is suppressed and driven at the frequency of the SA node. The SA node is then the actual pacemaker, while the AV node and all the lower parts of the conducting system are potential pacemakers.

If the SA node ceases to function, one of the potential pacemakers can take its place. The secondary pacemaker is the AV node with a spontaneous frequency of 40-60 beats per minute (the nodal rhythm). If the AV node should also fail, the conducting system in the ventricle can take over as a tertiary pacemaker, although the ventricular rhythm is only 25-40 beats per minute. This frequency is generally too low for the heart to operate as an effective pump.

Cardiac Mechanics

From a mechanical viewpoint, a complete heartbeat can be subdivided into four periods [7, 8]:

1) isovolumetric contraction,
2) ejection,
3) isovolumetric relaxation, and
4) filling.

Figure 11 shows the pressure changes in the pulmonary artery and in the aorta during the heart cycle along with the states of the valves in each of the four periods. Initially, the ventricles are filled with blood due to the preceding atrial contraction. In the first period,

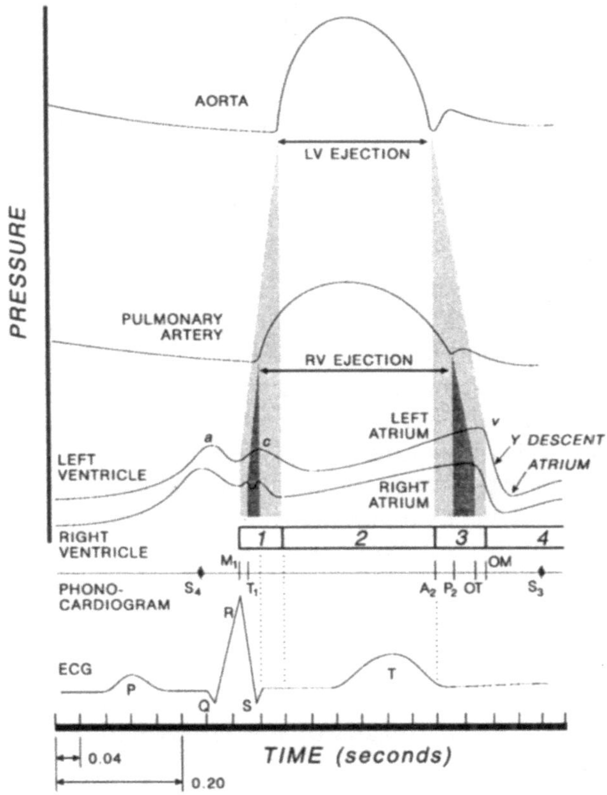

Figure 11. *Diagram of pressure changes in the pulmonary artery and in the aorta during the four periods of the cardiac cycle.*

excitation spreads through the ventricular myocardium causing the muscle cells to begin contracting and the pressure within the ventricle to rise. The valves between the atria and ventricles are closed because the ventricular pressure is higher than the atrial pressure. The semilunar valves in the outflow tract are also closed because the ventricular pressure does not yet equal the blood pressure in the aorta and in the pulmonary artery. During this period (see 1 in Figure 11), there is no change in volume of the heart as the tension in the heart wall increases. As soon as the pressure in the ventricles exceeds that in the large arteries, both semilunar valves open and about half of the ventricular blood volume is expelled into the vessels. The force exerted by the myocardium is so great that the ventricular pressure continues to rise during this ejection period (see 2 in Figure 11). Toward the end of contraction, the pressure in the ventricles decreases so that the blood pressure in the aorta and the pulmonary artery exceeds that in the ventricles and the semilunar valves close, terminating the ejection period and initiating the period of isovolumetric relaxation (see 3 in Figure 11). Although the myocardium is now relaxing, the volume of the ventricles is unchanged as long as the ventricular pressure remains above the atrial pressure, keeping the mitral and tricuspid valves closed. When the ventricular pressure falls below the atrial pressure, these valves open and the filling period begins as blood flows from the atria into the ventricles. Filling is mainly a passive process, that is, the venous pressure drives the blood into the ventricles with no need for atrial contraction. It is only at the end of the filling period that the atria contract, actively pumping a small additional volume of blood into the ventricles. The isovolumetric contraction and ejection periods together constitute the phase of the heartbeat known as *systole*; the phase comprising the isovolumetric relaxation and filling periods is known as *diastole*.

Adjustments of the Heartbeat

Under resting conditions, a systole expels about 70 ml of blood from each ventricle (the stroke volume). About the same amount (the reserve volume) remains in the ventricle at the end of the ejection period. The resting heart rate is about 70 beats per minute so that in each minute about 5 l of blood are pumped (the cardiac output, CO).

Under a workload, cardiac output according to the cardiovascular need is considerably increased. The heart rate can rise to more than 170 beats/min and the stroke volume can double so that during extreme physical exertion a CO of up to 25 l/min is possible. This adaptive change in the heartbeat is brought about chiefly by the influence of the autonomic nervous system. Its two components, sympathetic and parasympathetic, have opposing effects on the heart as illustrated in Figure 12. Sympathetic activity increases the amount of blood pumped, and parasympathetic activity decreases it. The parasympathetic component acts primarily on the SA and AV

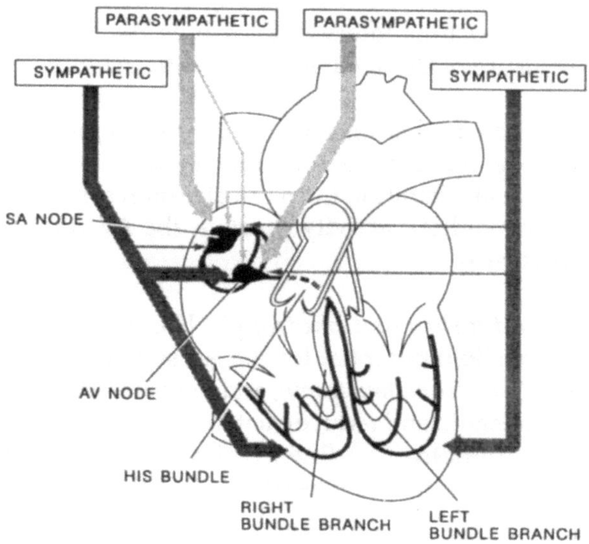

Figure 12. *Schematic description of the autonomic nervous system's two components, sympathetic and parasympathetic.*

nodes and exerts lesser control over the myocardium itself. The sympathetic influence on the heart, mediated by nerve fibers and also by the blood-borne hormone, adrenaline, affects the atrial and ventricular myocardium as well as the SA and AV nodes. The autonomic nervous system has four qualitatively different effects on the heart, as follows:

Influence on the heart rate (chronotropy): Sympathetic activation causes a distinct increase in heart rate (*positive chronotropic action*), while an increase in parasympathetic tone can lower the heart rate to the point of cardiac arrest (*negative chronotropic action*).

Influence on the strength of the heartbeat (inotropy): The strength of myocardial contraction is increased by sympathetic activity (*positive inotropic action*) and slightly decreased by parasympathetic activity (*negative inotropic action*). The sympathetic input acts to enhance the influx of calcium into the cell during the plateau phase, which increases the speed of contraction and the rate of rise of ventricular pressure. Therefore, the time between the onset of excitation of the ventricular myocardium and the opening of the semilunar valves is shortened. This systolic interval, known as the *pre-ejection period* (PEP), is a measure of the level of sympathetic activity.

Influence on conduction in the AV node (dromotropy): Sympathetic activity shortens the conduction time in the AV node (*positive dromotropy*), so that excitation from the atrium reaches the ventricle sooner. Parasympathetic activity shortens the AV conduction time (*negative dromotropy*).

Influence on excitability (bathmotropy): The sympathetic action is to lower the threshold potential for the AP so that the cell becomes more excitable (*positive bathmotropy*), and the parasympathetic action is to raise the threshold and reduce excitability (*negative bathmotropy*). However, the bathmotropic influence plays only a very minor role in the pumping function of the heart.

The heartbeat is affected not only by the autonomic nervous system but also by intracardiac mechanisms that cause temporary adjustments to changes in the amount of blood returning to the heart—that is, to differences in the degree of filling the ventricle at the end of diastole. The heart responds to an increase in end-diastolic volume by augmenting the stroke volume and to a decrease by reducing it. This intracardiac adjustment mechanism is called the Frank-Starling mechanism. For example, the venous return (the preload) can increase when a standing person lies down; the volume entering the ventricles rises because the amount of blood that was contained in the large veins of the legs (about 500 ml) now flows to the heart. With the Frank-Starling mechanism, the increased preload causes the strength of the heartbeat to increase so that the stroke volume becomes larger. Within a few beats, the extra volume has been pumped into the arteries. An increase in aortic pressure (the afterload) also indirectly enhances the strength of the heartbeat. In this case, it is a reduction in stroke volume, due to the elevated aortic pressure, that increases the end-diastolic volume; the Frank-Starling mechanism then ensures that the force of the heart's contraction rises sufficiently to bring the stroke volume back to normal, overcoming the higher pressure in the aorta.

Neural-Humoral Control of Circulation

The *autonomic nervous system* (ANS) regulates and coordinates the functions of the internal organs, matching their activity to the needs of the organism as a whole. The target sites in the body are many and varied. In addition to the heart, the lungs, and the blood vessels, the ANS controls organs involved in adjusting body temperature and excretion, metabolism, and reproduction. However, only the aspects of circulation and the actions of the ANS on the cardiovascular system are considered here [9].

The peripheral ANS comprises two independent functional units, the sympathetic and the parasympathetic systems. Peripheral sympathetic nerve pathways originate in the thoracic and lumbar spinal cord (Figure 13). The nerves to the target organs usually run side by side with the arteries. The peripheral sympathetic system can also transmit information to target organs via hormones. For instance, sympathetic fibers terminating in the interior of the adrenal gland (in the adrenal medulla) cause the release of adrenaline and noradrenaline into the bloodstream, which carries them to the target organs—particularly the heart and blood vessels. These hormones are responsible for crucial aspects of cardiovascular regulation.

The parasympathetic system originates in two regions: the brainstem (the part of the central nervous system between the spinal cord and cerebrum) and the lowest (sacral) part of the spinal cord. The nerve fibers from the brainstem run to their effector organs in the thorax and upper abdominal cavity within one of the large cranial nerves, the vagus. The parasympathetic fibers from the sacral spinal cord supply the organs in the lower abdomen.

The sympathetic and parasympathetic systems have opposite effects on the internal organs—that is, they are antagonists. Throughout the body, sympathetic activation enhances performance and creates a state of increased readiness to meet physical demands (*ergotropic response state*). The parasympathetic system promotes the organ's recovery and reconstitutive processes (*trophotropic response state*).

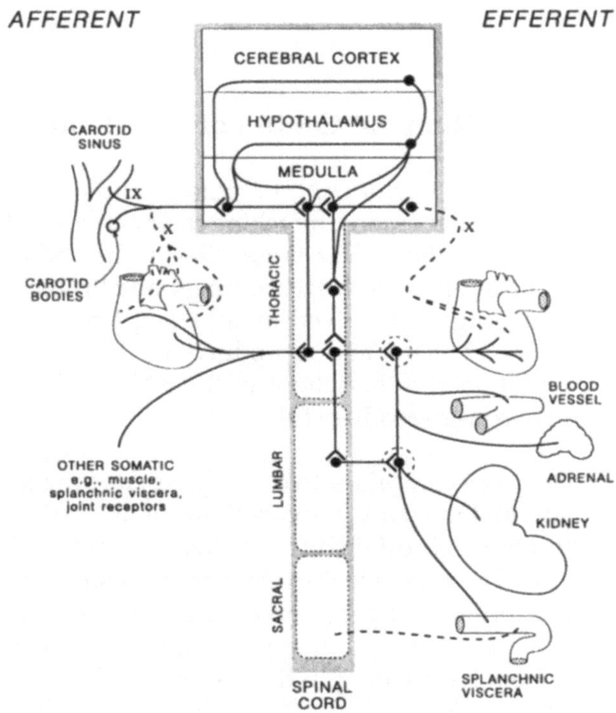

Figure 13. *Schematic diagram of the neural connections important to circulation control. Although the system is bilaterally symmetrical, in this schematic drawing, inputs (afferent fibers to the left) and outputs (sympathetic and parasympathetic efferent fibers to the right) are shown only on one side.*

The actual effects of sympathetic and parasympathetic activity on each individual organ are very diverse. Sympathetic activation can best be observed in the example of a stress situation. A specific means of increasing the body's readiness to respond is to improve the oxygen supply to the muscle. To accomplish this, increased activity of the sympathetic system expands the airways in the lungs, improves the pumping efficacy of the heart, and adjusts the circulatory system so that more blood is distributed to the muscles (by constricting the vessels in the digestive tract and skin, and dilating those in the muscles).

To assist recovery when the emergency has passed, the parasympathetic system reduces the heart's pumping, increases the flow of blood through the digestive organs, and enhances digestive processes through the secretion of digestive juices, more vigorous intestinal movements, etc.

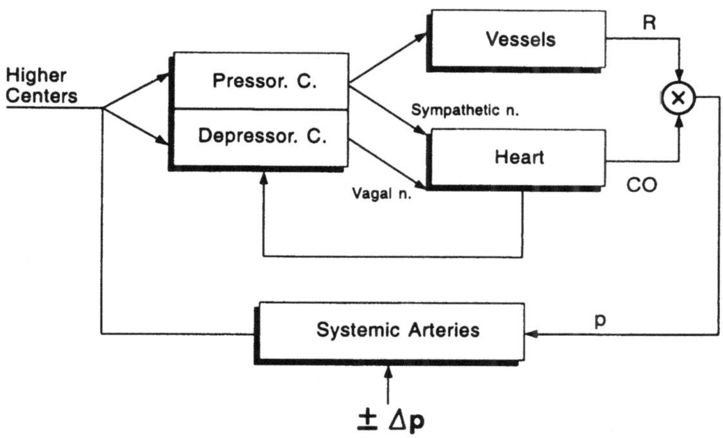

Figure 14. *Simplified block diagram of the nervous control of circulation.*

The antagonistic actions of the two autonomic subsystems are particularly clear in the pumping efficacy of the heart—increased by sympathetic and decreased by parasympathetic activation (Figure 14). Here the parasympathetic influence is exerted mainly on the heart's intrinsic pacemaker, the SA node, and on the element in the conducting system that links the atria and ventricles, the AV node. The parasympathetic system has little effect on the myocardium. Sympathetic activity directly affects the myocardium as well as the pacemaking and conducting system. Therefore, it plays an essential role in the development of new closed-loop principles in cardiac pacing.

Role of the Autonomic Nervous System in Cardiovascular Regulation

The ANS not only adjusts the state of the organ according to external requirements; it also plays a central role in maintaining a constant internal milieu, especially by its actions on the cardiovascular system. It keeps the pressure in the arteries constant, that is, the arterial mean pressure, to ensure that the organs are adequately perfused and receive all the oxygen and nutrients needed even under a workload. In so doing, the ANS compensates for changes in blood pressure caused by various external or internal factors, returning the pressure to its reference level.

This mechanism for maintaining constancy of the arterial mean pressure can be considered as a biological control system (Figures 13 and 15). Here the mean arterial pressure is the controlled variable. The controller itself is the circulatory center in the brain (in the medulla oblongata, the transition region between brain and spinal cord) as shown in Figure 15. The controlling variable in this system is complex, comprising both the change in caliber of the vessels (and, hence, in vascular resistance) and, primarily, the pumping efficacy of the heart (the cardiac output). Finally, various factors can disturb the system (*e.g.*, changes in blood volume, muscle activity, temperature, or position of the body) by altering—usually decreasing—the arterial blood pressure. The function of the ANS (sympathetic and parasympathetic) in this control system is extraordinarily important in transmitting information between the controller in the brain and the controlling variables, particularly cardiac output. Cardiac output is determined by heart rate and the volume of blood ejected by each ventricle per heartbeat—the stroke volume. As described above, sympathetic activity affects both the frequency (positive chronotropy) and the stroke volume (positive inotropy) of the heart, whereas the influence of the parasympathetic system is largely restricted to the heart rate. In addition to its action on the heart, the sympathetic system also alters the degree of vasodilation and, hence, the peripheral vascular resistance. Under normal conditions, this regulation of vessel diameter provides additional control of arterial blood pressure, control of the perfusion distribution (*e.g.*, between viscera, skin, and musculature) and control of venous return to the heart. When the blood pressure falls drastically, this mechanism is employed to raise the arterial pressure ensuring continued perfusion of the most important organs—the brain, heart, and lungs; that is, *centralization of the circulation in shock.*

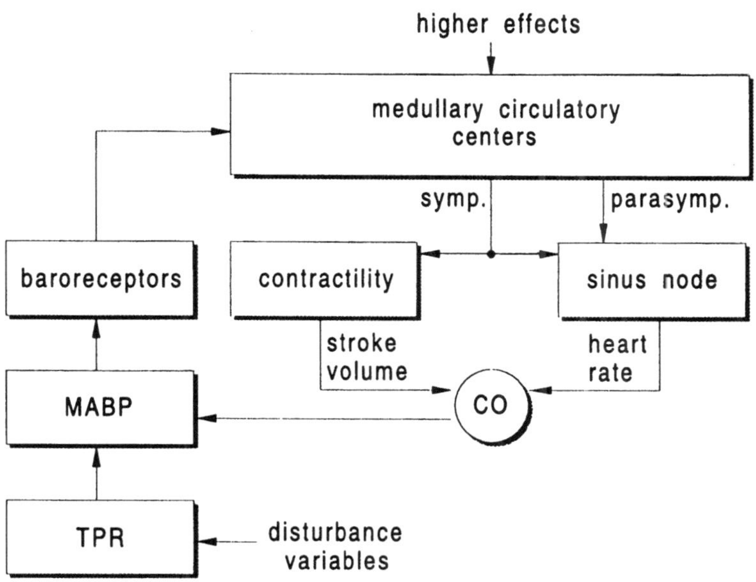

Figure 15. *Simplified block diagram of the autonomic nervous system in cardiovascular regulation.*

Circulatory Regulation in Patients with Disorders of the Heart's Pacemaker or Conducting System

When either the cardiac pacemaking or conducting system is impaired, the control system described above is no longer fully functional (Figure 16) [10, 11, 12]. A marked decrease in the spontaneous heart rate of these patients cannot be compensated even by an immediate increase in the strength of contraction, and, hence, in stroke volume. As a result, the blood pressure falls. If the intrinsic pacemaker fails completely, the heart cannot pump and the arterial blood pressure falls to ~15 mm Hg. The restriction of blood flow through the organs has particularly severe effects in the brain. A loss of consciousness can ensue, at least temporarily. Should the blood pressure not recover for many minutes, some organs can be permanently damaged and the patient may die—all as a result of malfunction of the pacemaking/conducting system.

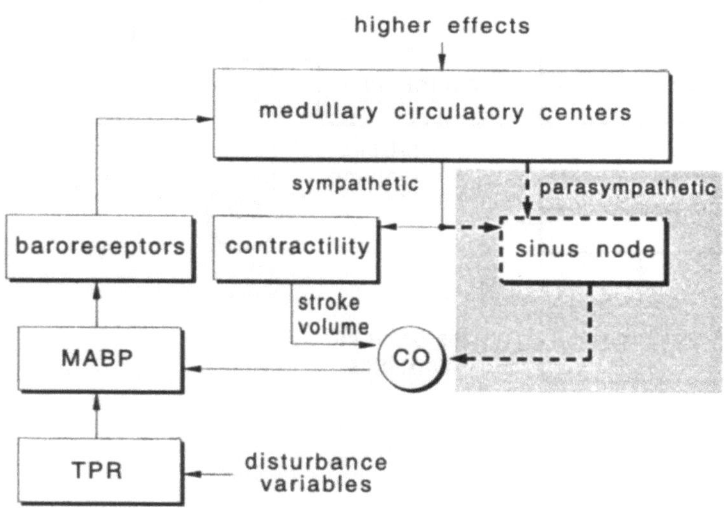

Figure 16. Simplified block diagram to describe chronotropic insufficiency, where the sympathetic and parasympathetic activation of the sinus node is impaired.

These patients may be treated by implantation of an artificial pacemaker that generates electrical pulses that stimulate the myocardium since the heart's own system cannot stimulate the myocardium. In the simplest case (Figure 17), the pacemaker triggers myocardial APs at a fixed, preset frequency. This, at least, prevents a life-threatening drop in blood pressure, but such a device cannot regulate the arterial blood pressure according to demand. The further development of pacemaker technology must, therefore, be directed toward creating an artificial substitute for natural pacemaking and conducting functions as **part of the blood pressure control system** so as to actively maintain physiological conditions. The most appealing solution to this problem is to continuously measure the arterial blood pressure and use this information to adjust the frequency of the pacemaker's stimulation impulses accordingly. This solution is impractical for two reasons:

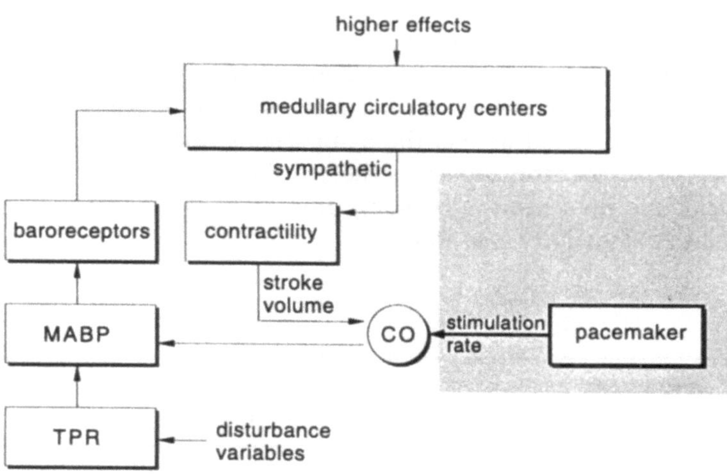

Figure 17. *Simplified block diagram of the cardiovascular control system to demonstrate myocardial AP at a fixed rate.*

1) *It is impractical to obtain such a measurement of arterial pressure since it requires the permanent placement of a catheter in a major artery, and*

2) *an artificial pacemaker of this type acts as an extra, independent controller in the blood pressure control system.*

Because the normal, biological controller (the circulatory center in the brain) remains intact and continues to regulate the force of myocardial contraction as well as the peripheral vascular resistance, unpredictable kinds of interference between the two controllers can be expected.

Since the natural circulatory center of these patients remains functional, and only part of the control system connecting the controller to the controlling variable "heart rate" is interrupted, the next best solution is the technical replacement of this segment of the control circuit. That is, a particular activity of the ANS is to be converted to the appropriate heart rate by technical means, as shown in Figure 18. Direct measurement of sympathetic or parasympathetic activity, expressed as the action-potential frequency in the nerves of the ANS, is impractical. Therefore, an indirect measure of ANS activity is required. As listed above, the sympathetic and parasympathetic influences affect three aspects of the heartbeat; and each of these can, in principle (in a healthy person), provide a signal that accurately reflects ANS activity (see Table I).

Table I
ANS activity of the heart

ANS Activity	*Signal*
chronotropic	duration of the RR and QT intervals in the ECG:
inotropic	maximal rate of pressure rise in the ventricles, duration of pre-ejection period (PEP);
dromotropic	duration of the PQ segment in the ECG.

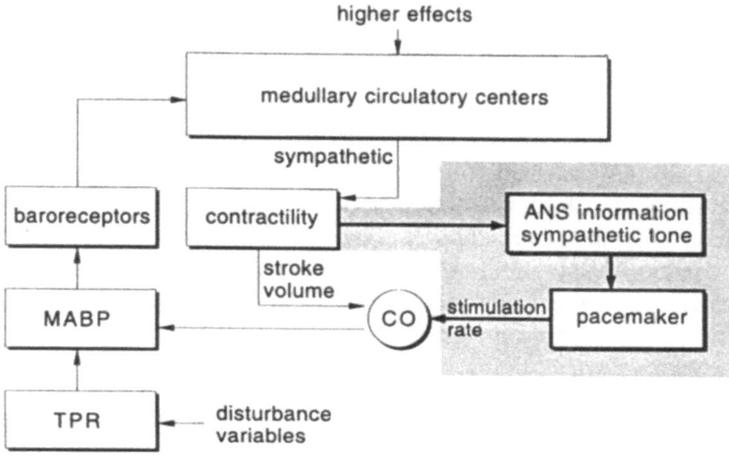

Figure 18. *Schematic description of closed-loop controlled pacing system based on the sympathetic activity of the ANS.*

Because of the impaired intrinsic pacemaking and conducting in patients who need an artificial pacemaker, the ECG features that reflect the chronotropic and dromotropic actions of the ANS are unavailable for use as signals to control the pacemaker. Only the variables associated with the inotropic action of the ANS are useful measures of the level of sympathetic and parasympathetic activity. The maximal rate of ventricular pressure increase, although it directly parallels the degree of inotropic ANS activity, is not a candidate due to the lack of an implantable sensor with long-term stability. However, systolic time intervals, ventricular inotropic parameters, and stroke volume can be monitored by intracardiac impedance measurement. PEP, the time between the onset of ventricular excitation and the opening of the semilunar valves, is, therefore, a good measure of ANS activity. Under sympathetic influence, its duration decreases and parasympathetic activity lengthens the PEP (Figure 19). Any signal representing sympathetic tone may be used in a characteristic manner to adjust the frequency of the stimuli generated by the pulse generator to the individual needs of the patient. Therefore, an ANS-controlled pacemaker bridges the gap in the cardiovascular control circuit, replacing the missing segment. The natural regulatory function of the circulatory center in the brain persists with the result that the heart rate is increased on demand. The development of an ANS-controlled pacemaker vividly illustrates that technical advances can be derived from knowledge of physiological control systems.

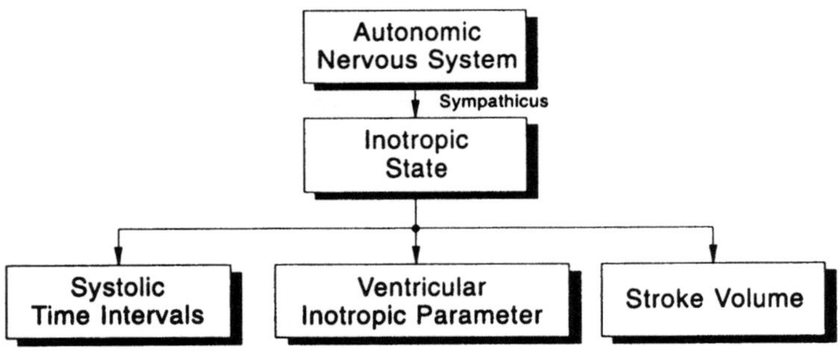

Figure 19. Schematic description of measurable parameters of the ANS activity.

MONITORING THE ELECTRICAL ACTIVITY OF THE HEART

The standard method of observing the electrical events in the heart is *electrocardiography* (ECG) and is routinely used for pacing evaluation and patient follow-up [13]. All together, the individual APs in the heart produce an overall time-varying voltage difference between the excited and unexcited parts during a contraction phase. This voltage signal can be recorded with electrodes attached to the surface of the body, amplified and displayed with a multichannel recorder. The recording can be either bipolar, in which the potential difference between electrodes at two active sites is measured, or unipolar, in which case a single recording electrode measures the potential at one site with respect to a reference electrode. The most common bipolar arrangement is the combination of three limb leads known as Einthoven's triangle. Unipolar recording usually employs Wilson's precordial leads attached at various positions on the chest.

A normal ECG recorded with bipolar electrodes positioned on the heart axis consists of a characteristic sequence of voltage fluctuations, including both slow waves and sharp deflections, with a maximal amplitude of about 1 mV. The elements of this voltage curve are associated with excitatory events in the individual parts of the heart as illustrated by the typical ECG in Figure 20. The feature known as the P-wave corresponds to atrial excitation, the QRS complex to excitation of the ventricular myocardium, and the T-wave to repolarization of the ventricles.

To explain the electrocardiogram in terms of the extracellular potential between excited and unexcited parts of the myocardium, this voltage difference can be represented as a dipole vector. The vector always points from the excited to the unexcited region, and its length corresponds to the magnitude of the momentary voltage difference.

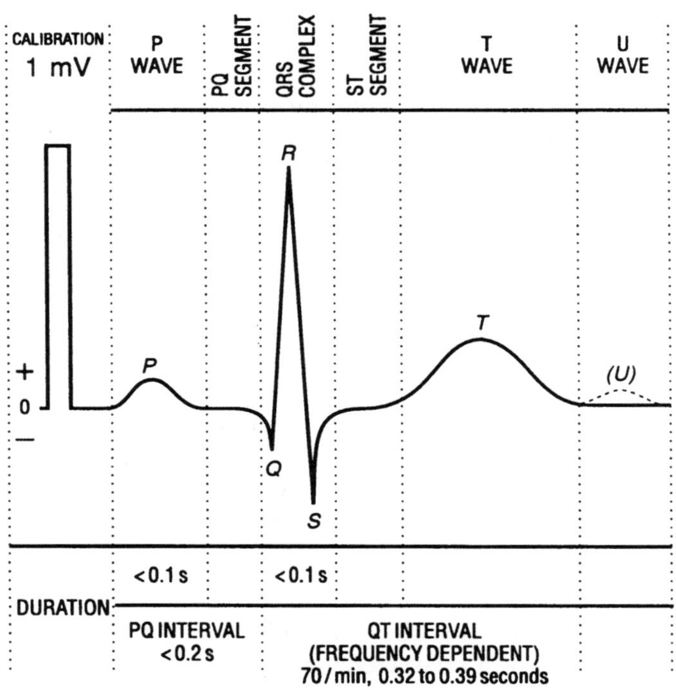

Figure 20. *Description of a normal ECG recorded with bipolar electrodes showing characteristic sequence of events.*

Normally, the dipole sweeps through a characteristic loop during each heartbeat. This loop can be recorded by *vector cardiography* (VCG) with two pairs of electrodes positioned on the chest so that the recording directions are perpendicular to one another. The electrodes are connected to the vertical and horizontal inputs of an oscilloscope so that the display on the screen represents the projection of the vector loop onto the plane defined by the two electrode pairs. By repeating the measurement in all three perpendicular planes of space, the entire vector loop can be reconstructed in three dimensions. Deviations from the normal shape of the vector loop give evidence of disturbances in the spread of excitation during the heartbeat.

The resting ECG and VCG are informative but additional information is obtained by recording them during physical activity. The main value of this procedure is that workload-dependent changes can indicate an inadequate supply of oxygen to the heart. As the patient operates a bicycle ergometer, the load is gradually increased until a predetermined heart rate is reached or until there are other reasons (*e.g.*, signs of ischemia in the ECG) for stopping the exercise test.

Long-term ECG recordings also assist the diagnosis of rhythm disturbances. For 24 to 48 hours, the patient wears a battery-powered multichannel recorder that stores the signals from several ECG leads as well as a time trace. The storage medium can be small magnetic tapes, tape cassettes, or semiconductor memory devices. The stored data is later analyzed in various relations (heart rate, rhythm disturbances, signs of ischemia, etc.). Although automatic analysis is increasingly common, it is essential for significant findings to be checked by an experienced operator with direct examination of a monitor display or printout.

The conventional ECG is recorded from the surface of the patient's body. However, techniques have become available for recording voltage directly from the myocardium with a catheter inserted into the heart. Such intracardiac recordings are particularly useful in the diagnosis of conduction disorders and serve as an indicator for proper selection of the type and mode of implantable pacemaker.

The most important for clinical purposes is the *His bundle electrogram* (HBE) obtained by introducing an electrode catheter through the right atrium and the tricuspid valve into the right ventricle so that its tip rests against the septum. The electrode is then in the immediate vicinity of the His bundle and can record the electrical activity in that structure and the AV node. This method is always combined with a surface ECG so that the timing of these intracardiac events with respect to the other elements in the heartbeat (atrial and ventricular excitation) can be observed. The HBE is especially informative about disorders such as delayed conduction (AV block) and nonphysiological conduction pathways (preexcitation syndrome). When combined with high-frequency atrial stimulation, the HBE assists evaluation of AV transmission under a workload and allows the reliability of SA pacemaking and conduction in the atrium to be tested. Finally, the reliable recording and analysis of a local intracardiac signal has become indispensable in clinical practice for controlling various temporary and permanent artificial pacemaker systems (demand pacemakers).

Pathophysiology

Under physiological conditions, APs are generated in the SA node by spontaneous depolarization of the cell membrane. The AP is then propagated over the atria, through the AV node, the His bundle, the bundle branches, and over the ventricular myocardium, where it produces a coordinated contraction of the muscle cells. The at-rest heart rate is about 60-80 beats per minute.

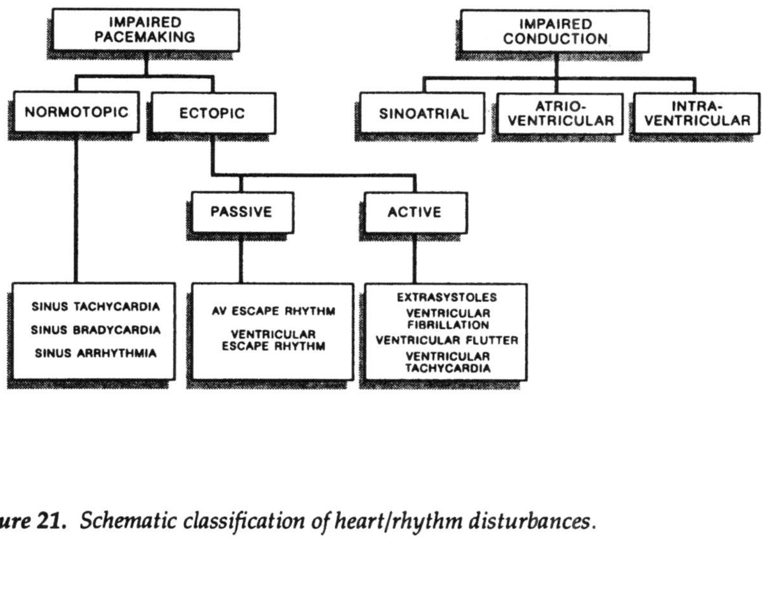

Figure 21. *Schematic classification of heart/rhythm disturbances.*

This normal course of events can be disrupted in many ways [14]. An increase of the resting heart rate to more than 100 beats/min is called *tachycardia,* a fall to less than 60 beats/min is *bradycardia,* and irregular timing of the beats is *arrhythmia.* These kinds of rhythm disturbances can be caused by malfunction at various sites in the pacemaking and conducting system. Figure 21 shows a schematic classification of heart-rhythm disturbances.

Disorders of the Pacemaking Function

In the normotopic situation, the rhythm of excitation is generated in the SA node, the primary pacemaker. However, the secondary and tertiary pacemakers in the conducting system can take over this role, and even cells in the working myocardium can function as pacemakers; these sites of rhythm generation are known as *ectopic centers* [15, 16].

Normotopic Rhythm Disturbances: In a normotopic disturbance of the heart rhythm, the malfunction lies in the SA node.

Sinus tachycardia can result from an increase in sympathetic tone—caused, for example, by mental agitation or as a reflex response to cardiac insufficiency or infarct, anemia, inflammation of the heart (myocarditis, pericarditis) or hyperactivity of the thyroid gland. Under physiological conditions, fever can also raise the heart rate by 10 beats/min for each 1°C increase in body temperature.

Sinus bradycardia often occurs under physiological conditions in athletes trained for high performance at endurance sports. Some athletes' hearts have a resting rate well below 50 beats/min. However, a reduced SA rate can also be a manifestation of increased parasympathetic tone, *i.e.*, during sleep. Pathological sinus bradycardia can be caused by brain damage, by thyroid hypoactivity, and by myocardial infarct. Complete failure of the SA node, producing cardiac arrest (*asystole*), can also result from sclerosis of the carotid artery. This artery contains baroreceptors that sense the blood pressure and, by signals in sensory nerve fibers, inform the cardiovascular center in the brain of its current level. Normally, a pressure increase triggers a reflex that lowers both the heart rate and the strength of contraction. Arteriosclerosis makes these baroreceptors so sensitive that a slight mechanical pressure on the carotid artery, *e.g.*, caused by looking backward while backing into a parking space, can interrupt SA node activity. The heart stops pumping, the brain does not receive enough blood, and the person becomes unconscious. This phenomenon is called hypersensitive carotid sinus.

Sinus arrhythmia under physiological conditions is common, especially in juveniles (*respiratory arrhythmia*). The heart rate rises during inspiration and falls during expiration. Pathological sinus arrhythmia is rare, mainly developing in old age, and indicates damage to the SA node (*sinus node syndrome*).

Ectopic Rhythm Disturbances: This abnormal rhythm is generated by an ectopic center—a pacemaker other than the SA node. Ectopic pacemaking can occur passively when a superordinate pacemaker fails or actively when an ectopic center independently generates extra APs although the actual pacemaker and conducting system is still intact.

Passive Ectopic Disorders: When the SA frequency decreases, it may be superseded by an "escape rhythm" generated by the next lower pacemaker in the conducting system. If the SA node becomes completely nonfunctional, or if conduction from the SA to the AV node is completely blocked, normally the pacemaker function will be taken over by a lower center. If the SA node fails as previously shown, it is replaced by the AV node so that the escape rhythm has a frequency between 40 and 60 beats/min (AV rhythm).

If the AV node also fails, the His bundle is the next in line (resting rate below 40 beats/min). As a last resort, the bundle branches can act as pacemakers (resting rate 20 to 35 beats/min). However, the frequency of this ventricular escape rhythm is usually too low to provide an adequate blood supply to the organism.

Occasionally, the ECG reveals a slow shifting of the pacemaker center from the SA to the AV node. This phenomenon is called "wandering pacemaker."

It often occurs when the SA node functions normally except for an occasional single heartbeat so that only one ectopic AP is generated. Such an escape systole is likely to occur when conduction is disturbed (see below: second degree AV block with Wenckebach periodicity).

Active Ectopic Disorders: An active ectopic center, usually a damaged part of the heart, adds extra APs to those of the normally functioning SA node and conduction system, altering the heart's normal basic rhythm. This most commonly results in *extrasystole*. Depending on the site of origin of the extrasystole, it is classified as ventricular or supraventricular (originating in the atrium). Monotopic extrasystoles are all generated at the same ectopic site; extrasystoles originating at various sites are called polytropic. Polytropic extrasystoles are an adverse sign for the patient since they are often the forerunners of ventricular tachycardia or fibrillation (see below). Extrasystole may appear in isolation but it is not uncommon to be repeated in a fixed relation to the normal rhythm. When every second heartbeat is an extrasystole, the pulse is known as *bigeminal;* when each normal beat is followed by two extrasystoles, it is a *trigeminal* pulse. Table II shows Lown's classification of *ventricular extrasystole* (VES) into classes 0 through V [17].

Table II
Classification of ventricular extrasystole (VES) according to Lown

Class	*Disorder*
0	No VES
Ia	Isolated monotopic VES with a frequency <30/h and <1/min
Ib	Isolated monotopic VES with a frequency <30/h and <1/min
II	Isolated monotopic VES with a frequency <30/h
IIIa	Polytropic VES
IIIb	Bigeminus
IVa	VES pairs or couplets
IVb	Bursts of VES and ventricular tachycardia (3 or more consecutive VES)
V	Early-arriving VES (R onto T phenomenon)

Extrasystoles of class V, which are generated so soon after a normal beat that parts of the myocardium have already regained their excitability, present a special danger. This situation can easily develop into ventricular fibrillation in which the heart is unable to continue pumping. Because repolarization does not simultaneously occur in the inner and outer layers of the myocardium, an extrasystole that arrives early is not uniformly propagated. Therefore, reentry can occur—the situation in which an AP travels continuously over closed paths of various lengths—always encountering excitable tissue.

Under normal conditions, reentry into a recently excited region is prevented because the cells are still refractory. But if the APs during a heartbeat are very brief (with correspondingly short refractory periods), or if there is a unidirectional conduction block in the muscle tissue, it can happen that when an AP returns to the starting point after its first passage over the myocardium, the cells there have recovered their excitability and permit reentry. The result can be prolonged ventricular tachycardia, ventricular flutter, or even fibrillation. During ventricular flutter, the heart contracts at a rate of 200 to 250 beats/min. The orderly spread of excitation is partially disrupted so that the heart fills only partially and cannot eject enough blood to sustain normal organ function. Still higher heart rates are called ventricular fibrillation. In this case, the contraction is entirely uncoordinated and pumping is impossible—which, without therapy, soon leads to death.

Such extreme tachycardia can also occur in the atrium—atrial flutter at frequencies of 200-350 beats/min, and atrial fibrillation at >350 beats/min. Since the spread of excitation from atria to ventricles is delayed at the AV node, this high frequency is not transmitted to the ventricle. During flutter, in most cases, a certain fraction of the atrial AP is blocked so that only every second or third AP reaches the ventricle. During fibrillation, the potentials are transmitted at irregular intervals so that absolute arrhythmia of the heartbeat results. Atrial fibrillation has little effect on the efficacy of pumping because the ventricle remains capable of coordinated contraction.

Disturbances of Conduction

Even with a functional pacemaker, the conduction of the AP can be partially impeded or completely prevented at various places. Such a conduction disorder is known as a *block*: an SA block in the pathway from the SA to the AV node, an AV block within the AV node, and an *intraventricular* block in the His bundle and bundle branches. Three degrees of severity are customarily distinguished, as shown in Table III.

Table III
Conduction block classification

Severity	*Symptom*
First degree	delayed conduction
Second degree	occasional failure of conduction (partial block)
Third degree	complete interruption of conduction (total block)

Because SA block affects conduction within the atrial myocardium, first-degree SA block is normally not detectable. In second-degree SA block, atrial conduction becomes intermittent so that the AV node occasionally generates escape systoles. A third-degree SA block cannot be distinguished from failure of the SA node as a pacemaker (sinus arrest). In this form of block, the AV node takes over as pacemaker. With the onset of the block, or if SA node failure is sudden, it commonly takes some time for the AV node to begin functioning as pacemaker. In the interim, the heart stops beating and the lack of oxygen in the brain leads to unconsciousness (Morgagni-Adams-Stokes attack). For these patients, the implantation of an artificial-demand pacemaker becomes the only choice of treatment.

AV block is the most common of the three forms. In first-degree AV block, conduction in the AV node is delayed but each atrial AP leads to a ventricular contraction. Usually there is no functional impairment. Second-degree AV block, in which not all atrial APs are transmitted, can be either *Type I* (Wenckebach) or *Type II* (Mobitz). In the more common *Type I*, transmission deteriorates gradually (the AV conduction time lengthens) until finally a potential does not reach the ventricle at all. In *Type II*, the conduction velocity in the AV node is normal but transmission to the ventricle fails at regular intervals. When transmission is completely interrupted in third-degree AV block, a secondary or tertiary pacemaker generates the ventricular rhythm (Figure 22). Again, if the block is of sudden onset, cardiac arrest and unconsciousness may occur before the

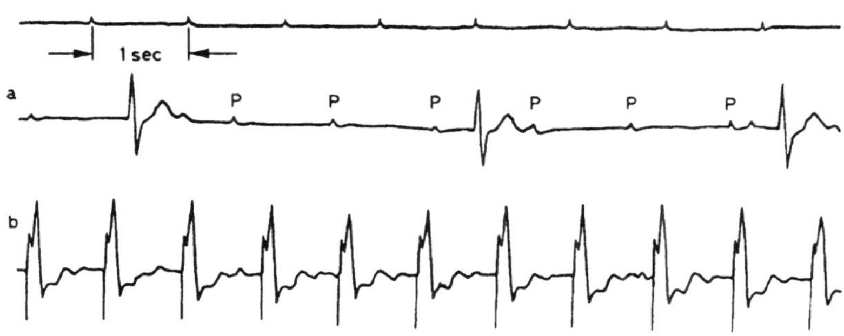

Figure 22. *ECG of third-degree block representing (a) the low ventricular rate and (b) demonstrating AV sequential ventricular pacing (VAT).*

substitute pacemaker becomes effective. A second or third degree AV block is usually associated with decreased pumping performance, so that although the blood supply is sufficient for resting conditions, the patient cannot support a physical workload.

Intraventricular block occurs when conduction in one or more branches of the His bundle is interrupted. The fault may lie in the right bundle branch (right-branch block), the entire left bundle branch (left-branch block), or the anterior or posterior division of the left branch (left-anterior or left-posterior hemiblock). A bifascicular block involves two branches. In any case, once excitation has reached the ventricle, and because it can propagate everywhere from cell to cell, the hemodynamic consequences of intraventricular block are usually very slight.

Not only can the normal conduction pathways be blocked, it is also possible for abnormal accessory routes for AV transmission to exist so that the AV node is not the only site at which excitation can cross the boundary between the two chambers. Because propagation is not delayed at these aberrant transmission sites, as it is in the AV node, ventricular contraction occurs prematurely, beginning at the base of the ventricle and spreading toward the apex of the heart. Therefore, this condition is known as the *preexcitation syndrome*. A preexcitation syndrome can often give rise to supraventricular tachycardia in which the excitation circles between atrium and ventricle, using both the normal conduction system and the accessory routes. This circling can be broken by ECG-triggered electrostimulation.

THERAPY FOR CARDIAC RHYTHM DISTURBANCES

The most urgent goal in the treatment of rhythm disorders should be to ensure a coordinated contraction of the heart, corresponding to the physiological sequence of events during the heartbeat. In many cases, this goal is unattainable so that one must strive for at least partial success by achieving the following smaller goals:

1) to ensure that the heart pumps effectively enough to keep the organism alive;
2) to alleviate the symptoms that result from the disturbed rhythm;
3) to protect the patient from complications that could develop from the existing disorder; and
4) to provide a satisfactory quality of life (*e.g.*, capacity for physical exertion).

This list makes clear that it is not necessary to treat every disturbance of cardiac rhythm. If therapy is required, however, apart from causal therapy (*i.e.*, elimination of the cause of disease) three possibilities are available: medication, electrotherapy and surgical procedures [18].

Medication is employed chiefly for disorders of rhythm generation (extrasystole, etc.). Three groups of drugs are in use, as follows:

1) Membrane-stabilizing antiarrhythmic drugs, which act by inhibiting rapid AP;
2) β-receptor blockers that reduce the sympathetic influence on the heart, thereby reducing heart rate, strength of contraction, and conduction velocity;
3) Calcium antagonists, which primarily inhibit slow AP.

As therapy for supraventricular tachycardia, atrial flutter, and atrial fibrillation, digitalis retards transmission from the atrium to the ventricle.

Treatment of arrhythmia by surgical procedures is now usually reserved for carefully selected special cases. However, surgery can be particularly useful as therapy for life-threatening tachycardia and for preexcitation syndrome. A detailed state-of-the-art is given by El-Sherif and Samet [19].

Electrotherapy in the Treatment of Cardiac Rhythm Disturbances

The treatment of conduction disorders such as SA or AV block, in particular, is a domain of electrotherapy. Artificial pacemakers are also useful in cases of bradycardia and asystole (*e.g.*, hypersensitive carotid sinus) and are increasingly employed to prevent reentry in the case of tachycardia. In principle, electrotherapy can be acute, temporary, or chronic.

Acute electrical treatment is indicated, *e.g.*, to stop ventricular fibrillation. Defibrillation is accomplished by applying a strong current pulse so that all the myocardial cells are excited at the same time. Therefore, they are all simultaneously refractory, reentry is prevented, and the SA node can initiate a coordinated contraction.

In temporary electrotherapy, contraction of the heart is induced by an artificial, extracorporeal pacemaker. The device produces rhythmic electrical pulses that are conducted to the right ventricle by an electrode catheter introduced through the veins. The pulses excite the ventricular myocardium near the electrode, and this excitation propagates from cell to cell over the entire heart wall. Such stimulation may be necessary when an organic problem (*e.g.*, a myocardial infarct) has abolished normal pacemaker function. An indication for use of this method is spontaneous cardiac arrest lasting for more than three seconds. When this indication no longer exists, the temporary pacemaker can easily be removed. However, if there is a persistent risk of asystole, chronic electrotherapy by permanent implantation of an artificial pacemaker must be considered.

A pacemaker to be implanted in a pocket of skin or muscle incorporates a small pulse generator and a battery within a single housing to which the electrodes can be connected. If it is a single-chamber procedure, the electrode is positioned in the right ventricle; electrical pulses from the generator then trigger rhythmic ventricular contractions. Dual-chamber pacemaker systems employ two electrodes, one of which is anchored in the right atrium, so that the atrium and ventricle can be stimulated in a sequential mode.

The different indications and technical solutions available in today's pacemakers can be described according to their different functions interacting with the heart's own electrical events. Referring to the recommendations of the British Pacing and Electrophysiology Group (BPEG) and the North American Society of Pacing and Electrophysiology (NASPE), the different modes of pacing can be summarized in Table IV below [20].

Table IV
The NASPE/BPEG Generic (NBG) Pacemaker Code

Position				
I	*II*	*III*	*IV*	*V*
Chamber(s) paced	Chamber(s) sensed	Response to sensing	Programmability rate modulation	Antitachyarrhythmia function(s)
0 = None A = Atrium	0 = None A = Atrium	0 = None T = Triggered	0 = None P = Simple programmable	0 = None P = Pacing (antitachyarrhythmia)
V = Ventricle	V = Ventricle	I = Inhibited	M = Multi-programmable	S = Shock
D = Dual (A + V)	D = Dual (A + V)	D = Dual (T + I)	C = Communicating R = Rate modulation	D = Dual (P + S)
	S = Single (A or V)	S = Single (A or V)		

Note: Positions I through III are used exclusively for antibradycardia functions.

Historically, the first pulse generators operated at a fixed rate, but later designs were introduced which would stimulate the heart on demand (VVI, AAI); that is, an electrical impulse is sent to the myocardium only if the heart has shown no spontaneous activity for a certain period of time. Demand pacemakers monitor the intrinsic activity of the atrium or the ventricle. With a dual-chamber pacemaker that has an additional atrial electrode the atrial activity can be monitored separately, and if loss of sinus rhythm occurs, according to the DDD principle, the two chambers will be stimulated. If conduction between the atria and ventricles is blocked, for example, but a physiological P-wave is still present and can be recorded in the long term by an atrial electrode, the atrial signal is used to time the stimulation of the ventricle with the reestablishment of the physiological contraction sequence. The atrial activity is sensed, and after a delay corresponding to the normal AV conduction time, the ventricle is paced so that the ventricular contraction it elicits occurs in the normal physiological relation to the atrial events. These pacemakers are, therefore, able to monitor electrical events and to independently control all chambers of the heart. These multiprogrammable pacemaker systems provide a great number of stimulating and sensing parameters that may be altered according to the patient's needs. Recent attempts have been made to use physiological parameters other than the heart's own activity to control cardiac stimulation and to adapt the pacing rate to the patient's hemodynamic needs. Table V summarizes the forms of control now available. Of these, the only systems in clinical use are those controlled by temperature, respiratory parameters, physical activity (muscle contraction, vibration), or those using a change in QT time as the control variable.

Table V
Open- and closed-loop control principles for rate-adaptive systems

Triggered by the atrium	VDD, DDD, VAT	Closed-loop control
Other physiological parameters for open- or closed-loop rate control	Minute volume, respiratory rate	Open-loop control
	QT time	
	Body activity	
	Motion energy	
	Blood temperature (CVT)	
	pH value	
	Central venous oxygen saturation	
	Ventricular Inotropic Parameter (VIP) Pre-Ejection Period (PEP) Stroke Volume (SV)	Closed-loop control

Figure 23 summarizes the systemic control of cardiac output for a patient with a cardiac pacemaker. The lower shaded portion of the diagram includes normal physiological control processes. The regulated cardiac actions are stroke volume and heart rate to deliver CO

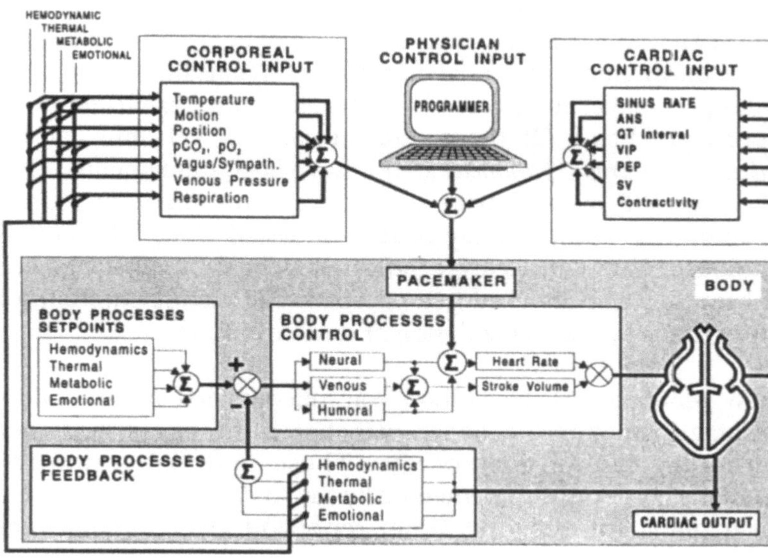

Figure 23. Cardiovascular control process including different rate-responsive pacemaker concepts.

to the systemic circulation. The systemic circulation is closely managed by numerous feedback control processes that are collectively listed in the lower box as hemodynamic, thermal, metabolic, emotional, and homeostatic processes. Deviations from set values [center left box], which requires adjustments in cardiac output are effected via pathways listed in the center box. All of the effector signals impinge on the heart and contribute to modifying HR and stroke volume. The upper unshaded portion of the diagram represents potential control opportunities for rate-adaptive pacemakers. There are three major control classes:

1) The rate may be reprogrammed by the physician [center];
2) The rate may be adjusted by the pacemaker based on an open-loop corporeal signal that is correlated to cardiovascular demand [upper left box];

3) The rate may be adjusted based on an intracardiac signal that reflects changes in the intrinsic control signals to the heart [upper right box].

The DDD pacemaker with access to the sinus rate is a classic example of an intracardial control signal. In the absence of suitable atrial performance, other intracardiac signals may be used to obtain rate control information. Potential intracardiac signals that may be used by the pacemaker to restore chronotropic control are listed in the upper-right box. Of these, the measure of QT interval and *Ventricular Inotropic Parameter* (VIP) have been implemented clinically. The VIP measurement is based on intracardiac impedance measurements and is closely related to *Systolic Time Interval* (STI) techniques. VIP is an accurate indicator of the sympathetic tone. It has been successfully used in a pacemaker to restore chronotropic control to the systemic regulation of cardiac output.

A new generation of closed-loop, controlled pacing systems taking advantage of monitoring the sympathetic tone has established a new frontier in pacing technology. These multiprogrammable pacemaker systems, supplemented by the ANS system of rate adaptation, may, for the first time, represent real physiological stimulators with electrical frequency excitatory events that correspond to the natural cardiovascular control principles. This is the major focus of the following chapters.

PACEMAKER TECHNOLOGY

To summarize the preceding chapters, the sequence of contractions in the heart is controlled by autonomous mechanisms that are subject to extracardiac regulative influences. Through sympathetic innervation, these mechanisms permit inotropic and chronotropic adaptation of cardiac output to physical workload. The automatic action begins at the SA node. If its impulse formation ceases, *e.g.*, an SA blockage, or if there is a total AV block, the remaining parts of the conduction system are capable of spontaneous excitation. However, as the distance of the secondary site from the SA node increases, the spontaneous depolarization rate decreases and the workload-dependent variation of the heart rate is compromised so that the cardiac output is exclusively regulated through changes in stroke volume. This compensation mechanism presupposes a healthy ventricular myocardium, capable of adapting the stroke volume. In the event that this adaptation does not compensate for the absence of increase in heart rate, the bradycardia symptoms can be successfully treated by implanting a cardiac pacemaker [21-23]. Normally, heart rate makes the largest contribution to the workload-dependent adaptation of the cardiac output. This function can be maintained by either reestablishing the physiological excitation through atrium-controlled pacing or by an adaptation of the pacing rate through control and regulatory techniques using corporeal, cardiovascular, hemodynamic, and metabolic parameters [24-32].

The multidisciplinary nature of the technical task is derived from the diverse components of the physiological, physical, electrochemical, and material sciences. The engineering and scientific contributions not only entail signal processing, control engineering, and microelectronics, but also incorporate material sciences and process engineering. As with the solution of other problems in biomedical engineering, the development of cardiac pacemakers critically depends on mastery of basic technologies. The reliability and safety of the technical therapeutic device requires comprehensive quality control measures that must include all the system components.

As can be seen in Figure 24, the component tasks of pacemaker technology include:

1) the electrode and its interface with the myocardium,
2) microelectronics and energy source,
3) the housing and insulating materials, and
4) bidirectional telemetry (programming unit).

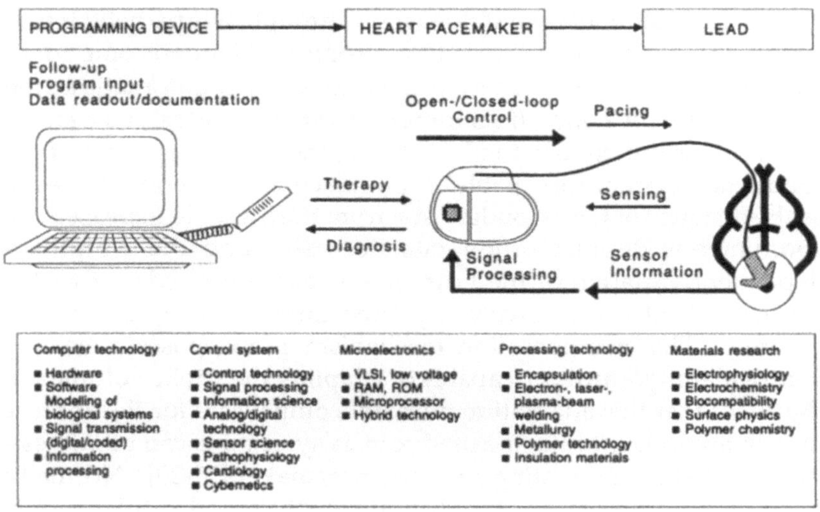

Figure 24. *Multidisciplinary nature of engineering and scientific contributions to pacemaker technology.*

The processes at the electrode are characterized by the electrophysiological and electrochemical conditions at the electrode/myocardium phase-boundary [33]. Pacemaker design with low-power consumption electronics makes stringent demands on microelectronics. The microelectronics must not only serve for signal processing but must also meet the control requirements of the device. Achieving a pacemaker lifetime matched to the life expectancy of the patient primarily depends on the power supply [34]. Experience has shown that other aspects such as long-term stability of the insulation materials, minimizing foreign body reactions (biocompatibility), and resistance to corrosion also have a considerable influence on the trouble-free operation of the pacemaker. As pacing devices have become more complex, bidirectional communication is required to manage the signal processing and the pacing modes of the device. The principle of the pacemaker is essentially based on a timer with a switching stage, the timer being controlled by cardiac activity. The energy necessary for artificial pacing is transferred to the myocardium from a storage capacitor. The subsequent repolarization process reestablishes charge neutrality and, thus, minimizes irreversible chemical reactions and a rise of the pacing threshold [35, 36].

In today's pacemakers, interference of the artificial excitation with the heart's intrinsic rhythm is avoided by continuously monitoring the intracardiac EGM. If cardiac activity remains absent during the escape interval, the pacemaker takes over and, thus, prevents a drop of the heart rate or cardiac arrest. Therefore, artificial pacing is provided only on demand. The demand pacing principle forms the basis of all modern single- and dual-chamber pacemakers.

The development of electrotherapy for cardiac rhythm disturbances is closely connected to our understanding of pathophysiology, the pathogenesis of impulse formation, impulse conduction disturbances, and the technical progress in circuit, battery, and electrode technology. Technological progress has made it possible to manufacture pacemakers that, in practice, realize most of the existing therapeutic concepts. Improved functionality is achieved in a reasonable volume while guaranteeing an operating time matched to the life expectancy of the patient.

Implantable Pacemakers History, Current Status, and Trends

As previously indicated, the first artificial pacemaker was developed by Elmquist and implanted by Senning in 1958 [37]. Electrostimulation soon became the therapy of choice for bradycardia arrhythmias. The implantable pacemaker was made possible by the invention of the transistor. The circuit in the first pacemakers made use of a simple RC element to control the pacing rate. The pacemaker did not monitor the intrinsic cardiac activity and, therefore, the asynchronous mode of operation (VOO) resulted in possible interference between the artificial pacemaker and the heart's spontaneous rhythm.

The next step in the mid 1960's, was the development of triggered (VVT) and of inhibited (VVI, AAI) pacemakers. These devices required the addition of an ECG amplifier and an R- or P-wave detection circuit respectively. But these pacemakers were built using discrete transistors, which limited the desired functionality. In the early 1970s, the incorporation of integrated circuits permitted the design of programmable units. Those devices allowed non-invasive postoperative adjustment of many pacing parameters. The programmability greatly facilitated patient management throughout the service life of the device. At about the same time, the introduction of the lithium iodine battery extended the operational life of the device.

Parallel to the efforts to enhance single-chamber devices, "physiological pacemakers" were developed. The first devices operated in the VAT mode, pacing the ventricle following a detected atrial electrical event. For patients with AV block but intact sinus node function, these devices were a decisive step toward an improved quality of life. The next logical step was to design devices that offered atrial and ventricular demand pacing combined with P-wave triggered ventricular pacing (DDD).

Today, most advanced pacemakers combine DDD pacing with many features that greatly increase the indication range of the device and that facilitate the postoperative management of the patient. Non-invasive, two-way communication between the implant and an external programmer provides access to these features. They include:

- programmability of the control and the pacing parameters,
- external trigger capability for both chambers,
- transmission of operating data, such as the battery voltage, the electrode resistance, or the results of a self-test,
- automatically performed tests of the pacing threshold, storage of patient data,

- acquisition, real-time transmission, processing, and storage of physiological data for the purpose of evaluating the therapeutic effectiveness of the implant,
- acquisition and processing of physiological data that complements the atrial and ventricular intracardiac EGM to enhance the chronotropic response, *e.g.*, in the event of a sick sinus syndrome.

Despite the increase in complexity, today's devices are smaller, are more reliable, and offer an increased service life compared to the earlier devices [38]. Technological and physiological aspects of artificial pacing, as well as the current state of development, will be illustrated and examples of pacemakers that are currently in use will be presented. This assessment will be supplemented with an overview of the next generation of pacemakers.

Multiprogrammable Single-Chamber Pacemaker

The use of single-chamber pacemakers prevails in antibradycardia pacing. Depending on the indication, they are used for pacing either the atrium (AAI, M) or the ventricle (VVI, M). The multiprogrammability permits the adaptation of the implant to a form of electrotherapy that is optimized with respect to indications. It offers the advantage of postoperative noninvasive adjustment of pacing therapy. In addition, this feature effectively reduces the multiplicity of pacemaker types with consequent economic advantages. Table VI summarizes the programmable parameters and their therapeutic and diagnostic significance.

Programmability of the pacing-pulse parameters not only improves the efficacy of pacing, *e.g.*, in connection with changes of the pacing threshold, but also achieves effective energy savings and, thus, an extension of the service life. By providing for noninvasive adjustment of the sensitivity of the cardiac signal detection channels and permitting either a unipolar or a bipolar electrode configuration, protection is provided against extra- or intra-corporeal noise potentials. Programming the rate and the refractory period is useful for optimizing the important cardiological parameters of the artificial pacing system to the requirements of the therapy. Rate programming especially favors the patient's own heart rhythm if pacing is needed only occasionally. A hysteresis setting serves to avoid interference between artificial pacing and an intrinsic rhythm when the two rates are nearly the same [39].

Table VI
Clinical significance of programmable parameters

Parameter	*Diagnostic Benefit*	*Therapeutic Benefit*
Basic rate	Testing of sensing and pacing performance; ECG analysis of spontaneous rhythm	Hemodynamic optimization; arrhythmia suppression
Hysteresis rate	Hemodynamic studies	Providing priority for spontaneous rhythm
Pulse energy (pulse duration, pulse amplitude)	Determination of stimulation threshold and safety margin; electrode pacing performance studies	Efficient energy utilization; prevention/correction of exit block and muscle stimulation; prolongation of service time
Sensitivity	Determination of sensing threshold and safety margin; signal and noise amplitude evaluation	Prevention/correction of entrance block: myopotentials, EMI
Electrode polarity	Study of polarity-dependent sensing and pacing performance	Prevention of muscle stimulation and myopotential inhibition
Refractory period	Analysis of oversensing phenomena	Prevention of T-wave and polarization potential sensing

Technical Achievement for the Multiprogrammable Single-Chamber Pacemaker

The block circuit diagram (Figure 25) provides an overall view of the operational principle of the multiprogrammable single-chamber pacemaker. The input and output amplifiers of the pacemaker are connected to the myocardium through the electrode. The impulse energy is adjusted via the amplitude and duration of the pacing pulse. Like the sensitivity of the input amplifier, these variables are controlled digitally from the program memory. The central, crystal-controlled timer and counter are used for the time sequencing of all control processes, such as pacing rate, refractory period, hysteresis, and data transfer. Bidirectional inductive telemetry is used to exchange data with the implanted unit by means of an external programming unit. Programming takes place via a coil, a receiving amplifier, a decoder, a controller, and the register in which the temporary and permanent programs are stored. An encoder and a program output stage make available the actual pacing and control parameters, the operating parameters (battery current, battery voltage, internal battery resistance, electrode resistance, patient information, etc.), as well as intracardiac EGM signals and markers [40].

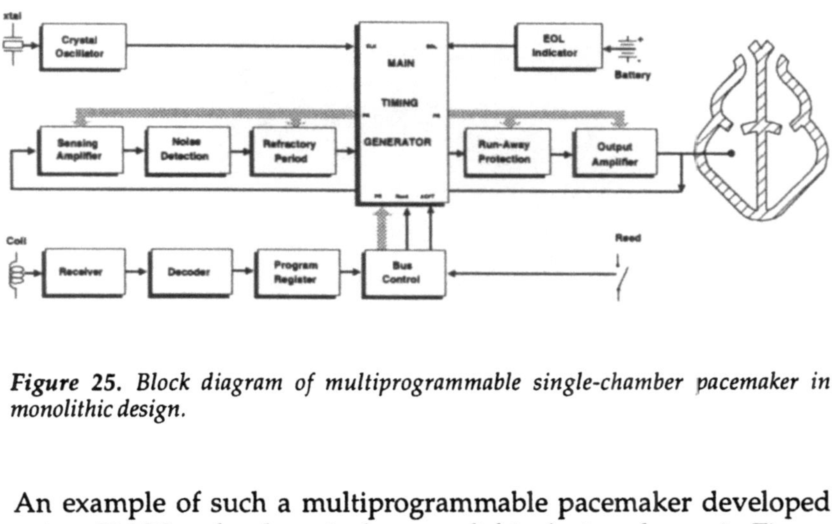

Figure 25. *Block diagram of multiprogrammable single-chamber pacemaker in monolithic design.*

An example of such a multiprogrammable pacemaker developed using CMOS technology is the monolithic design shown in Figure 26. All analog and digital functions are integrated on a silicon chip 4 mm x 4 mm in size. The control signals are amplified in a two-stage band-pass amplifier with eight sensitivity ranges. The pace pulse amplitude is generated by means of a voltage multiplier and can be adjusted between 2.4 and 9.6 V. The topographic arrangement shows the position of the signal-, program-, and output-amplifier, as well as the position of the memory, the oscillator, the reference voltage, the various digital components, and the programmable logic array (PLA). 3 μm silicon-gate technology is used. On an area of 16 mm^2, about 5,000 gates and transistors perform the pacemaker functions. When inhibited, the circuit consumes 5 μA operating current; under a standard program with a pacing rate of 70 min^{-1} and an output voltage of 4.8 V, the circuit consumes about 15 μA. The operating voltage range of the integrated circuit, 1.5 V to 3.0 V, is matched to the lithium battery technology. Bidirectional telemetry in conjunction with a programming unit forms the interface between the pacemaker and the physician. By means of this unit, the user is guided through the programming process.The parameters and the settings are tested for consistency and for patient safety prior to transmission. The implant can be interrogated to ascertain the pacemaker type and the active stimulation program, and an automated pacing threshold test can also be performed.

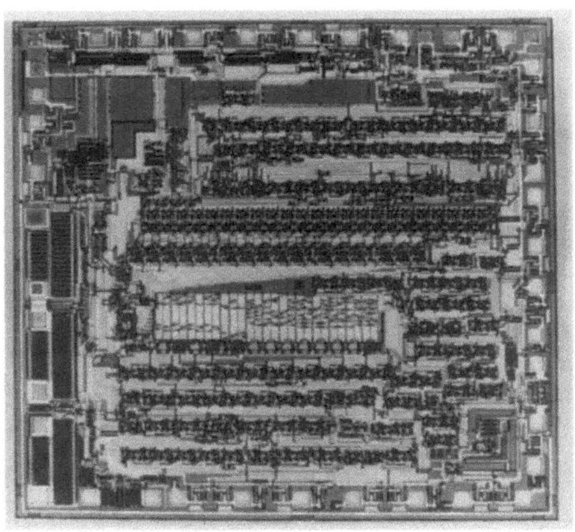

Figure 26. *Monolithic design of a single-chamber pacemaker using 3μm-Si-gate technology (Biotronik Nanos).*

Figure 27 shows the assembled unit. Hybrid technology is used to establish the connections between the monolithic circuit and the passive components of the pacemaker electronics [41, 42]. The hybrid is manufactured using thick-film technology. The integrated circuit is hermetically encapsulated in a chip carrier. A coil situated under the hybrid substrate is used for bidirectional telemetry. The lithium battery is located in the left part of the housing. The pacemaker housing consists of a two-part titanium capsule that is corrosion-proof for long periods and is hermetically sealed by means of laser- or electron-beam welding. The cast epoxy-resin head with a connecting block and a hermetically sealed feedthrough forms the electrode connection block. Further miniaturization of the implantable unit was achieved by using chip-and-wire technology (Figure 28). The result is a very small implantable device that has full complex functionality that provides high reliability and long-term stability.

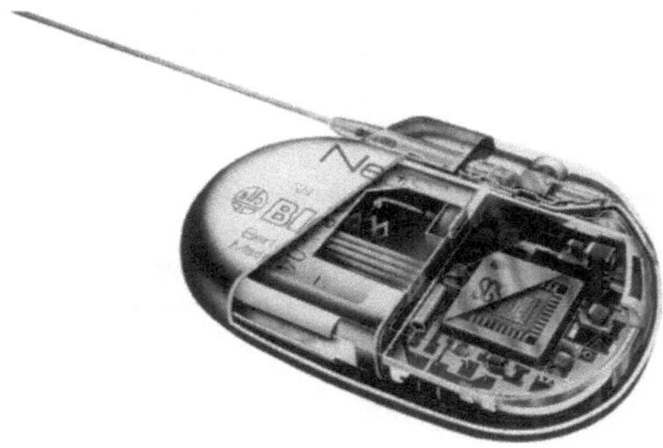

Figure 27. *Multiprogrammable single-chamber pacemaker, assembled unit (Biotronik Neos).*

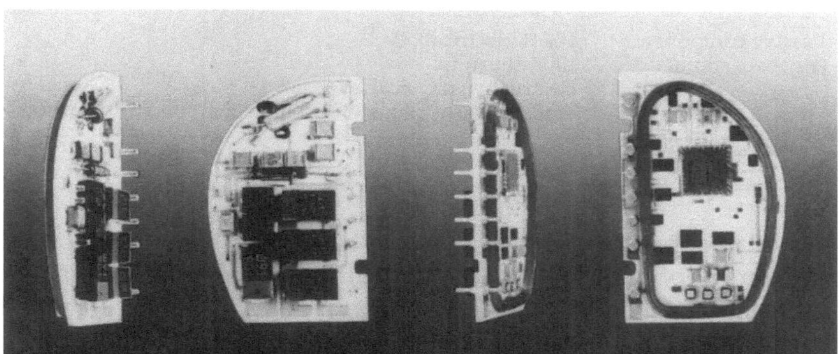

Figure 28. *Hybrid circuitry of multiprogrammable pacemaker using chip-and-wire technology (Biotronik Nanos).*

The high quality of these monolithic circuits and of their hybrid structure has been clinically proven in more than 500,000 implants. Reliability of $\lambda = 10^{-7}h^{-1}$ with a confidence level of 90% for the hybridized overall circuit and $\lambda = 10^{-9}h^{-1}$ for IC's and passive components has been achieved [43]. In pacemaker technology, reliability cannot be realized by the usual methods of redundancy of critical circuits and components since this would entail increased power consumption. Due to the volume limitations on battery capacity, the operating-time requirements can only be met if the power consumption of the circuit is low. The quality requirements can only be fulfilled by a high degree of integration, good design practices, and the consistent application of control measures in the various phases of manufacture. These include a 100% final testing of the components, semi-finished goods, and the finished product. The quality control measures that have been established for the manufacturing of pacemakers are listed in Table VII.

Table VII
Reliability Through Quality Control

Component type	*Test/Inspection procedures*
LSI, VLSI, transistors, diodes	Component qualification 100 % electrical wafer test 100 % SEM metallization analysis (MIL-STD-883/2018) 100 % visual inspection of the chip (MIL-STD-883/2010) Metallization and bondability Traceability (diffusion and metallization lots) Continuous process control Group B and Group C testing
Passive components (*e.g.*, capacitors)	100 % electrical test 100 % burn-in Destructive physical analysis Traceability of all materials in the manufacturing process

The burn-in procedure on a 100% basis has proven to be an important instrument of quality control. According to experience, the failure probability of the electronic components is given by a trough-shaped curve. The failure rate is relatively high in the beginning, then remains at a relatively low level for a long time, and then rises again only at the end of the lifetime. The burn-in serves the purpose of advancing the device past the initial period of high failure probability. It is clear that the reliability of the implant and the safety of the patient require a profound understanding of all technical and medical details of the therapeutic device.

Multiprogrammable Dual-Chamber Pacemaker

With intact SA impulse formation and conduction, but in the presence of AV conduction disturbances, electronic bridging of the AV block through P-wave-controlled pacing of the ventricles (VAT) is the obvious choice. Abnormal excitations, however, such as prematurely occurring supraventricular or retrogradely conducted ventricular extrasystole, ectopic ventricular rhythms, etc., severely restrict the broad application of VAT pacing. Pacemaker-induced parasystoles can be prevented only by additional control functions for ventricular pacing. Consequently, dual-chamber pacemakers with an atrial and a ventricular control circuit have been developed

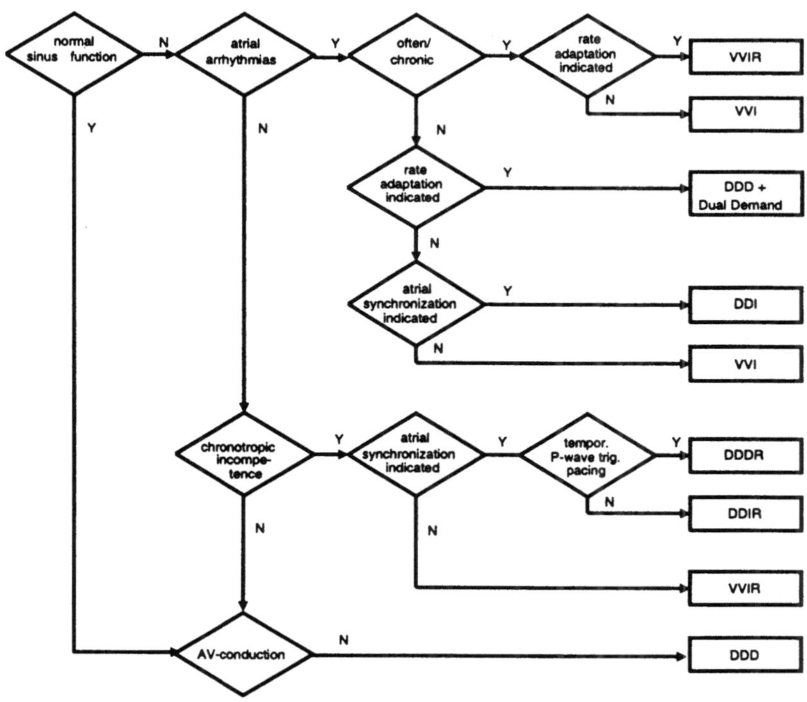

Figure 29. Logic diagram of relationship between rhythm disturbances and therapeutic pacing mode.

(DDD). These devices provide physiological pacing of the atria and ventricles under nearly all conditions of disturbance in impulse formation and conduction. However, in the case of sick sinus syndrome, the rate cannot be physiologically adapted to stress conditions (Figure 29).

As with a single-chamber unit, multiprogrammability of all control and pacing parameters allows adaptation of the pacemaker to the individual-electrophysiological and hemodynamic requirements and facilitates the management of the patient throughout the service time of the device. In a universal dual-chamber pacemaker, the atrial or ventricular pacing and control can be selectively switched off. Thus, this pacemaker allows the programmability of all known single- and dual-chamber pacing modes (AAI, VVI, AOO, VOO, DVI, VDD, VAT, DAD, DDI; see Table VIII [44]).

Table VIII
Programmable Parameters for Dual-Chamber Pacemakers

Parameter	Diagnostic Benefit	*Therapeutic Benefit*
Pacing modes DDD with and without Dual Demand Modality	Evaluation of atrial arrhythmia tracking	Physiologic pacing in high degree AV block (both modes) even in patients with intermittent atrial arrhythmias (Dual Demand)
Pacing mode DDI	Hemodynamic studies	AV-synchronous pacing in the absence of spontaneous atrial activity
Pacing modes with triggered response	Sensing tests; performance of series drug testing and hemodynamic studies (in combination with chest wall stimulation); retrograde conduction test	Acute therapy of tachyarrhythmias using chest wall stimulation
Pacing mode DDD with AVT modality	Electrophysiological studies	DDD pacing with the additional benefit of preventing supraventricular reentry tachycardia

(table continued next page)

Table VIII
Programmable Parameters for Dual-Chamber Pacemakers
(continued)

Parameter	*Diagnostic Benefit*	*Therapeutic Benefit*
AV delay pace and sense, Dynamic AV delay	Hemodynamic studies; testing for retrograde P-wave sensing	Hemodynamic optimization; compensation of latency time; prevention of pacemaker mediated tachycardia by (automatic) prolongation of the post-ventricular atrial refractory period
Safety AV delay	Crosstalk evaluation	Prevention of inappropriate inhibition of ventricular pulse in event of crosstalk
Blanking time	Crosstalk evaluation	Prevention of crosstalk
Atrial refractory period	Testing for retrograde conduction initiating pacemaker mediated tachycardia	Prevention of pacemaker mediated tachycardia
External pulse triggering by coded signals	Pulse triggering independent of pacemaker timing cycles for electrophysiological studies	Acute therapy of tachyarrhythmias
Temporary programming	Test programs for follow-up	Optimization of pacing therapy
Measurement of battery and lead parameters	Information on current consumption and remaining service time	Optimization of pacing therapy
Event counter/rate histogram	Testing for effectiveness of drug therapy	Adaptation of drug therapy; optimization of pacemaker and drug therapy
Automatic IEGM storage	Arrhythmia analysis; performance analysis of pacing system	Optimization of pacemaker and drug therapy
IEGM-transmission with marker signals	Arrhythmia analysis; pacing system performance analysis; pacemaker ECG interpretation	Optimization of pacemaker and drug therapy

In a DDD pacemaker, the ventricles are synchronized with the atria by technical means in analogy to the natural excitation sequence. Detection and pacing are time-synchronized with the depolarization events in the atria [45]. The AV delay, as well as the atrial and ventricular refractory periods, control the pacing sequence. Automatic adaptation of the AV delay to the heart rate permits optimization of the hemodynamic contribution of the atrium toward filling the ventricle (Figure 30). The same holds true for the shortening of the AV delay after a detected P-wave (AV delay fallback) to compensate for the latency time between the atrial pulse and its response.

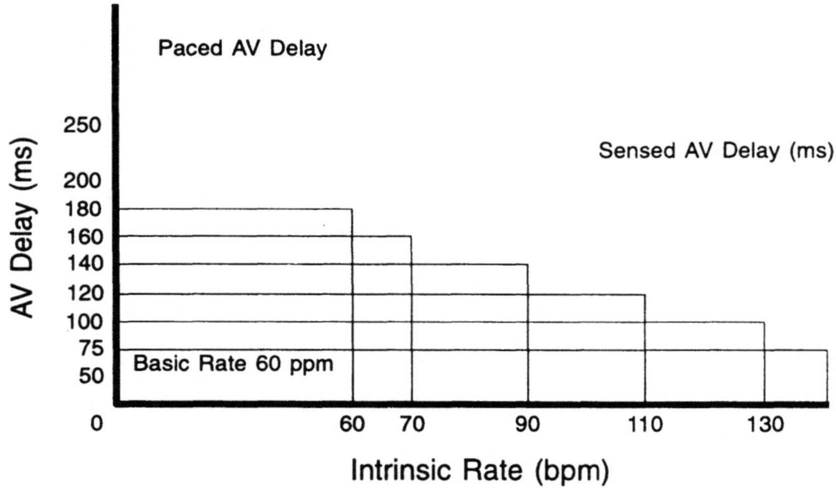

Figure 30. Dynamic AV delay; the AV delay is dependent on the sinus rate.

The occurrence of arrhythmias, such as the consequence of ventricular extrasystole or abnormal conduction of excitation, can be prevented by selecting suitable control periods. For the majority of clinically significant rhythm disturbances, this permits automatic adaptation of pacing. For the treatment of certain supraventricular reentry tachycardia, special control programs offer the possibility of shifting the depolarization of the myocardium ahead of the rhythm cycle, thereby preventing or terminating the tachycardia.

Technical Achievement of the Dual-Chamber Pacemaker

Like the single-chamber pacemaker, the universal DDD pacemaker consists of analog and digital subassemblies. The main structural difference is that the dual-chamber pacemaker requires two parallel channels for processing of the intracardiac EGMs and for pacing (Figure 31). As an example, a hybrid circuit of a multiprogrammable

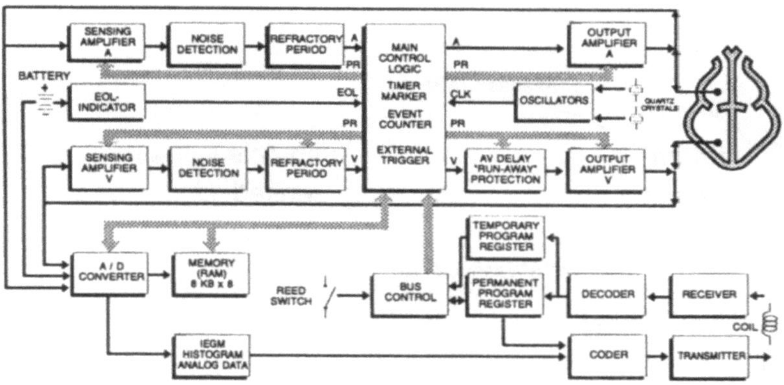

Figure 31. Block diagram of a multiprogrammable dual-chamber pacemaker (DDDM) permitting IEGM analysis and containing a patient data memory (Biotronik Diplos).

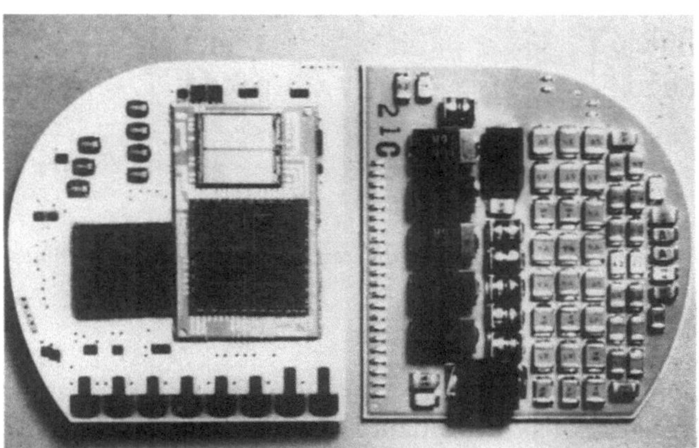

Figure 32. Hybrid circuit of multiprogrammable dual-chamber pacemaker, front and back sides, showing modular design concept (Biotronik Diplos).

dual-chamber pacemaker is shown in Figure 32. The multi-stage band-pass amplifiers for the acquisition of the intracardiac atrial and ventricular signals, together with the analog data-processing circuits, the voltage multiplier, the output stage, and the telemetry unit, are collected together on one *integrated circuit* (IC). The digital signal processing and the control unit are likewise designed monolithically. In conjunction with the separate memory chip (RAM), it forms the central logic and data acquisition unit. The capacitors and the input protection diodes are assembled on the back side of the hybrid

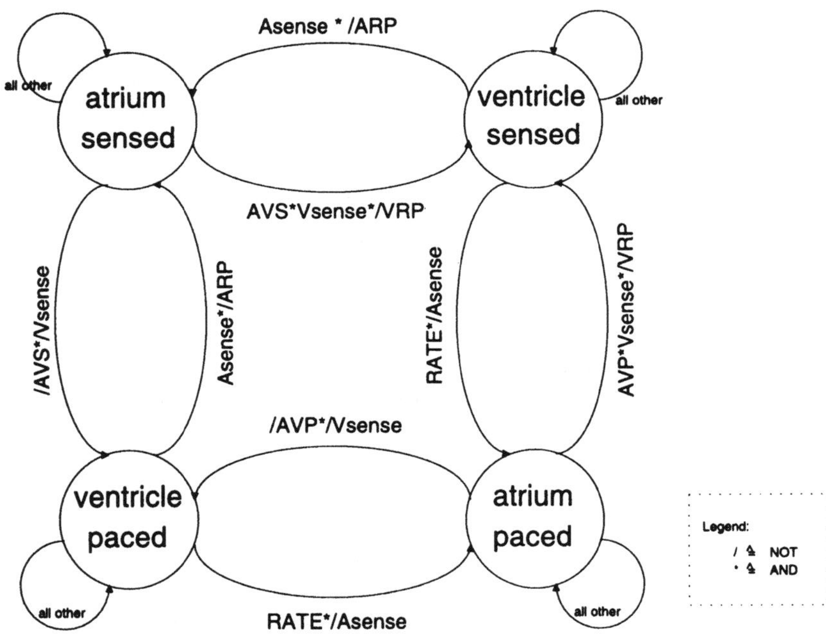

Figure 33. *Simplified state and flow diagram of a pacemaker programmed to DDD mode.*

circuit. The technical task of reestablishing the natural excitation process consists of synchronizing the ventricles with the atria. The control logic circuitry evaluates signals from the atrium and the ventricle, and ensures pacing in accordance with need [46]. In the dual-chamber mode, states "sensed" and "paced" can be formally defined for both chambers so that four basic states with their transitions can be distinguished (Figure 33). The state machine changes states depending on the heart's intrinsic activity and the program parameters, thereby affecting the signals described in Figure 34.

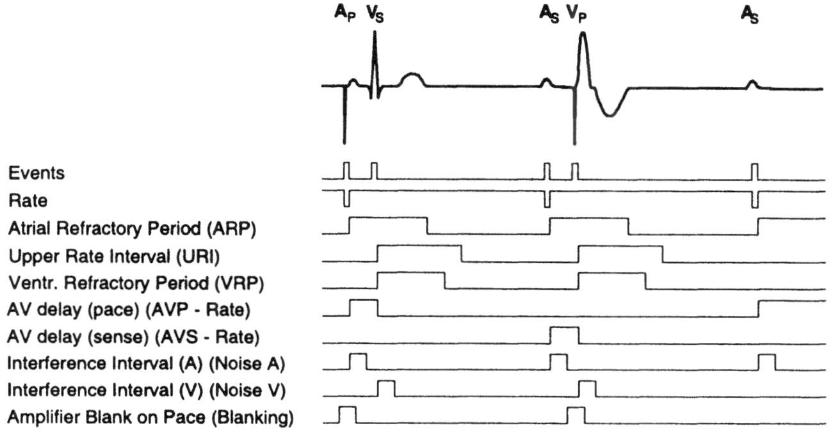

Figure 34. *Correlation between ECG signals and pacemaker timing.*

The state sequencing is performed by the control unit (controller). The controller is a state machine, the essential components of which are the state register and the random logic block (Figure 35). The state register stores the current state of the system and also conducts this information to the logic block. The latter, depending on the current state and on its other input signals, operates a collection of programmable timers (timer file) and, if necessary, changes the state upon the next cycle of the system clock. A timer that has been started in this way will put the state machine into another state after the time has elapsed and/or will trigger other timers through the state machine. Furthermore, the logic block of the state machine generates signals for the output register that controls the analog circuit parts in accordance with its programming. The various execution sequences of the state machine are selected through the mode-control register. For example, in the DDD mode of operation, the basic execution sequence is shown in Figure 33 with the time behavior described in Figure 34.

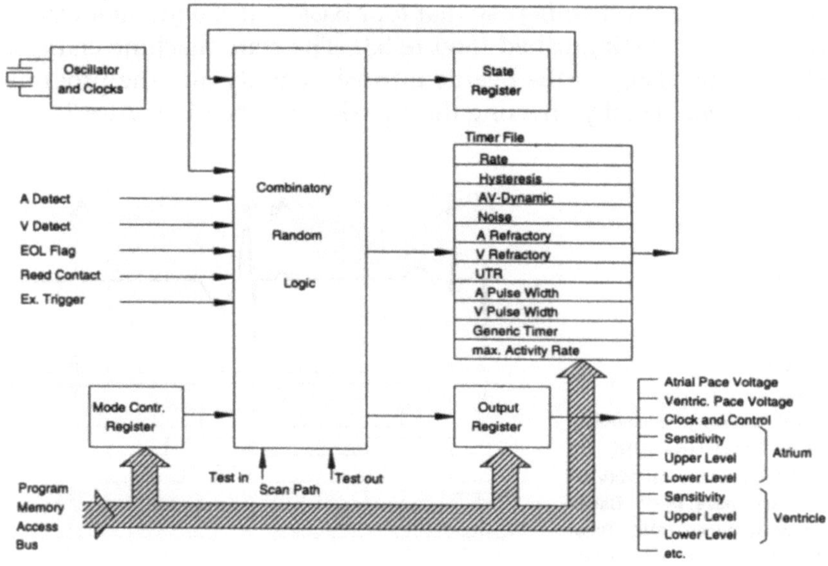

Figure 35. *Block diagram of digital control functions.*

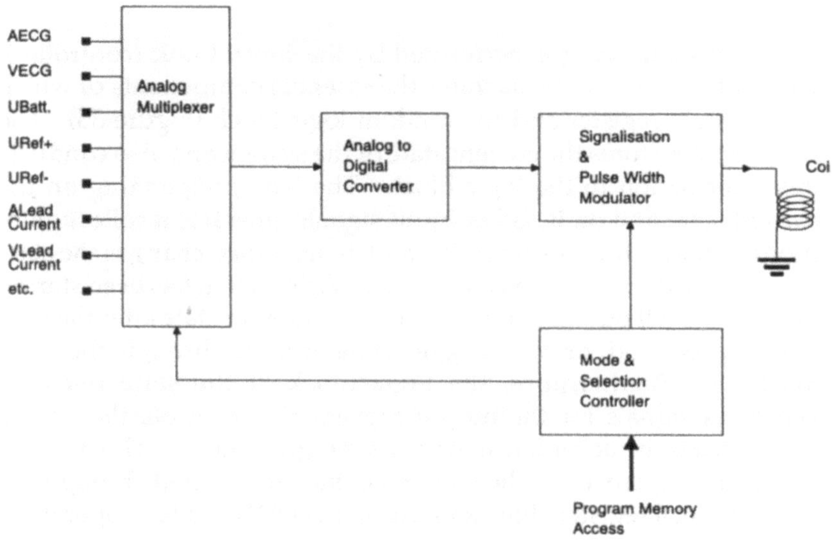

Figure 36. *Block diagram of analog telemetry circuit.*

In addition to dual-chamber pacing and multiprogrammability, the unit also offers rhythm monitoring through internal analysis and transmission of intracardiac electrograms (Figure 36). Thus, data are available to evaluate the therapeutic effectiveness of the artificial conduction system, and these data can be retrieved at any time from the implant. Arrhythmias are analyzed in real time; in addition, up to three atrial and ventricular electrograms can be stored according

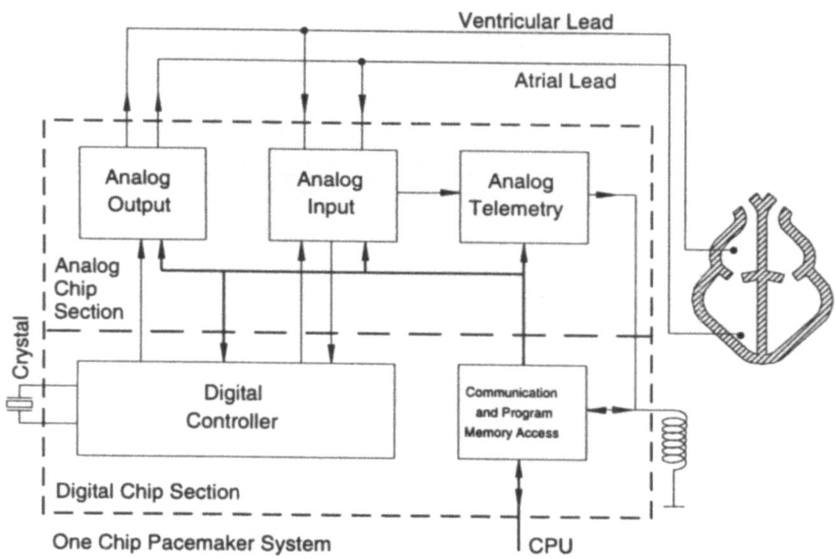

Figure 37. Simplified block diagram of the universal dual-chamber pacemaker.

to preselectable criteria. Several event counters are provided to monitor the pacing and sensing activity, and are used for rhythm diagnosis. Detection of intracardiac electrograms, such as the occurrence of pace and sense events, can be stored in the form of a rate histogram. For example, such a histogram provides information concerning the effectiveness of antiarrhythmic drug therapy. The pacemaker includes the feature of being able to serve as an electrophysiological measurement system. By means of the programming unit, a noninvasive connection to, and synchronization with, an external stimulator can be established. This and the plurality of programmable control and pace parameters conveys the complexity of the system (see Figure 37).

Using a new integration technique, the complex functions of a modern dual-chamber pacemaker can now be implemented on a single chip. This further miniaturizes the pacemaker and allows more efficient production (Figure 38). The reduction of components especially applies to the analog input amplifiers and the output

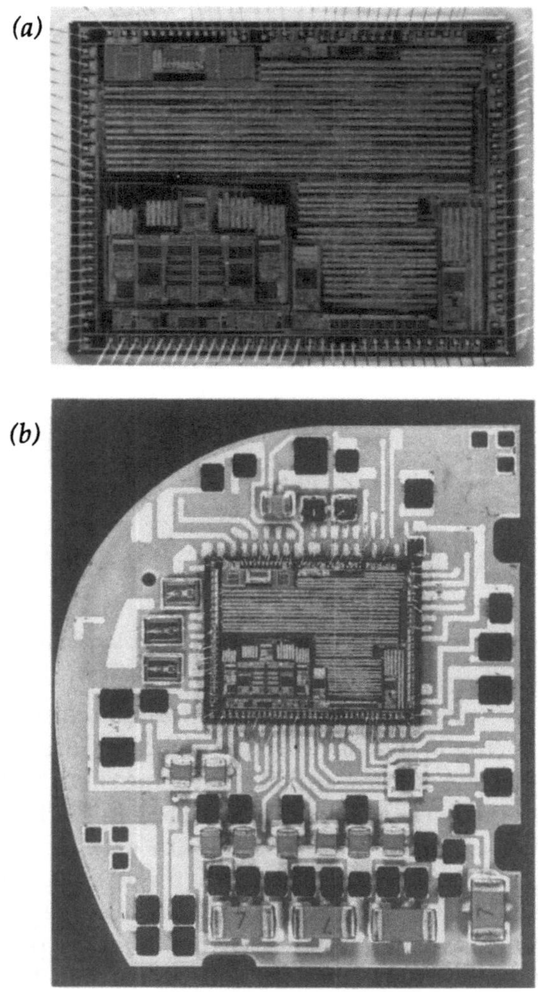

Figure 38. *(a) monolithic pacemaker chip and (b) hybrid circuit of a universal dual-chamber pacemaker (Biotronik Physios).*

stages. As a result, pacing amplitude and control sensitivity can be adjusted over a wider range and capture can be verified. The same applies to the supplementation of the dual-chamber principle by an additional physiological sensor that makes possible an important extension of the electrotherapy to include the chronotropic insufficiency associated with sick sinus syndrome.

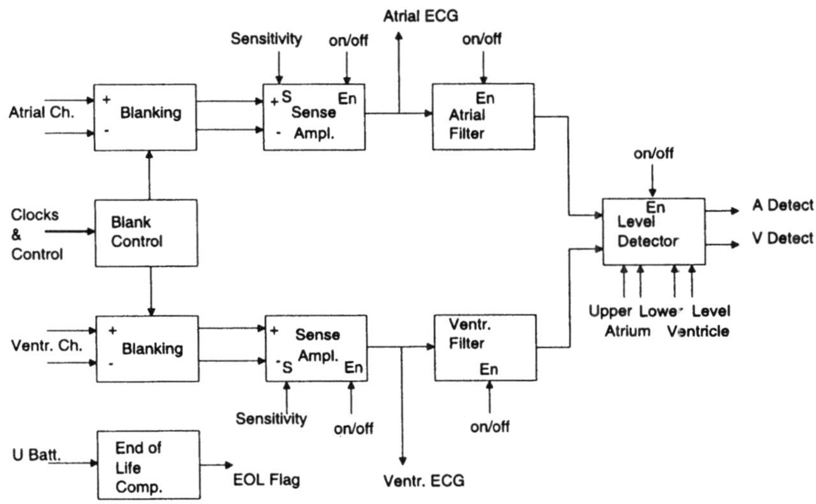

Figure 39. *Analog input circuit.*

Atrial and ventricular control are achieved using multi-stage band-pass amplifiers (Figure 39). These amplifiers, together with the analog data acquisition circuitry, the output amplifiers, and the telemetry unit, make up a significant part of the chip's area. The digital data-processing unit for the control and the signal analysis unit form the remainder of the integrated circuit. Signals are processed in the input circuit by multiplexing the "level detector" (Figure 39) for the atrial and ventricular channels. The same principle is used to generate the pacing voltage for the output stage (Figure 40). This results in a corresponding increase in circuit integration. The detection threshold and the pacing voltage can be independently programmed for the atrial and the ventricular channels. Sixteen detector thresholds and 128 pacing voltages are available.

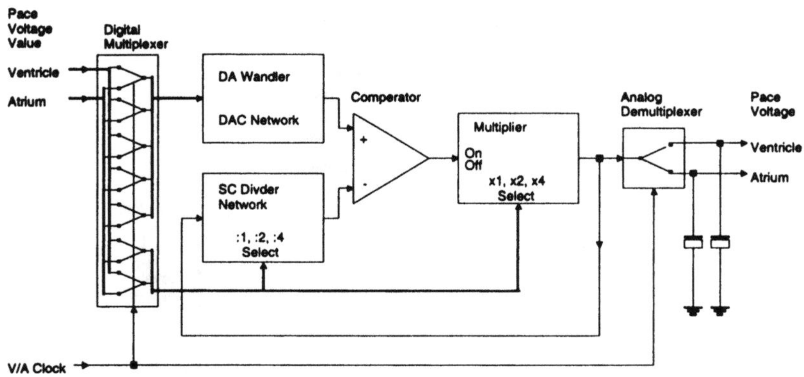

Figure 40. Analog output circuit.

All analog blocks are designed in *switched capacitor* (SC) technology [47-49]. Unlike standard operational amplifiers, the amplifiers used for the SC circuit of this chip do not have a voltage output but instead have a current output with an idealized transfer characteristic. They are known as *operational transductance amplifiers* (OTA), and have special advantages for low-power circuits in combination with SC networks. The reason for this is that no current surges occur when driving capacitive loads. The maximum current output of an OTA is limited by its design.

Figure 41 shows a combination of OTAs and switched capacitors that serve as input amplifiers of the intracardiac EGM channels. The first stage consists of a differential amplifier; the second stage takes care of the programmable adjustment of the amplification, which is determined by the ratio C_1/C_2 of the input capacitance to the feedback capacitance. The multiphase drive of the transfer gates, which are symbolized as switches, keeps in equilibrium the stages that operate dynamically by virtue of their structure. The four capacitors in the feedback branch of the second stage can be switched in a parallel manner. Their binary weighting permits a 16-step sensitivity setting of the input amplifiers of the pacemaker.

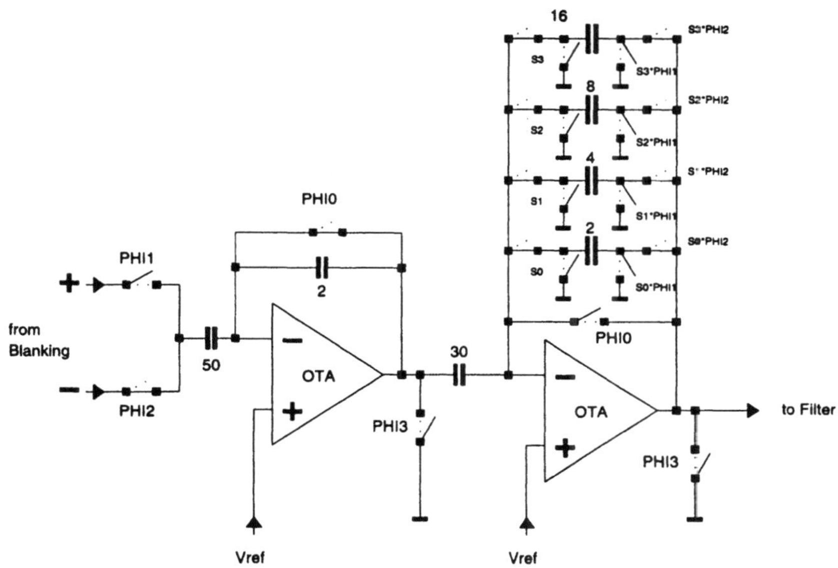

Figure 41. Detection amplifier in switched capacitor technology.

The dynamic circuit technique, characterized by a combination of OTAs and SC networks, is consistently used for the remaining analog circuit blocks. Especially emphasized are the filters of the input circuit, the AD converter for the analog telemetry unit, and the DA converter in the analog output circuit. The filters have four high-pass poles (80 dB/decade) and two low-pass poles (40 dB/decade). The center frequency is 70 Hz for the atrial channel and 40 Hz for the ventricular channel. The AD converter has a resolution of 6 bits and operates according to the principle of successive approximation with automatic zero-point adjustment and measurement-range matching for weak and strong signals (for ECG signals or supply voltages). The DA converter has five binary weighted capacitors that can be switched in parallel, similar to those in the last feedback branch in the input amplifier.

The pacemaker offers not only a large number of programmable control and pacing parameters, but also the possibility of being used as an electrophysiological measurement system. With the aid of the programmer, the system permits linkage and synchronization with an external stimulator. The digital circuit of the pacemaker consists of two blocks, an execution control, and a communications interface. These are built in standard cell logic. For the first time, it is now possible to create the layout of this digital circuit completely automatically, using modern computer-aided design (CAD) software.

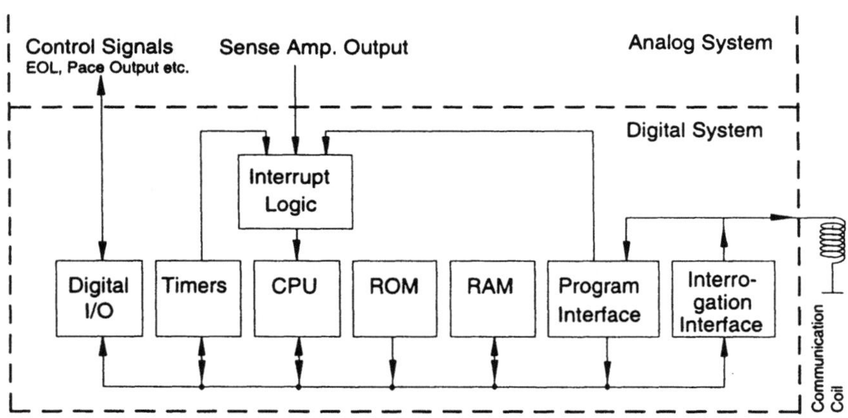

Figure 42. Microprocessor-controlled digital circuit of a pacemaker.

Operation of a dual-chamber pacemaker follows the principles given in Figure 42. However, the functionality of the execution control unit is much more extensive. The state machine has nearly 100 possible states. In conjunction with the output register, it allows more pacemaker functions to be introduced. These comprise automatic adaptation of sensitivities, thresholds, and pace impulse amplitudes with efficiency checks. Together with the timers, additional functions can be implemented, such as dynamic, rate-dependent AV delays. When developing the combinatorial logic of the state machine, its complexity was taken into account, and software tools were used not only for simulation but also for logic synthesis. This greatly shortened the development time. The circuit can be created automatically from a syntactical description of the state machine. Because the input description is automatically tested for consistency, manual intervention is no longer necessary. The occurrence of errors is thus reduced. The calculation is followed by a node analysis of the gate logic. This yields test patterns that guarantee a 100% test coverage of the logic block and that are set up via the test connections (scan path test) during the chip test.

A special protocol takes care of orderly access to the program memory and to the registers and timers of the system from multiple sides. Besides the telemetry coil connection for a wireless link to the programming unit, there is another interface available to connect a microprocessor for system expansions (Figure 42). Because of the process selected and because of the small transistor geometries (channel lengths about 1.75 μm), the chip size can be held within reasonable limits. Double-layer metallization is used to link the semiconductor structures. The idling current is about 6 μA. With atrial and ventricular pacing at a frequency of 60 min^{-1}, about 30 μA are drawn from the lithium battery (2.5 Ah). This yields an expected operating time in excess of 8 years. Despite the monolithic solution, the hybrid technology retains its role as a reliable connection technology for the wiring of the IC with non-integrated discrete components (quartz oscillators, tantalum capacitors, coils for transmission, reed contacts, etc.).

Controlled and Regulated Pacemakers

The development of future pacemakers aims at providing additional benefits to the patient. The pacemaker will not only sustain the patient's life but will also improve the quality of life [30]. Cardiac output is mainly influenced by physical exercise, metabolism, temperature (environmental as well as body temperature), emotion (or psychological stress), and posture. Increase in cardiac output is normally accomplished by an increase in both cardiac rate (from the resting rate of about 70 bpm to about 200 bpm) and stroke volume (from the resting value of about 70 ml to about 120 ml) [2].

For patients treated with pacemakers that stimulate the heart with a fixed rate, adjustment of cardiac output to different demands can only be achieved by an increase in stroke volume. Regulation is, therefore, not performed in the physiological way but is limited to a much lower maximum of cardiac output due to the restricted range for stroke volume. At present, there are different methods for the adaptation of the pacing rate in clinical use [30, 50]. Up to now, none of the systems has achieved a state comparable with the quality of physiological control. Therefore, the development of a rate-adaptive pacing system is still a challenging goal for interdisciplinary cooperation between physiologists, biomedical engineers, physicians, and the pacemaker industry. Analysis of the physiological control system offers the possibility of overcoming some fundamental problems.

CONTROL ASPECTS OF CARDIAC OUTPUT ADJUSTMENT

It must be taken into account that even the detailed description of a physiological system and its behavior can only be a simplified model of the reality. Nevertheless, analysis and modelling are very efficient and powerful methods to a better understanding of complex physiological systems and their behavior, and are essential in the design of those pacing systems using a physiological control mechanism to maintain homeostasis.

When blood is pumped into the peripheral circulation, the most relevant value is the *mean arterial blood pressure* (MABP) defined as the effective pressure for a whole cardiac cycle. Blood flow through the peripheral circulation is proportional to MABP and inversely proportional to the total peripheral resistance. The same relation is valid for the output of the left ventricle since cardiac output must be equal to the blood flow through the peripheral resistance.

Cardiac output is controlled in such a way that MABP is nearly kept constant even if the peripheral resistance changes considerably over a wide range. With regard to systemic hemodynamic regulation, cardiac output is the controlling quantity whereas MABP is the controlled quantity. All quantities that affect the controlled quantity, mainly via the peripheral resistance, can be called disturbance quantities. Changes of the peripheral resistance can be local or systemic. Carbon dioxide mainly affects the cerebral circulation whereas the catecholamines have systemic effects. Physical exercise influences the resistances in the working muscles. Heavy physical activity can raise regional muscular blood flow from 0.02 ml/g/min to 1.3 ml/g/min [51] or (in global values) from 1.2 l/min to 22.0 l/min [52]. This increase in blood flow is caused by a corresponding decrease in flow resistance whereas MABP is only slightly increased [53] in a way that is typical for a proportional controller.

It has been shown that even a significant reduction in the peripheral resistance causes an increase in cardiac output inversely proportional to the change in resistance [54]. This behavior can only be explained if MABP is the controlled quantity in a closed-loop system. Other evidence for MABP as controlled quantity is the fact that artificial stimulation of the heart with increasing frequency and constant peripheral resistance results in a decrease of stroke volume in such a way that cardiac output and, consequently, MABP remains constant [55, 56].

The primary functions of the circulation system have been briefly discussed in the preceding chapters. In the following section, the regulation aspects of **two** different mechanisms available for the physiological adjustment of cardiac output are presented.

Intracardiac mechanism (*i.e.*, the Frank-Starling mechanism) relates to the intrinsic ability of the heart to adapt itself to increasing loads of inflowing blood (preload) or to increasing output pressure (afterload). Under physiological conditions and with changes in the peripheral resistance, this mechanism is of minor relevance for cardiac output control. However, it supports the adjustment of the cardiac output of both ventricles to each other. Furthermore, this principle seems to be of major relevance for the adaptation provoked by changes in posture (orthostatic effects on the preload of the right heart) or caused by resistance variations in the pulmonary circulation (afterload for the right and preload for the left heart).

Cardiac Innervation. As a result of the cardiac innervation by the autonomic nervous system, sympathetic effects cause an increase in heart rate by increasing the frequency of impulse generation in the sino-atrial node (positive chronotropy) as well as in the force and speed of myocardial contraction (positive inotropy). The vagus, the main representative of the parasympathetic nervous system, significantly reduces the rate of spontaneous depolarization in the sino-atrial node (negative chronotropy); however, it does not noticeably affect the contractility of the ventricular muscle. The effect of both components of the ANS on the AV node (dromotropy) can be ignored for this consideration. Together, both nervous systems determine the heart rate under resting conditions. Complete inhibition of the parasympathetic nerves allows the heart rate to rise from its normal value (72 bpm) to about 160 bpm, whereas complete inhibition of sympathetic influences on the SA node allows the heart rate to decrease from its normal value (72 bpm) to about 60 bpm [2].

The effect of the sympathetic nervous system is transmitted to cellular receptors by norepinephrine whereas the transmitter of the parasympathetic nervous system is acetylcholine. The catecholamines, norepinephrine and epinephrine, are also secreted by the adrenal medulla following sympathetic stimulation. The two hormones are transported in the circulation. Their effect in the heart is comparable to the direct sympathetic influence. In the peripheral circulation, the catecholamines especially increase the flow resistance in skeletal muscle. Release of the catecholamines is considerably enhanced during emotional stress situations.

If the peripheral resistance begins to change and, thereby, causes a deviation in MABP, this deviation is detected by pressure-sensitive receptors. Mechanoreceptors for blood pressure monitoring are found in:

- the systemic arterial tree at specific sites,
- the pulmonary trunk and its division in the right and left pulmonary arteries,
- the atrio-caval junctions and the pulmonary veno-atrial junctions, and in
- the atria and ventricles.

Of major importance for blood pressure control are the systemic arterial receptors; of minor importance are those in the atria and ventricles. The systemic arterial mechanoreceptors are called baroreceptors. They are located in the carotid sinus, the aortic arch, the root of the right subclavian artery, and in the junction of the thyroid artery with the common carotid artery. The signals of the receptors are transmitted to the medullary cardiovascular center. The center receives not only signals from the baroreceptors but also from other receptors. This information is transformed by the center into signals that are sent via the efferent pathways of the ANS to the sino-atrial node and the myocardium. The effect is such that the MABP is kept constant with only slight deviations that are typically found in proportionally operating controllers. This negative feedback ensures stability of the whole system.

The physiological mechanism described here is capable of responding adequately to acute changes in the hemodynamic situation. For the long-term regulation of cardiac output, other mechanisms must be considered [57, 58]. In patients with pacemakers of fixed pulse rate (*e.g.*, sick sinus syndrome patients), only the sympathetic influence on inotropy is still effective. The sympathetic tone enables the ventricles to adjust the stroke volume. This effect, possibly supported by the Frank-Starling mechanism, results in an increase in cardiac output to a limited extent, which is significantly lower than in healthy people.

Excessive increase of stroke volume, however, can include the risk of over-exertion of the myocardial contractility reserve. There is clear evidence, however, that in spite of chronotropic insufficiency control signals concerning the hemodynamic situation are sent from the medullary cardiovascular center and received at the heart. This means that the signal for the adjustment of the controlling quantity is still detectable in the action of the heart.

Different Strategies of Rate Adaptation

From control theory, it is well known that different strategies may be utilized in order to eliminate the effect of disturbance variables on the controlled quantity. Some of these strategies will be briefly explained in the following. Figure 43 shows a schematic diagram that is based on the relevant physiological control circuit. All variables that affect the mean arterial blood pressure, with the exception of stroke volume and heart rate, are called disturbance variables. Most of these disturbance variables influence the peripheral resistance. Only blood volume (*e.g.*, changed by blood loss) and orthostasis affect directly arterial blood pressure. Orthostasis that depends on posture influences both blood distribution within the circulation, with a direct effect on blood pressure, and to some extent,

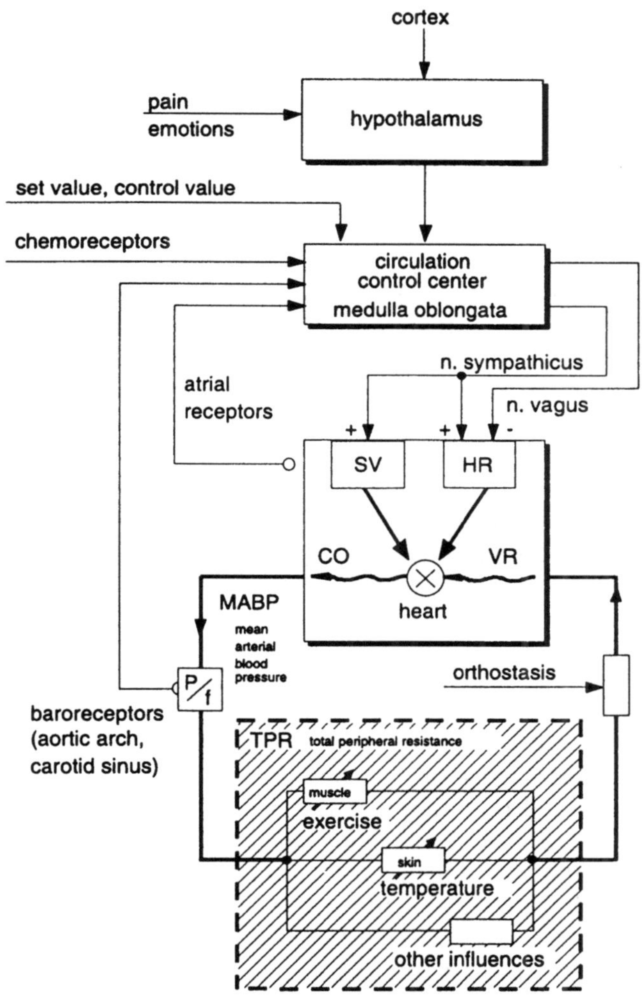

***Figure 43.** Schematic diagram of the physiological closed-loop control system for mean arterial blood pressure (MABP). The controlling quantity is cardiac output CO = SV x HR with SV = stroke volume and HR = heart rate, respectively. For steady state, CO = VR with VR = venous return. All parameters with the exception of CO that affect MABP are disturbance variables. Most of these disturbance variables affect MABP by their effect on the total peripheral resistance (TPR). The control loop is closed through the pathway from the baroreceptors via the medullary circulation control center with its nervous connections (n. sympathicus, n. vagus) to the sinoatrial node (rate) and the myocardial muscle (contractility).*

also peripheral resistance. Additionally, peripheral circulation is supplied with sympathetic signals causing vasoconstriction.

Forward Propagation of Disturbance Variables

If the effect of any disturbance variable on the controlled quantity (*i.e.*, cardiac output) is exactly known, a direct pathway from the sensor via a controlling system (the pacemaker with integrated transfer characteristics) to the controlled system (the heart) can be used to compensate the influence of the disturbance variable. It is obvious that not only must the static transfer characteristic be considered, but also dynamic behavior in order to avoid positive feedback that may result in instability.

The most relevant disturbance variable is muscular exercise. There is a strong and linear relation between workload and heart rate [53]. This relation depends on individual features that must be considered for adequate rate adaptation. However, the direct and continuous measurement of workload is not possible during everyday activities. Hence, strong effort is directed to the indirect measurement of physical activity. One of these indirect methods is the measurement of vibration or motion energy related with physical activity. Pacemakers that make use of this strategy are, at present, the most frequently implanted rate-adaptive systems [50, 59]. However, this solution does not represent physiological closed-loop control. Only that disturbance variable which is measured can be considered. With a sensor measuring physical activity, it is not possible to also take into account, for example, the effect of environmental temperature or emotional stress. **In principle, all** relevant disturbance variables must be considered to overcome these problems. The outputs of the sensor elements must be integrated and combined so that their effect on the pacing rate approaches the physiological dependency. Due to insufficient information, this may not serve the intended purpose.

Open-Loop Control Using Corporeal, Non-Cardiac State Variables

Some systemic or corporeal non-cardiac state variables are affected by the same disturbance variables that cause the heart to increase its output, as demonstrated in Figure 44. These variables can be utilized for rate adaptation. The most relevant of these is the central venous

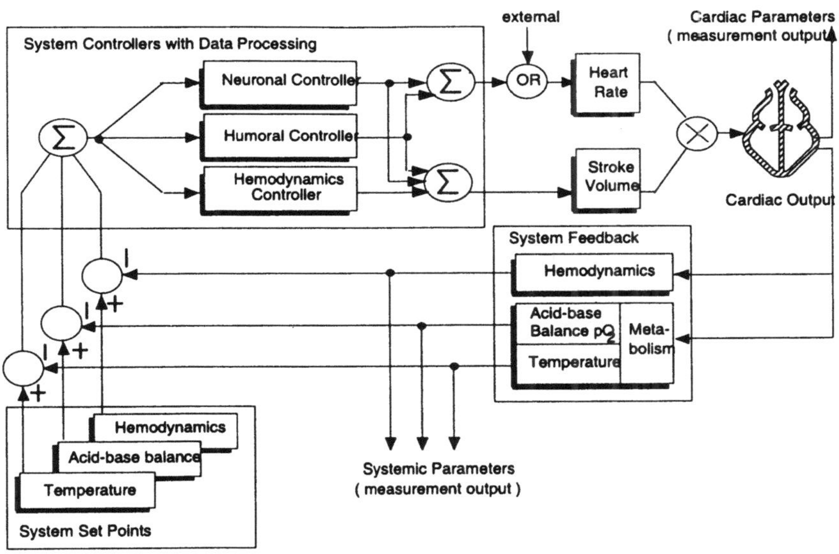

Figure 44. Schematic presentation to illustrate the interrelationship between cardiac output and systemic non-cardiac state variables.

oxygen saturation (decreased by muscular activity) [60] and the central venous blood temperature (increased by muscular activity and fever) [61, 62]. Respiratory variables, *e.g.*, oxygen uptake and ventilation, are related to metabolism associated with muscular activity and body temperature. Oxygen uptake is perfectly correlated with heart rate during physical exercise [63]. No parameter, however, adequately reflects in an integrated manner all disturbance variables. As a consequence, it would be necessary to monitor different systemic variables for rate adaptation, a solution that does not satisfy the integrating purpose of rate adaptation in a physiological way.

When control is achieved using an open-loop control strategy, the relationship between the measured variables and heart rate must be well understood. In physiological systems, this understanding must include assessment of the forward propagation of the disturbance signal. If the disturbance variable is dependent upon cardiac output, then the risk of positive feedback must be carefully evaluated.

The only variable in this group that allows establishment of a closed-loop control system is MABP, the controlled quantity. However, even if the measurement of blood pressure with a sensor implanted in the aorta could be achieved, another difficulty has still to be solved: the transformation of the pressure signal into the adequate pacing rate. The static transfer function of the physiological baroreceptors is non-linear, showing an S-shaped form. The receptors respond extremely rapidly to changes in arterial pressure. The number of output impulses increases greatly during systole and decreases during diastole. Furthermore, the receptors respond much more to a rising pressure than to a stationary pressure and also respond less to a falling pressure. Such a dynamic sensitivity is described as proportional-differential behavior. Hence, in order to ideally compensate the failure of the physiological system with the supporting system it might be necessary to simulate this control characteristic with the technical system, *i.e.*, the sensor and the pacing system. This approach to a closed-loop system, however, also considers other influences on the medullary circulation center (*e.g.*, signals received from the chemoreceptors, higher influences like pain, emotion) by their hemodynamic effects.

Rate Adaptation Using Cardiac Parameters

Some cardiac parameters are potential candidates for rate adaptation [64, 65]. The ideal candidate would be a parameter that reflects both the sympathetic and parasympathetic effect on cardiac output. Such a parameter is not available or can only be obtained indirectly.

The parasympathetic activity is mainly reflected in the sino-atrial rate and, thus, cannot be sensed in patients with chronotropic insufficiency. However, sympathetic activity is reflected even in those patients in different cardiac parameters (Figure 45).

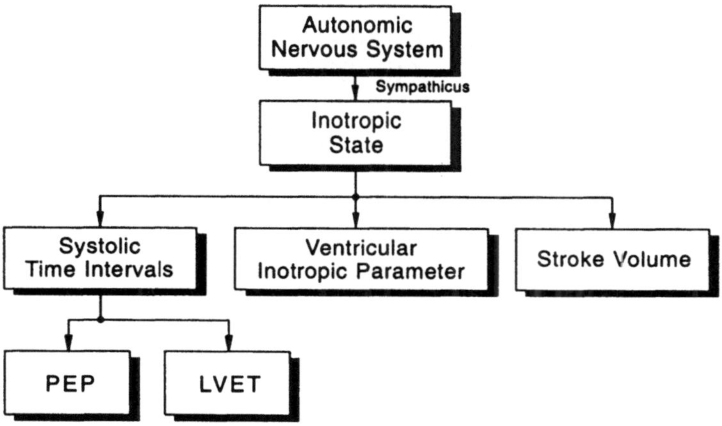

Figure 45. *Parameters containing ANS information.*

The sympathetic influence is reflected in the contractility of the ventricular muscle [31]. Contractility can be estimated by the time derivative of pressure (dp/dt) measured during the isovolumetric phase following the filling phase and preceding the ejection phase. The preferable site of measurement should be within the left ventricle. It should be remembered that contractility also depends on the intercellular calcium concentration. Factors which modify extracellulor or intercellular calcium may mimic the sympathetic influence.

In patients with chronotropic insufficiency, increasing the stroke volume is the only way possible to increase cardiac output. This increase in stroke volume is caused by the effect of sympathetic tone on myocardial contractility but, presumably, it is also supported by the Frank-Starling mechanism. Using the stroke volume, an algorithm can be employed to control heart rate in such a way that the actual increase in stroke volume is much lower than it would be without the rate adaptation. Stroke volume can be measured by impedance plethysmography [66-74]. However, although this method approaches the behavior of the physiological system by utilizing a balanced increase in both rate and stroke volume, it still provides open-loop control in regard to the ANS. The ANS information is also expressed in systolic time intervals that comprise the whole systole starting with the onset of ventricular electrical activity (indicated by the Q-wave of the ECG or, in pacemaker patients, by the stimulus). The pre-ejection phase (PEP) includes the delay required for the electromechanical coupling as well as the isovolumetric contraction phase. When the outlet valves open, the left ventricular ejection time (LVET) begins and continues as long as the valves remain open (see Figure 46).

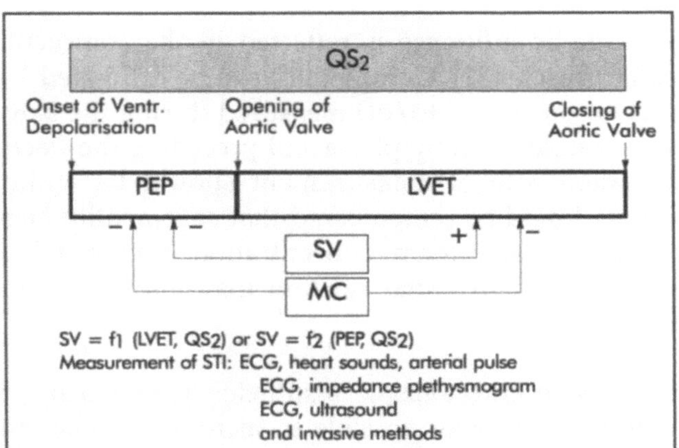

Figure 46. *Systolic time intervals.*

However, rate itself affects LVET so that the higher the rate the shorter the ejection time. A provoked increase in heart rate derived from a decrease in ejection time can further decrease the ejection time thereby causing positive feedback and the risk of overloading the ventricular muscle.

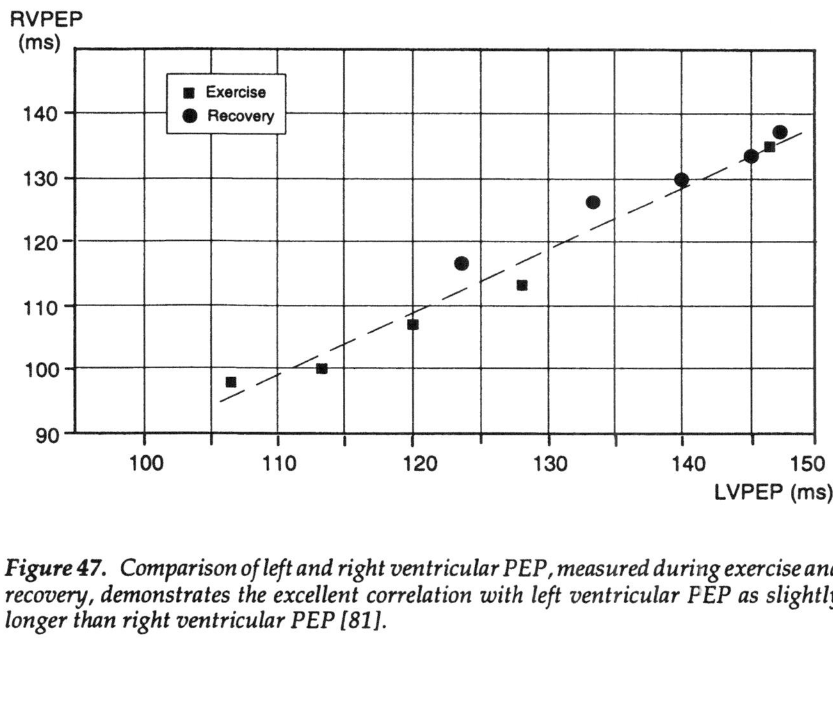

Figure 47. *Comparison of left and right ventricular PEP, measured during exercise and recovery, demonstrates the excellent correlation with left ventricular PEP as slightly longer than right ventricular PEP [81].*

Much attention has been given to the determination of the *pre-ejection phase* (PEP) [64, 65, 75-78]. The ejection phase of the left ventricle begins when the intraventricular pressure exceeds the aortic pressure thereby opening the aortic valve. Consequently, the PEP parameter is strongly related to contractility and similar arguments concerning the effect of calcium ions might by valid. Although PEP in the right ventricle is slightly shorter than in the left ventricle, both pre-ejection periods respond in a similar manner (Figure 47) [79]. PEP is not affected by heart rate. There is no risk of positive feedback [80].

The exact determination of PEP, however, requires methods that are not applicable in pacemaker technology. Therefore, a parameter PEP*, which can be derived from intraventricular conductance measurement, was introduced (see also the following chapters).

The *ventricular inotropic parameter* (VIP) is based on intraventricular conductance measurement. It can be determined from the conductance signal using a specific signal processing algorithm. Therefore, VIP is the parameter that most reliably represents ANS information according to the following criteria:

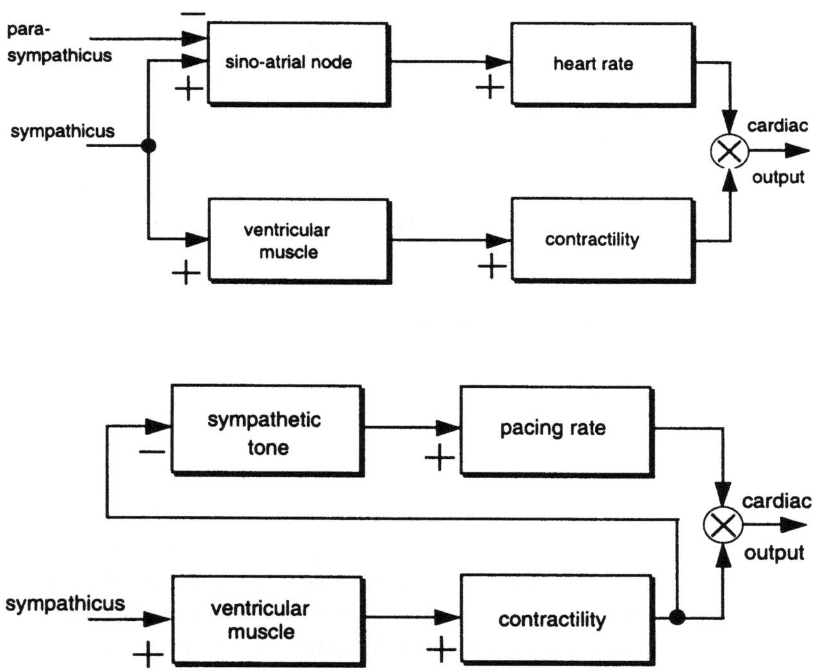

Figure 48. Sympathetic control of the normal heart (upper portion); in chronotropic insufficiency, the contractility, i.e., VIP, is used to reestablish physiological rate adaptation.

- it should be detectable at the moment the outlet valves are opening, but must be far enough away from the end of the systole to avoid any risk of positive feedback;

- it must be reliably identifiable under different conditions for the same patient (*e.g.*, rest, physical exercise);

- the same algorithm must be applicable in the majority of patients.

Parameters such as VIP or PEP representing contractility make possible a closed-loop control system for rate adaptation. This principle aims at the restoration of the physiological system and makes use of all components of the physiological system that are still functional within the range of recompensation (Figure 48). It uses the baroreceptors for the measurement of MABP, the medullary circulation center for the signal processing, and the transmitting process of the sympathetic nervous system to myocardial muscle. It bridges (only) the interruption between the autonomic nervous system and the SA frequency generator. Therefore, it can be considered a physiological approach although it does not take into account the parasympathetic tone.

Since, with the exception of the loss of parasympathetic tone in chronotropic insufficiency, the heart receives all physiological control information, no additional information is necessary. Even in the case of orthostatic load, which is of special interest since the change of posture with its hemodynamic effects can be performed in a very short time, the heart is supplied with all relevant information (1) by the sympathetic activity via the effect on the peripheral resistance and (2) by the change in preload due to pooling of blood in the lower extremities when posture is changed from supine to upright position. Under physiological conditions, the change from supine to upright position can cause a change in peripheral resistance of 30% and a reduction in central blood volume due to pooling by 400 ml.

As a consequence, heart rate is increased by about 30% and stroke volume is decreased by 40%, resulting in a decrease of cardiac output by about 25%. However, the controlled quantity, MABP, is only slightly decreased when the steady state has again been established [52].

In the following chapters, different pacemaker designs employing each of these rate-adaptation strategies will be presented.

STATUS OF THE APPLICATION OF CORPOREAL CONTROL PARAMETERS

Rate-adaptive pacemakers using different corporeal control signals have been developed. To achieve rate adaptation based on a sensor signal, a suitable signal transducer and a suitably robust algorithm must be available. Table IX summarizes the available sensor technology necessary to implement one of these adaptive strategies. The same high quality requirements imposed on the pacemaker also apply to the permanently implanted regulatory system. A stable and reliable measurement function must be maintained for many years. The sensor must fulfill the requirement of implantation and the same must apply to the power consumption. Advances in micro-

Table IX
Physiological Parameters for Pacemakers Providing Open- and Closed-Loop Control

Control Parameter	*Sensor*	*Evaluation Parameter*
P-wave	Atrial lead	Ideal closed loop
Respiration	Intrathoracic impedance	Metabolic open loop
QT-time	Stimulation electrode	Open loop
Activity	Piezoelectric sensor	Non-metabolic open loop
Motion	Piezoelectric sensor	Metabolic open loop
O_2-saturation	Optical transducer	Metabolic open loop
Blood temperature	Thermistor electrode	Metabolic open loop
PEP/VIP	Unipolar impedance electrode	ANS closed loop
SV, EDV	Multipolar impedance electrode	ANS closed loop

electronics permit the use of transducers in which the measurement and control variables are linked in a nonlinear fashion. As discussed below, there are numerous transducers that meet these requirements and may be incorporated into a rate-adaptive pacemaker [82].

Advances in technology permit a new definition of the medical-clinical requirements imposed on a pacemaker and, at the same time, open up a new arena of medical indications and therapies. In addition to eliminating a life-threatening state, the therapy focus has expanded to include improving physiological performance and further improving the quality of the patient's life. The patient is not the only beneficiary of microelectronics; the physician reaps the benefits of improved patient management, improved diagnostic capabilities, and greater therapeutic flexibility. More powerful support systems have been developed to assist in managing greater therapeutic flexibility. How the advances in technology have been applied to achieve high-performance, rate-adaptive pacemakers will be demonstrated with two examples.

Motion Energy as a Control Parameter

Thoracic motion is used to control a rate-adaptive pacemaker. A number of refinements of the principle have been made. They include improving the sensor to better detect motion associated with physical work and improving the algorithm to better discriminate between valid and invalid signals [59, 83]. An important improvement was to change the sensor from a frequency detector to an energy detector, in response to analyses showing that a direct proportionality exists between motion energy and the chronotropic heart rate. In many forms of exercise, activity of the large muscles accounts for the most important component of total body work and oxygen consumption. A measurement of the relative motions of the body mass therefore suffices for an estimate of whole body ergometry [84-86]. Considering the body as a system of coupled mass as shown in Figure 49, movement during exercise can be monitored by a vibration sensor located in the pacemaker implanted at the upper thoracic site.

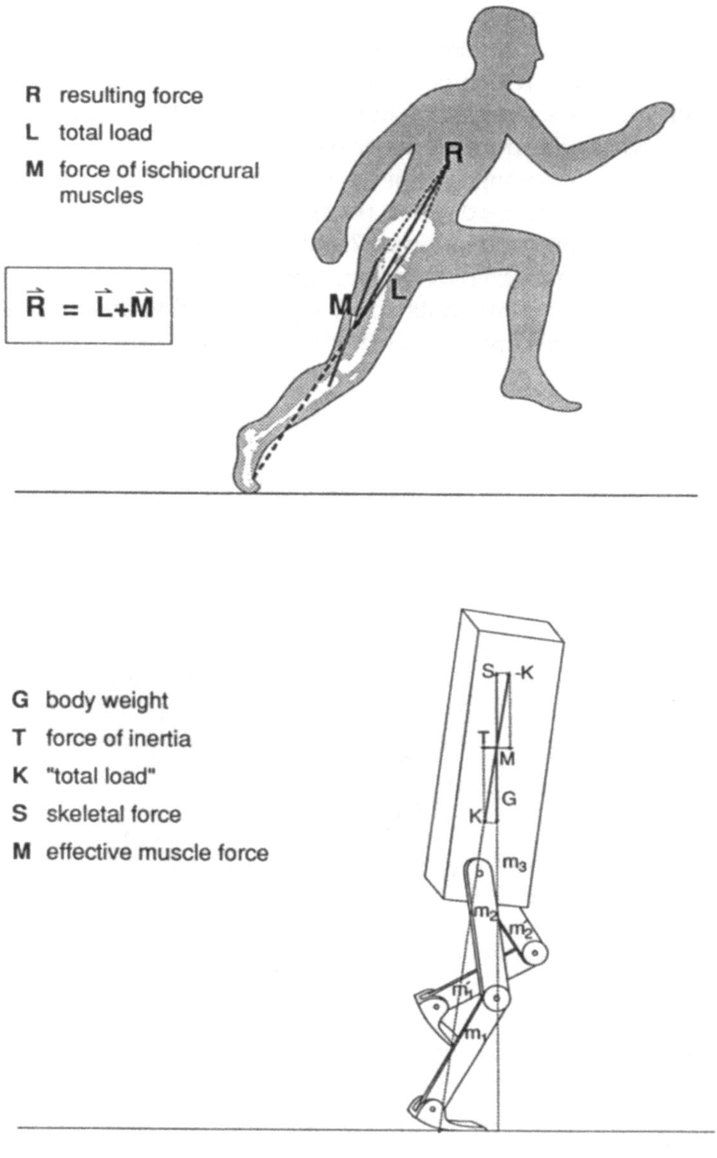

Figure 49. *Model of the interaction of forces as a result of movement of the large muscles under exercise.* UPPER FIGURE: *vertical axis: compulsive force (–) of ground. Lateral axis: pendulum motion (+) of trunk. Sagittal axis: displacement (+) of the center of mass.* LOWER FIGURE: *demonstration of the different contributing momenta detected by the motion sensor in the upper trunk.*

The following correlative chain is the basis for a motion-energy rate-adaptive pacemaker. Body motion is correlated with large-muscle activity, which is correlated with muscle O_2 consumption, which increases circulatory demand and must be supported by an increased cardiac output. To provide the higher cardiac output, the sensor-driven pacing rate is increased. The effectiveness of the motion-energy rate-adaptive pacemaker rests on the strength and selectivity of the correlations [87, 88]. The strength of the correlation is dependent on signal detection, signal processing, and the control algorithm. The constraints apply equally well to all corporeal control variables that are weak correlates of actual cardiac demand [89].

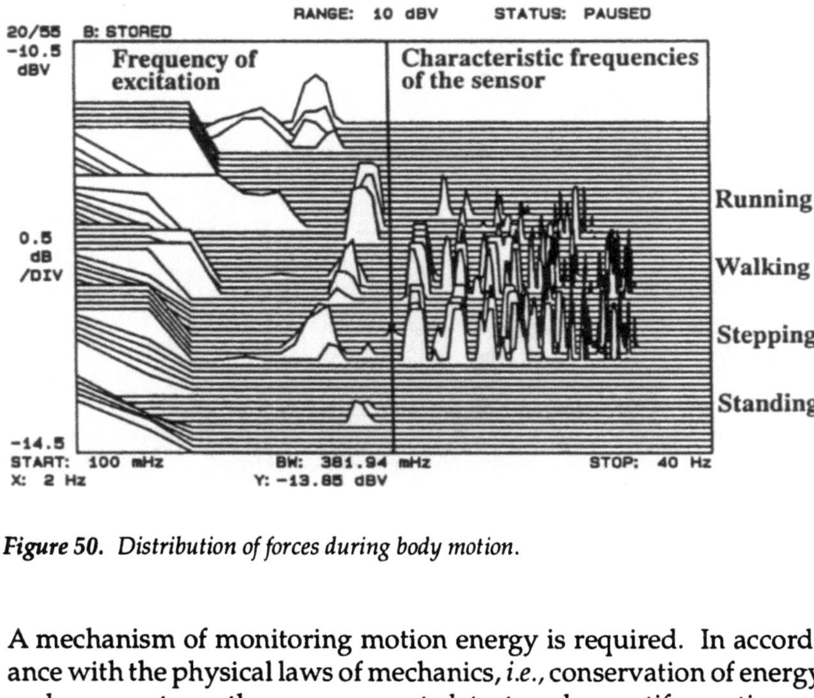

Figure 50. Distribution of forces during body motion.

A mechanism of monitoring motion energy is required. In accordance with the physical laws of mechanics, *i.e.*, conservation of energy and momentum, the sensor must detect and quantify motion or acceleration. Placing the motion sensor in the pacemaker housing allows it to detect motions in the largest mass of the body associated with the largest expenditures of energy. Pacemaker acceleration or thoracic tissue vibrations can be detected and used as indicators of energy expenditure by the patient. The sensor must be very stable and, as shown in Figure 50, allows the monitoring in the spectral range of motion-related signals. A sensor based on the piezoelectric effect meets these criteria and provides the advantage of no power requirement. This monitoring of thoracic acceleration for ergometry in sports medicine has been used for a long period of time [85]. The same principle can be extended to the control of rate-adaptive pacemakers.

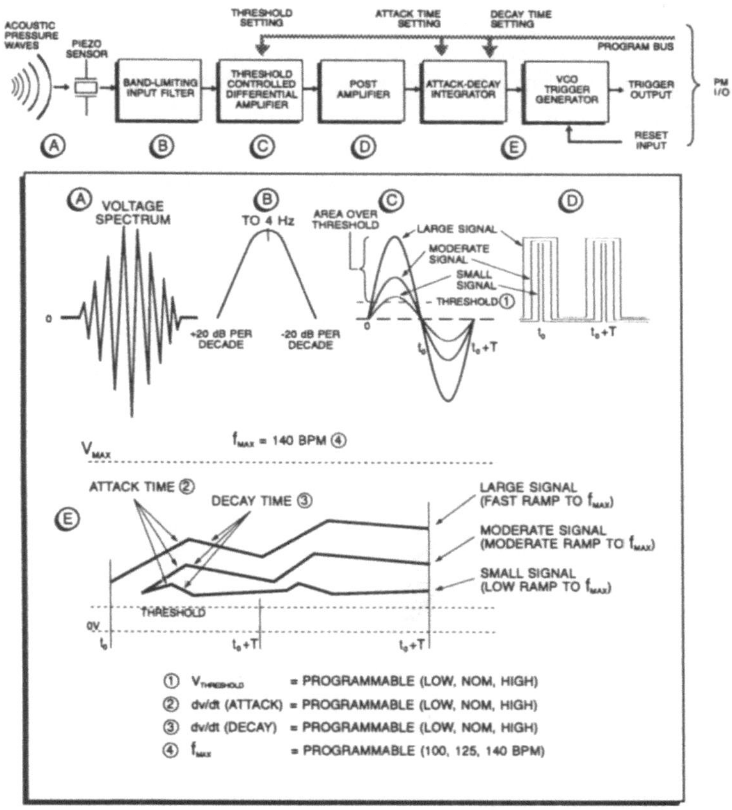

Figure 51. *Processing of corporeal motion signals gained by a piezoelectric transducer.*

In designing a sensor circuit, one must consider that the signal to be measured is dominated by frequencies below the resonant frequency of the piezoelectric transducer. Measurements made in this mode are proportional to acceleration. Therefore, the detected signal must be integrated to obtain a signal which is proportional to the motion energy. Figure 51 shows that the circuit for processing the transducer signal to achieve a rate-adaptive pacemaker consists of seven stages. The primary acceleration signal from the transducer must be filtered to extract only those frequencies that represent gross thoracic accelerations. The input filter has the form of a second order band-pass filter. The high pass characteristics are determined by the capacitive impedance of the piezoelectric sensor itself and the input impedance of the circuit. The low pass characteristics are determined by the subsequent amplifier. Since spectral filtering of the signal is not sufficient to achieve total energy specificity, a level-controlled amplifier provides rejection of low amplitude motions. The filtered acceleration signal passes through the integrator stage. The resulting integrated signal is an energy correlate that can be used to control the pacing rate via a *voltage-controlled oscillator* (VCO).

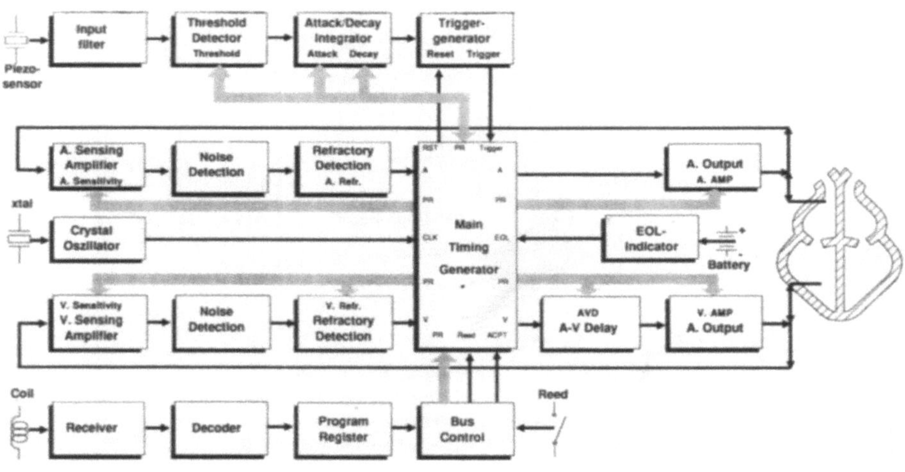

Figure 52. Block diagram of an implantable dual-chamber pacemaker with motion-controlled rate adaptation (Biotronik Ergos).

The technical design of the sensor and integrator circuitry permits single- and dual-chamber pacing. The combination of P-wave synchronous and sensor-controlled operating modes offers the advantage of a broad range of therapeutic strategies (Figure 52). The technical design has been implemented in a hybrid circuit shown in

Figure 53. The effectiveness of the sensor integration algorithm for the motion-energy-based, rate-adaptive pacemaker is demonstrated in Figure 54. The pacing rate is closely correlated with the exercise workload.

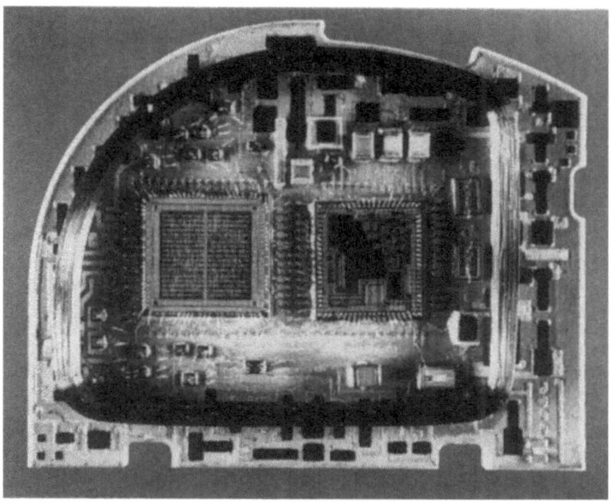

Figure 53. *Hybrid circuit of the motion-controlled pacemaker (Biotronik Ergos).*

Figure 54. *Motion-controlled pacing: sensor signal and metabolic rate during exercise (left); exercise-dependent pacing rate during exercise (right).*

A rate-adaptive pacemaker must provide additional programmable features to tailor the device performance to the patient's clinical requirements. Useful programmable features are sensor sensitivity, maximum pacing rates, and response times. Programming the sensor sensitivity allows parameter adjustment to the unique biomechanical and anatomical characteristics of the patient. The maximum pacing rate is programmed to specify a pacing range consistent with the patient's medical history, cardiac reserves, and exercise tolerance. Programming the response times will determine

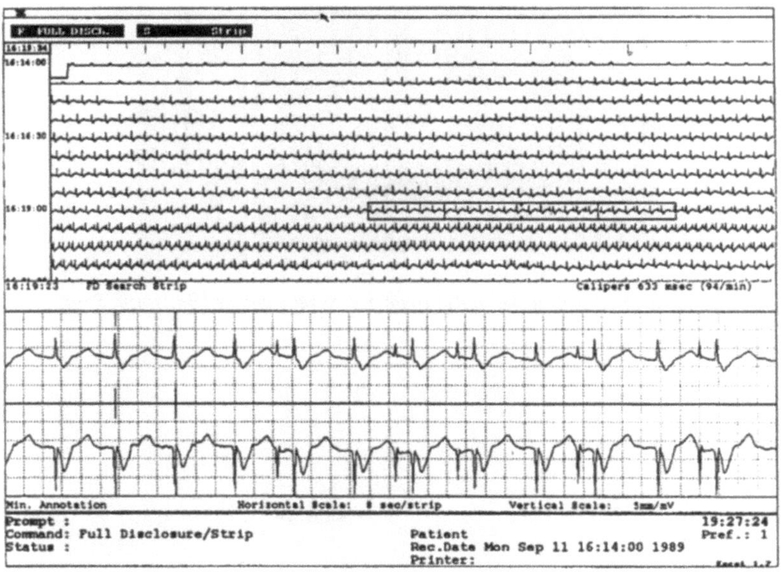

Figure 55. Holter-ECG exercise; the interaction of the natural sinus rate (VAT) and artificial motion-controlled rate (DDD, CR) demonstrate the appropriate function of the Biotronik Ergos pacemaker.

how rapidly the pacing rate is to increase at the onset of physical activity and how slowly it will return to the resting rate upon cessation of physical activity. The response times can be adjusted to keep the patient free of symptoms during the transition period between resting and active states.

With both sensor sensitivity and response times programmable, therapeutic pacing rates can be adapted to low workloads. Thus, the sensor-controlled pacemaker can be made to respond when the patient sits up or stands up. This rapid response behavior is particularly valuable for patients with an orthostatic circulatory insufficiency. The motion-energy pacemaker provides these patients with an adequate increase in HR and, thus, CO and prevents the hypotensive episodes associated with standing up.

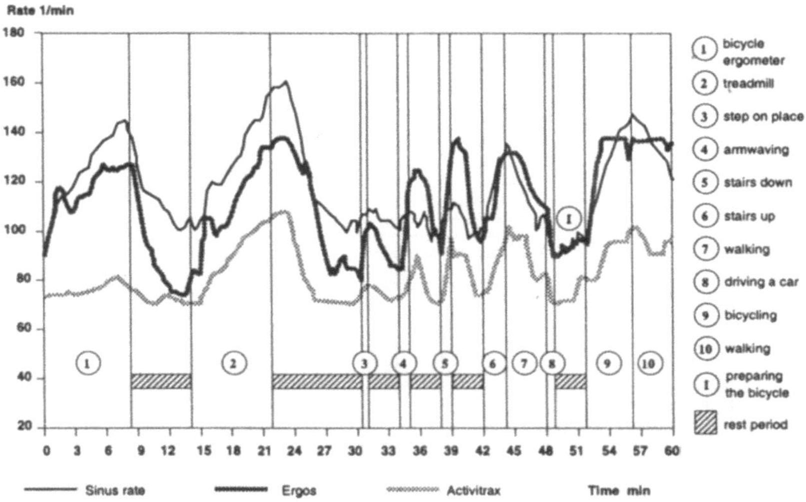

Figure 56. *Comparative study of different mechanical control principles for rate adaptation.*

The motion-energy, rate-adaptive pacemaker is effective in providing a pacing rate responsive to the patient's needs under a wide range of conditions (see Figure 55). However, motion energy is not the only algorithm that might be used in conjunction with a piezoelectric accelerometer. Motion frequency may also be used. Using the motion frequency algorithm, the pacing rate is controlled by the frequency with which the sensor signal exceeds a threshold value. The frequency-based algorithm does not perform well in normal daily activities such as climbing stairs or hills. When ascending stairs, elderly patients typically decrease their stepping cadence while increasing their level of exertion. In these conditions, the frequency-based algorithm inappropriately decreases pacing rate.

Figure 56 summarizes a comparative study in which motion sensor pacemakers were bandaged extra-corporeally to the thorax of test subjects. The pacing control algorithm based on motion energy exhibits better specificity than the one based on motion frequency. With a minimum of deviation, the motion energy device paces at rates very similar to the natural sinus rate [90].

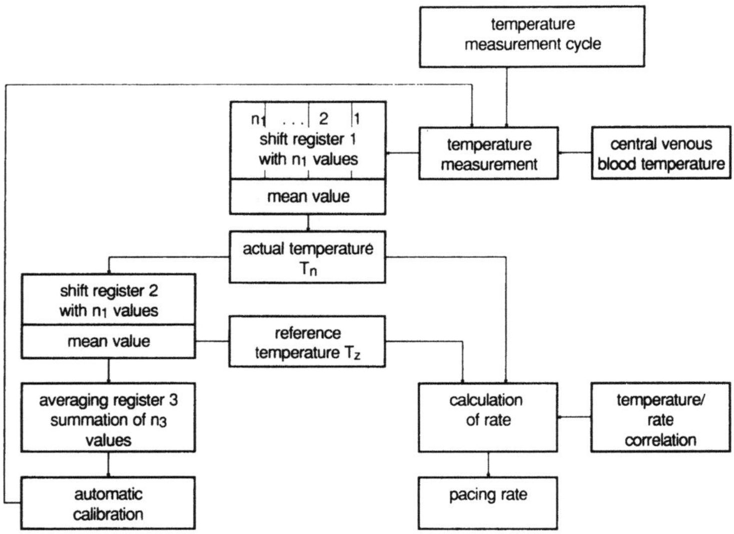

Figure 57. *Flow chart of software that uses central venous temperature for rate-adaptation.*

Central Venous Temperature as Corporeal Control Parameter

The application of *central venous temperature* (CVT) to the control of pacing frequency goes back to the work of Csapo [91]. There is a linear relationship, also confirmed by later studies, between the CVT, the workload, and the chronotropic heart rate. Multi-stage averaging permits a differentiation between workload-independent circadian temperature fluctuations and temperature changes of a workload-related origin (Figure 57). In this way, both classes of signals can be used for physiological rate adaptation to metabolic needs, especially during night rest [92]. A rapid rise of the pacing rate at the beginning of exercise can be achieved by relating the workload-specific temperature rise to a continually updated reference value that reflects workload-independent, slow-temperature variations [93, 94].

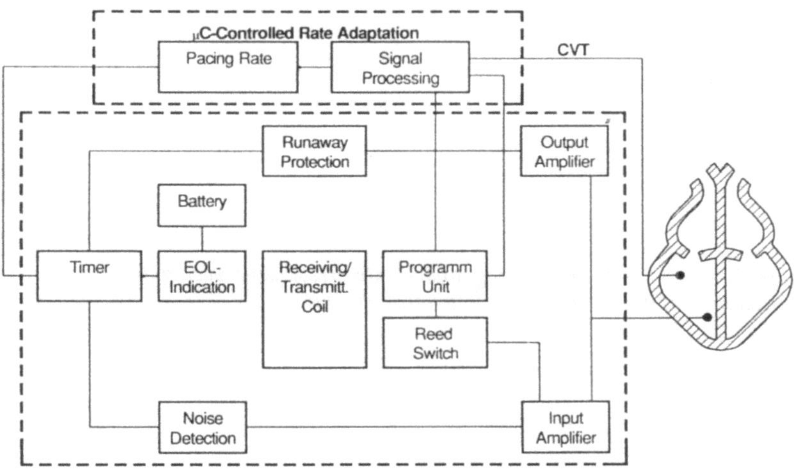

Figure 58. *Block diagram of an implantable pacemaker with central venous temperature to control the stimulation rate (Biotronik Thermos 02).*

The technical solution represented in the block diagram (Figure 58) makes use of a single-chip microcomputer that includes a microprocessor, timer and counter functions, system read/write memory (RAM, 112 byte), and program memory (ROM, 2 kbyte). An external data memory (RAM, 8 kbyte x 8) is used to store all programmed control parameters and to manage patient and diagnostic measurement data. These data are especially important since, after being interrogated via the programming unit, they permit the adjustment of the control parameters to optimize pacing therapy. As a special feature, the Biotronik Thermos offers rate and temperature monitoring. The CVT together with the pacing frequency behavior over a 24 h period can be documented graphically (Figure 59). These

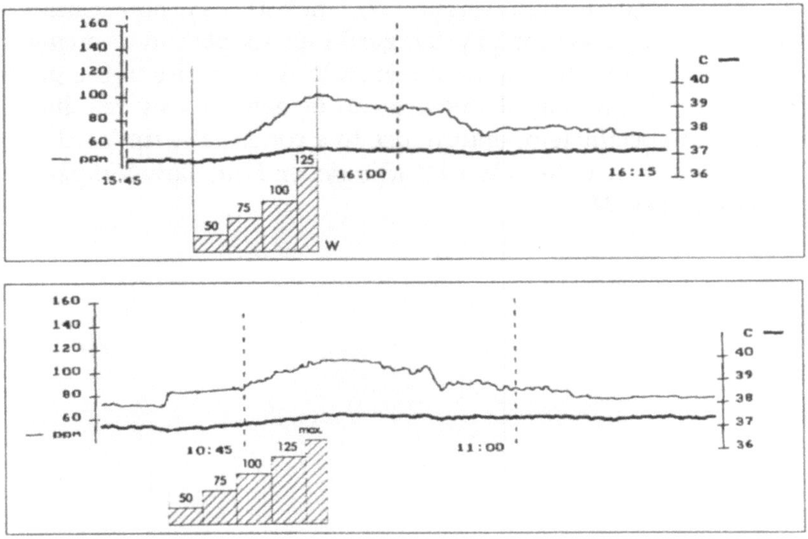

Figure 59. *Heart rate monitoring under rate- responsive pacing during exercise telemetered from the memory of the implanted pacemaker.*

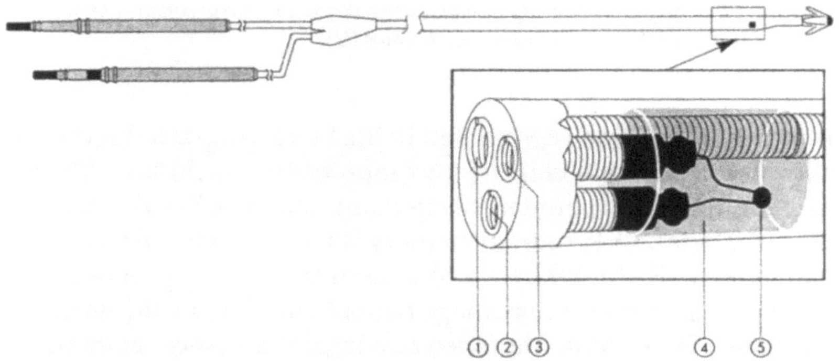

Figure 60. *Design concept of Biotronik pacing lead with integrated thermistor for the measurement of the central venous blood temperature.*

diagnostic data are processed by the external programming unit allowing optimal adjustment of the implanted unit. The intraventricular sensor consists of a thermistor that is integrated into the pacing electrode. The microelectronic design consists of a custom analog integrated circuit with a digital interface for temperature measurement that permits a measurement resolution of 0.025 C (Figure 60). The pacemaker-specific functions of the multichip design are prodvided by a separate integrated circuit serving also as a backup VVI,M pacemaker.

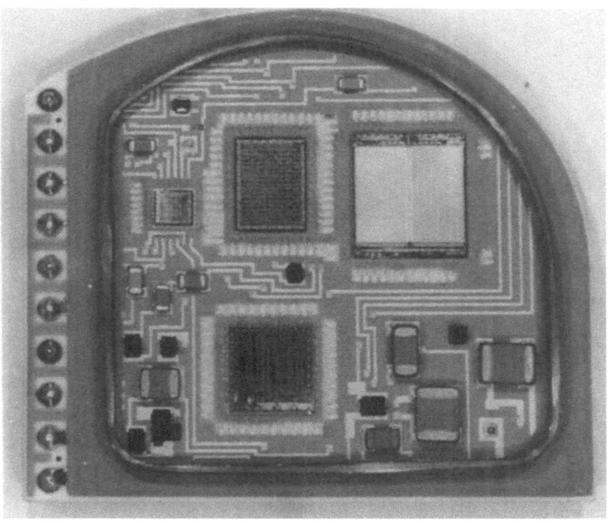

Figure 61. *Hybrid circuit of a central venous temperature-controlled, rate-adaptive pacemaker consisting of a customized IC for signal processing, an independent pacemaker circuit, and an 8kbyte x 8 RAM. The microprocessor mounted in chip-carrier technology is located on the rear side of the pacemaker (Biotronik Thermos 02).*

The individual integrated circuits and the non-integrated discrete components (quartz oscillators, tantalum capacitors, coils for transmission, reed contacts, etc.) are interconnected using a modular design allowing the use of well-proven components providing quick development under strict reliability requirements [95]. Figure 61 shows one side of the hybrid circuit with its components. The rate is adapted according to an algorithm that is based on linear workload-dependent temperature changes of the central venous blood, where different temperature/rate responses can be chosen. The rise of the pacing rate under stress and its decay in the following recovery phase can be adjusted to the individual needs of the patient (Figure 62). Figure 63 shows an example of a temperature/rate curve during exercise with a subsequent recovery phase.

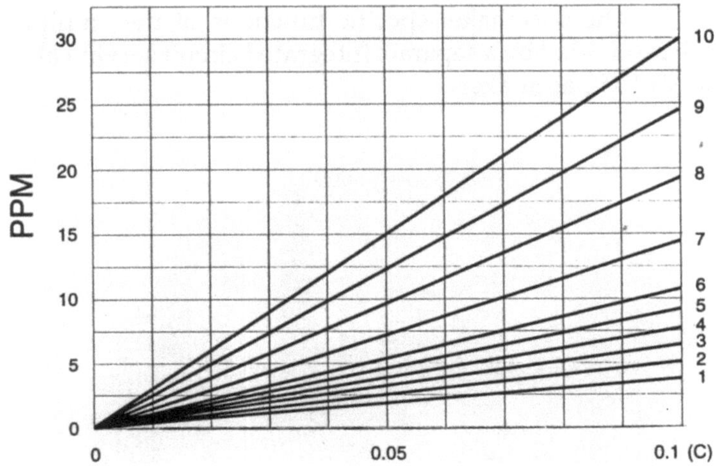

Figure 62. Programmable temperature/rate response characteristics.

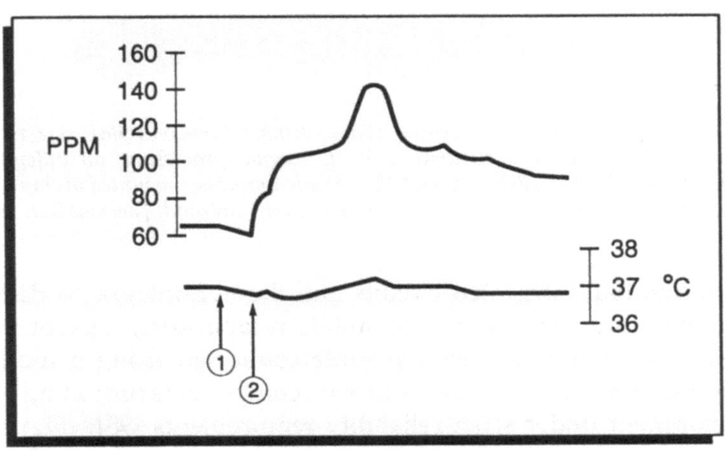

Figure 63. Time dependence of central venous temperature and pacing rate at beginning of exercise.

During the development of the device, it proved advantageous that the temperature-to-workload behavior could be simulated by a simplified computer model based on the circulation model described by Guyton and Coleman [36]. Figure 64 shows that the calculated data agree with the clinical results.

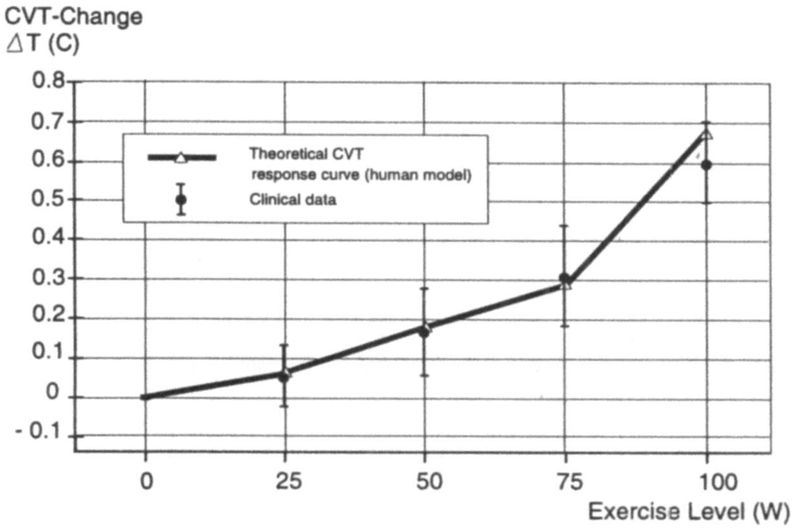

Figure 64. *Mean temperature increase during exercise: comparison of measured and theoretical values.*

Clinical experience confirms the workload specificity of the CVT for open-loop rate adaptation [62, 96-100]. Temperature-controlled rate adaptation proves physiological for a wide range of workloads. Clinical results demonstrate that the principle of the temperature-controlled rate adaptation is suitable for atrial as well as for ventricular stimulation. Since the introduction of this system, the question has been discussed whether the mixture of the venous block returning to the heart via vena cava superior and vena cava inferior into the atrium would be sufficiently accurate for adequate temperature measurement to be used for rate adjustment. Clinical findings provide the identity of atrial and ventricular CVT as represented in Figure 65. In this case, with an implanted **atrial** CVT-controlled pacing system, the Biotronik Thermos, a second thermistor catheter was used in the ventricular position to monitor the intraventricular temperature data simultaneously. The exercise test using a bicycle

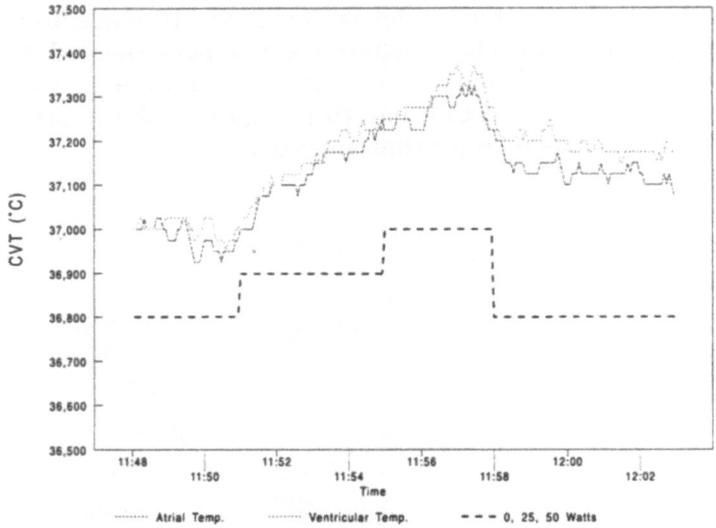

Figure 65. *Simultaneous record of atrial and ventricular temperature data during exercise test (interrogated pacemaker data).*

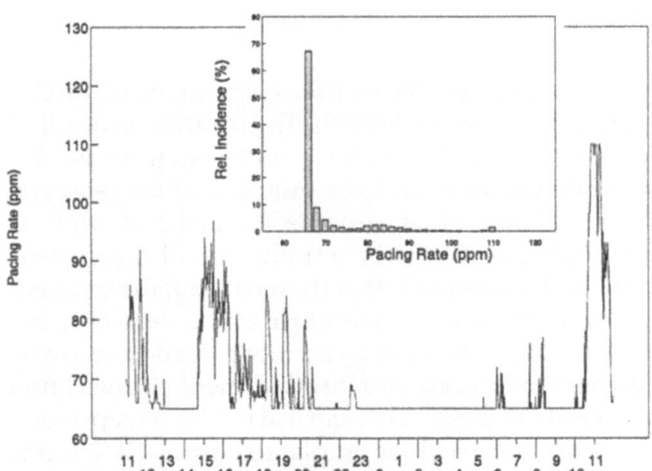

Figure 66. *A 24 h rate profile as interrogated from the implanted pacemaker. The inset shows the rate histogram from this record.*

ergometer demonstrated no difference in the timely courses of CVT in both chambers; response time as well as values of the CVT changes are identical within the statistical fluctuations. Using standard programmer software, the stored data of the pacemaker can be interrogated and saved as a data file. A typical example of a rate trend monitoring is given in Figure 66, which shows the time course of pacing rate, permitting analysis of daily activities. An additional histogram allows a detailed analysis of exercise-induced heart rate. Another 24 h trend monitoring of a CVT-controlled pacemaker in AAI mode provides evidence for the existence of circadian changes. The data obtained demonstrate that the stimulation rate with regard to activity and metabolism is comparable to the heart rate of a normal, healthy individual (see Figure 67). Taking into account the workload-related physiological and metabolic considerations, the advantage of CVT control can be summarized as follows:

1) physiological adaptation of the rate to the individual needs of the patient under physical stress,
2) a reduction of the rate during rest by night, controlled by the circadian temperature variations,
3) adaptation of the rate in case of fever, and
4) rapid adaptation of the pacing frequency at workloads above 50 W.

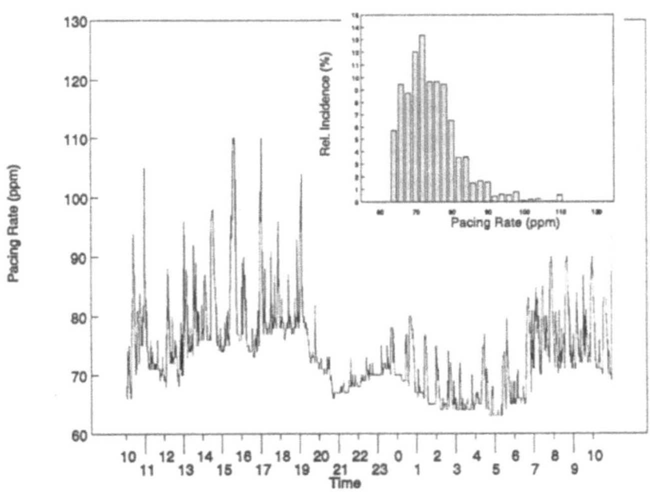

Figure 67. *Typical rate profile demonstrating a physiological basic rate (implanted pacemaker interrogation). Inset: rate histogram from this record.*

When including the thermoregulatory aspects of circulation, rate control based on central venous blood temperature shows a regulated behavior that leads to an increased system stability. The central venous blood temperature rises less under rate-adaptive pacing than it does under the same workload without rate adaptation. This corresponds to a negative feedback effect of the pacing rate on the central venous blood temperature [61].

CARDIAC CONTROL PARAMETERS

In comparison to corporeal control parameters, cardiac control parameters offer the general advantage that the sensed variable contains systemic control information [101]. A particularly useful control signal available via a cardiac parameter is the sympathetic tone, which is independent of the heart rate [53, 102]. This signal is an efferent control signal from the ANS to regulate cardiac output.

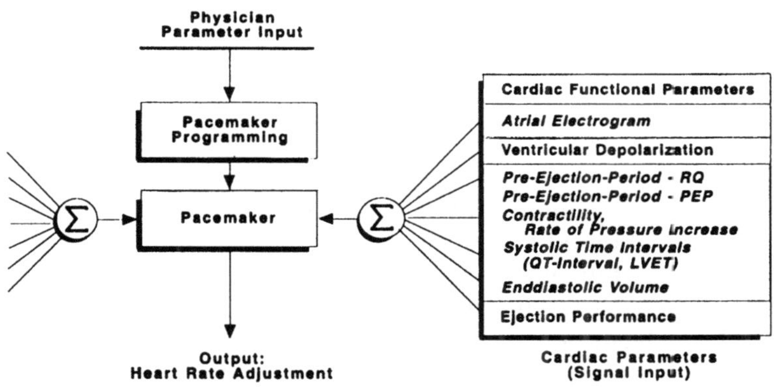

Figure 68. Cardiac control parameters for rate adaptation.

Cardiac parameters that contain sympathetic control information include atrial sinus rate, ventricular stroke volume, end-diastolic volume, end-systolic volume, and the systolic time intervals as well as the de- and re-polarization potential (QT) (Figure 68). The independence from the heart rate is not strictly met by the QT time interval [103]. The QT interval decreases not only with a rise of the sympathetic tone but also with a rise of the heart rate. Consequently, to avoid instabilities and possible positive feedback conditions, a QT interval control algorithm must detect instability and correct for it.

Principles of Intracardiac Conductance Measurement

The determination of ventricular mechanics from intracardiac conductance measurements can be extended to the detection of the inotropic response as an indicator of cardiovascular demands to serve as a control input for the reestablishment of chronotropy. Historically, transthoracic impedance, introduced by Kubicek [104], serves as a basis for today's understanding of blood volume displacement in the thorax. With the advent of powerful data processing, these principles have been extended to impedance tomography [105]. The application of these principles to intracardiac impedance measurements was initiated by Geddes [71] in 1966, followed by Salo, Baan, Valentinuzzi, Boheim and others [68, 70, 106-108]. The principle was extended to intracardiac plethysmography of the left and right ventricles. In addition to measuring ejection volume, intracardiac impedance was used to monitor systolic time intervals [64, 76, 78].

Impedance plethysmography is based on the difference in specific conductivity of blood and cardiac tissue. The observed impedance changes during the cardiac cycle can be attributed, in part, to changes in the fractional volume of blood and tissue near the measuring electrodes. The impedance changes are also influenced by anatomical and mechanical factors, including electrode location, catheter movement, and the kinetics of contraction.

The determination of ventricular volume from intracardiac impedance measurements requires knowledge of the relationship between chamber morphology (both size and shape) and the electrical potential distribution within the chamber [71-74, 80, 109, 110]. The direct application of Ohm's law to the analysis of intracardiac impedance measurements is not appropriate. A more detailed theoretical analysis based on Maxwell's equations must be considered for the relationship between volume and potential distribution.

Basic Aspects of Conductance

The heart's ability to transport blood is directly related to volume changes of its ventricles. The difference in the conductivity of blood and surrounding tissue or muscle allows assessment of volume changes through impedance measurement. To understand volume and geometry-related impedance changes, a simulation method is needed to describe the anatomical ventricular geometries (see Figure 69).

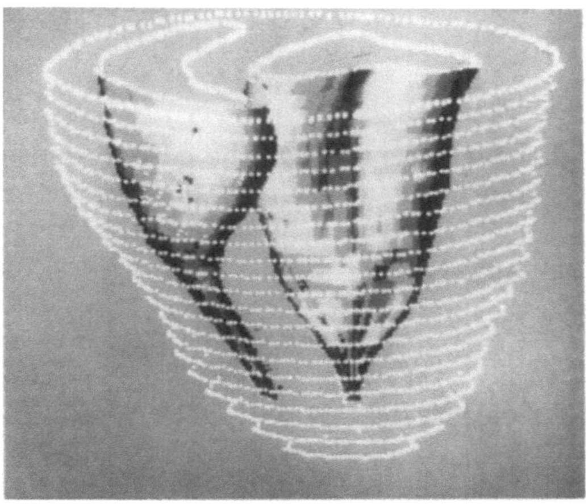

Figure 69. Cardiac chamber geometry, obtained by magnetic resonance imaging, demonstrates the palmate shape of the right ventricle.

Theoretical studies of unipolar electrode impedance as a relevant indicator of volume displacement involving the application of Gauss' theorem, Maxwell III, and Ohm's law for the right ventricular volume have resulted in the Poisson equation defining the resulting potential at any location in the ventricle as a function of induced current.

Assuming a volume V, having a conductivity σ and a charge density qv, the total current across one of its surrounding faces S is

$$I = \iint j \cdot dS$$

A change in current through the entire volume arises from a change in charge density inside the volume

$$I = \iint j \cdot dS = -\frac{d}{dt} \iiint q_v \cdot dV = -\iiint \frac{d}{dt} q_v \cdot dV$$

Applying Gauss' theorem and equating inner terms, the continuity equation is defined as follows:

$$\nabla j = -\frac{d}{dt} q_v$$

which states that the divergence of the current density in a given volume is caused by a change in charge density.

Application of Ohm's law

$$j = \sigma \cdot E$$

The definition of electric field

$$E = -\nabla \Phi$$

and Maxwell III

$$\nabla \times E = 0$$

results in

$$\sigma \cdot \nabla^2 \Phi(x,y,z) - \nabla \cdot j = 0$$

Calculation of $\nabla^2 \Phi(x,y,z)$ is achieved by expansion in a third degree Taylor polynomial. Results are the second derivatives of the electrical potential and are approximated by using the difference in potential to adjacent points, resulting in

$$\frac{\partial^2}{\partial x^2}\Phi(x,y,z) = \frac{\Phi(x+h,y,z)-\Phi(x,y,z)+\Phi(x-h,y,z)-\Phi(x,y,z)}{h^2} - \frac{h^2}{12}\cdot\frac{\partial^4}{\partial x^4}\Phi(\xi,y,z)$$

Integration of Maxwell III over a cubic element of volume h^3, and applying the definition of conductance $G = \frac{I}{U}$

results in Poisson's Equation

$$-(G_k^+ + G_k^- + G_l^+ + G_l^- + G_m^+ + G_m^-)\cdot U_{k,l,m} + (G_k^+\cdot U_{k+1,l,m} + G_k^-\cdot U_{k-1,l,m}) + (G_l^+\cdot U_{k,l+1,m} + G_l^-\cdot U_{k,l-1,m}) + (G_m^+\cdot U_{k,l,m+1} + G_m^-\cdot U_{k,l,m-1}) + I_{k,l,m} = 0$$

which describes the resulting voltage at a point $\Phi_{k,l,m}$ as a result of $I_{k,l,m}$ and the impressed current at this point.

Geometric modeling of the right ventricle (RV) presents a great challenge. An accurate model has utility not only for imaging applications, but for analysis of RV mechanics. Due to its complex crescentic shape, different types of models have been used to predict its absolute volume [111]. When considering a particular model, it is important that it accurately predicts both the geometry and geometric changes during the cardiac cycle.

Based on heart imaging observations, the RV takes on a crescentic shape in cross section. A model of the RV's geometry that fits its crescentic nature is shown in Figure 70; it approximates the cross-sectional shape of the chamber by two intersecting circles. In addition, the long axis curvature was found to take on a parabolic shape. The derived model predicts right ventricular volume changes when compared to actual volume changes during contraction with a high degree of correlation. Two basic conclusions are made:

1) The horizontal cross-sectional area is best represented by two intersecting circles of differing radii defined in the x-y plane.

2) The long axis septal curvature is best represented by a parabola.

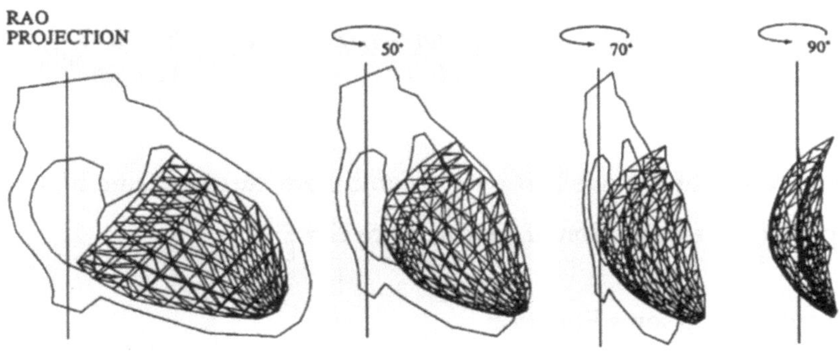

Figure 70. Model of the right ventricular volume from right anterior oblique projection.

Anisotropic Impedance

In order to determine the relationship between impedance-derived volume, and actual ventricular volume, studies were carried out in canine hearts *in situ* in a manner similar to Geddes [71]. After the animal was sacrificed, each of the major vessels was ligated with the impedance catheter in position and the chamber was emptied [112]. Known volumes of blood were introduced into the ventricle while determining the impedance. These studies apply to the passive filling of the ventricle. Analysis using cylindrical equations produced invalid results in the right ventricle, and simple models of resistivity of the right ventricle are inappropriate [113, 114]. Using finite-element analysis over the right ventricle, the volume can be described; however, unipolar as well as multipolar electrodes do not accurately allow the measurement of right ventricular volume. Many factors during contraction directly and indirectly influence the impedance. In this regard, considerations were given to the following items:

- Contraction Axis
- Geometry and Symmetry
- Supine Position: Preload
- Afterload and Hypertension
- Valve Insufficiencies and Regurgitation
- Myopatia

As an example of the relative influence of geometry on intracardiac impedance, changes of the following parameters are significant [115]:

1) right ventricular cross-section area
2) left ventricular volume
3) atrial volume
4) pacemaker and electrode placement

Based on theoretical considerations, right and left ventricular volume changes are evident in the right ventricular impedance. These phenomena are verified by clinical studies [116]. It is clear that unipolar impedance measurements do not allow direct and indirect volume measurements. However, contractile processes, as long as they are reflecting geometric changes, are detectable. Therefore, the pacemaker electrode serves three functions: First, as an actuator to conduct electrical energy from the pulse generator to the myocardium; second, as a sensor to detect intracardiac potentials (IEG) to control the pacemaker; and third, the same electrode serves for intracardiac conductance measurements recognizing the myocardial response that is evoked by a pacing impulse or detecting intrinsic rhythm, thus providing ANS information [117].

Employment of PEP for Rate Adaptation

In patients with artificially stimulated heart action, the time interval between the onset of ventricular electrical activity and the beginning of the ejection period is defined as the beginning of PEP. PEP includes the electromechanical delay and the isovolumetric contraction phase. Electromechanical delay is the time required for the process of excitation-contraction coupling. During this time, calcium ions are released from the intracellular stores (the sarcoplasmic reticulum) and diffused to their site of action at the actin filaments, resulting in cross-bridges between actin and myosin filaments as a condition to start the contraction. The isovolumetric contraction phase is the time interval between the beginning of the contraction and the moment the ventricular muscle has developed sufficient force to open the valves by the pressure difference on both sides. The beginning of the ejection phase is indicated by the first reduction in intraventricular volume. It has been shown that an increase in sympathetic activity augments the transmembrane inflow of calcium ions, thereby raising the availability of calcium ions in the sarcoplasmic reticulum and enhancing the development of contractile force. As a consequence, the time necessary to open the valves is shortened. It has now been shown that a unipolar intracardiac electrode arrangement is satisfactory to detect the beginning of the ejection phase. For this purpose, impedance is measured between the tip of the conventional pacing electrode and its subcutaneous counterelectrode.

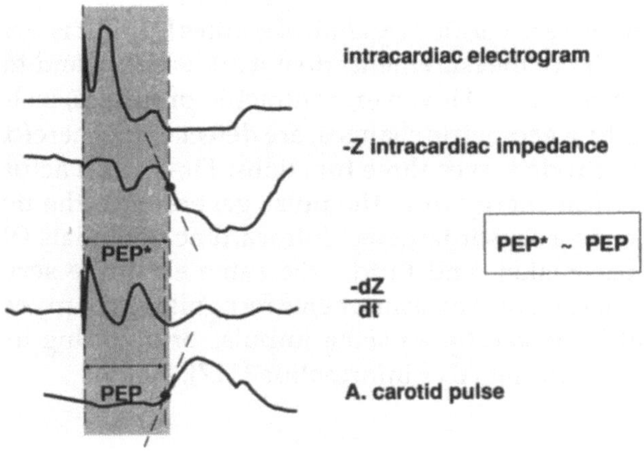

Figure 71. *Demonstration of the principle of how PEP* is determined by using the right ventricular impedance.*

A change in intraventricular geometry during contraction is associated with an increase in impedance. The first derivative of the impedance signal with high sensitivity may be interpreted as volume change. The first maximum of this curve after the stimulus pulse indicates the end of the pre-ejection period. This time interval between the stimulus pulse and the maximum of the first derivative of the impedance measured in the right ventricle is called PEP* in order to differentiate it from the original PEP (Figure 71). It has been shown in clinical examinations that PEP* can be measured with sufficient accuracy to use it for pacing rate adaptation. Like PEP, PEP* is independent of the pacing rate so that positive feedback with the risk of instability is avoided.

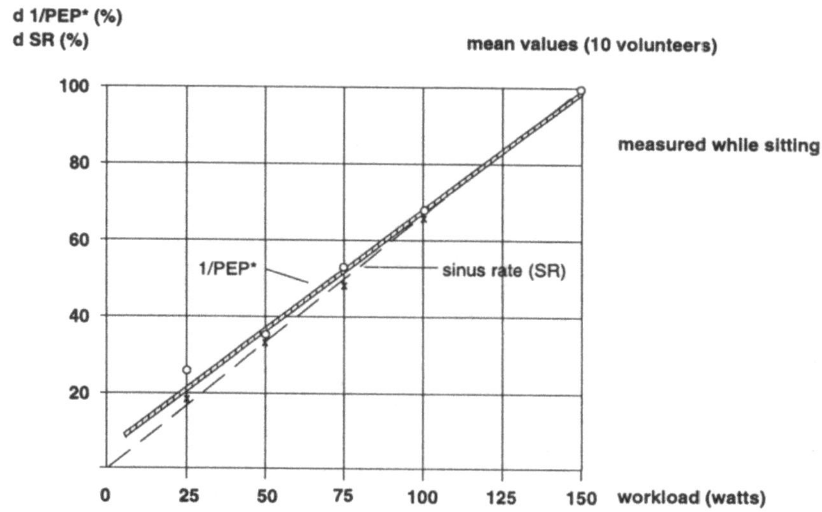

Figure 72. *Correlation between the percent change of 1/PEP* and sinus rate (SR), respectively, compared with the workload. Measurement was performed with the volunteers in a sitting position.*

As shown in Figure 72, the change in the reciprocal of PEP* referenced to the resting value is linearly related to the workload. The same is true for the sinus rate. This means that the reciprocal of PEP* perfectly relates to sinus rate. This is also demonstrated by the nearly perfect sinus rate approximation compared with the measured sinus rate during different kinds of posture and exercise (Figure 73).

For the reliable determination of PEP*, a sliding measurement window is used. Obviously, PEP* changes only slightly from one heartbeat to the next. Therefore, a window can be set during which the impedance signal indicating the beginning of the ejection phase is to be expected. With this method, current consumption can be reduced and, at the same time, erroneous signals can be eliminated. When no signal indicating the beginning of the ejection period has been detected in the detection window for one heartbeat, PEP* is assumed to be of the same length as in the preceding heart action. However, the width of the window is enlarged for the next measurement to avoid loss of PEP* detection with subsequent disengagement from the closed-loop control of heart rate.

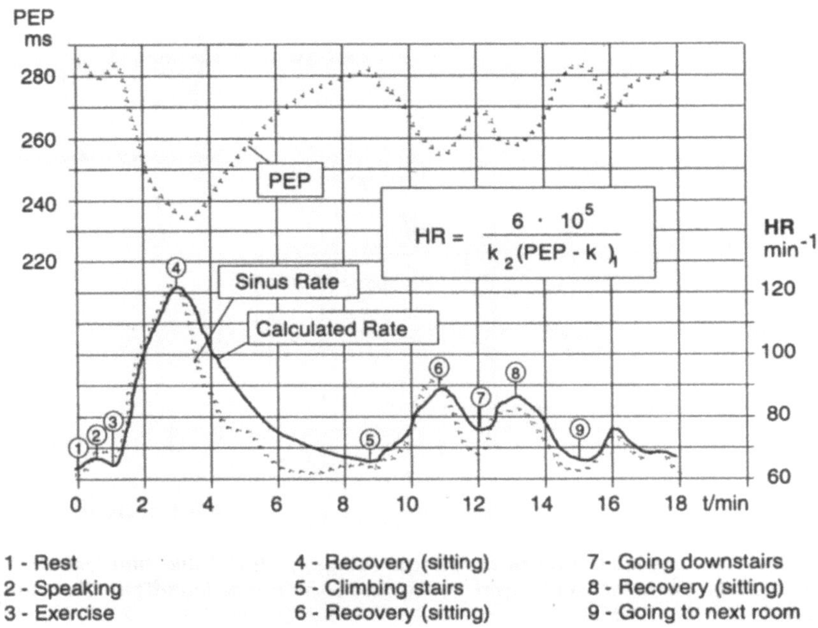

Figure 73. *Measured PEP and heart rate as compared with rate calculated with the indicated relation during rest, exercise, and recovery.*

Improving the PEP Control: Concept of VIP

The intraventricular conductance signal primarily contains mechanical motion, geometry, and wall tension changes near the pacing tip. By placing the electrode near the apex of the right ventricle, it records information from both the right and left ventricle contractile process. As shown above, the VIP is sensitive to time shifts in the conductance waveform due to the shortening of the isometric contraction time interval with increasing sympathetic tone, thereby providing the basis for extracting the inotropic status. Contractility is manifest in ventricular contraction and is directly related to conductivity, as shown above. The ANS information can easily be detected from the intracardiac impedance signal, which is generated within approximately 0.5 cm^3 of the pacing tip. The displacement information can be derived by signal processing means to extract time-based information for rate control. A typical example of an intracardiac impedance recording is shown in Figure 74. However, it was obtained under supine and resting conditions during pacemaker replacement. It shows systolic and diastolic influence on the waveform. The early ventricular systolic time marks can be identified by the derivative of the impedance amplitude to determine the pre-ejection period PEP* [65].

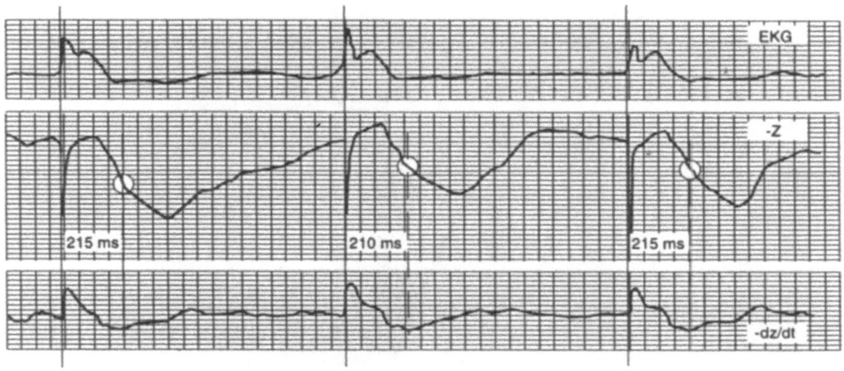

Figure 74. *Registration of right ventricular impedance measured via an implanted unipolar pacing lead.*

The initial clinical results using the principle in a fully implanted pacemaker were good and demonstrated the successful extraction of the ANS information. However, the patient-to-patient variations restricted the applicability of the PEP* algorithm to individual calibration requirements leading to excessive exercise exposure and complicated programming procedures.

Based on clinical data obtained, 28 different algorithms were systematically evaluated for the ability to extract the ANS information with the highest specificity, accuracy, and precision. This included standard signal processing, fast Fourier analysis (FFT), exploratory statistical analysis such as auto- and cross-correlations, and factor analysis. The algorithm which proved to be superior by all criteria was the Regional Effective Slope Quality (RQ). The method detects changes in the contraction velocity during early systole by detecting the exercise-related temporal shift in the conduction waveform. This principle can be demonstrated by using the measured waveforms of a 72-year-old male under resting conditions (see Figure 75). Each curve represents the average of over 50 cardiac cycles. During exercise, the features of the curve occur earlier and the second

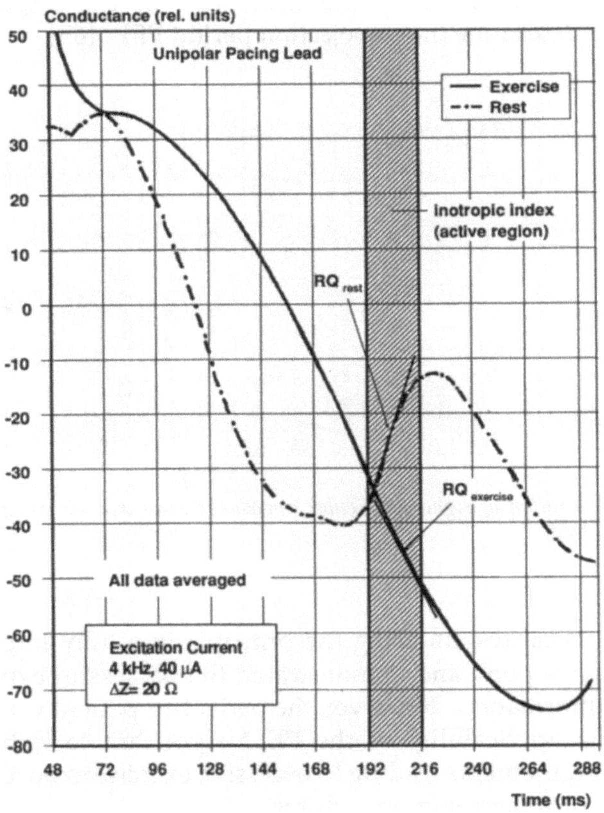

Figure 75. *Typical shapes of conductance waveforms under rest and exercise.*

conductance wave frequently fuses with the earlier conductance feature on the curve. The features shift in a manner parallel to the shortening of the isometric contraction under the same conditions. This shift in the conductance waveform is consistent with exercise changes in the timing of the phases of the cardiac cycle. By comparing the changes in the intracardiac conductance curve during rest and exercise, it is easy to identify the principal morphology which changes with PEP. As shown in Figure 76, the features of the intracardiac waveform, which change with exercise, are associated with the pulmonic and aortic valve openings and are consistent with the definition of PEP. The time period over which conductance changes are measured delineates the *Region of Interest* (ROI). ROI may be optimized to the specific conditions of the patient's heart.

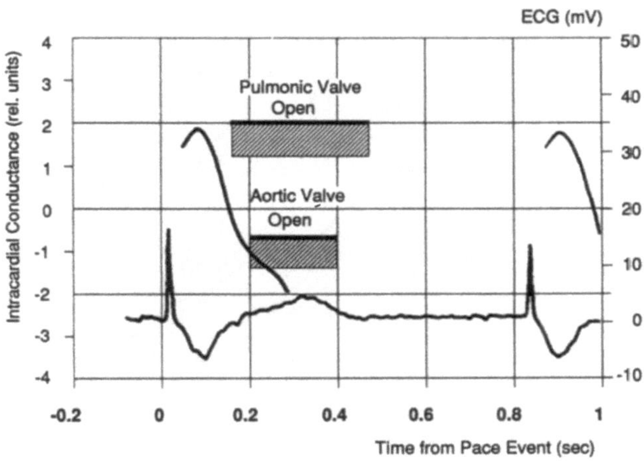

Figure 76. *Intracardiac conductance time (patient with resting rate of 70 ppm).*

The efficacy of the algorithm was verified by comparison of the detected values with the autonomic sinus rhythms in patients suffering AV block against the regulated stimulation rate. Ventricular Inotropic Parameter (VIP), a measure of sympathetic tone, provides the adjustment of pacing rate during physical exercise to individual requirements. It is defined in such a manner that it becomes zero for resting conditions and one for the maximum level of exercise that is appropriate to the age and physical fitness of the individual. Morphology analysis of the recorded conductance demonstrates the influence of the left ventricle under exercise, moving the portion of the curve associated with the early ejection time of the left ventricle to earlier time points as expected, responding to exercise-related increase in inotropy. This shift may also be detected by the PEP* algorithm.

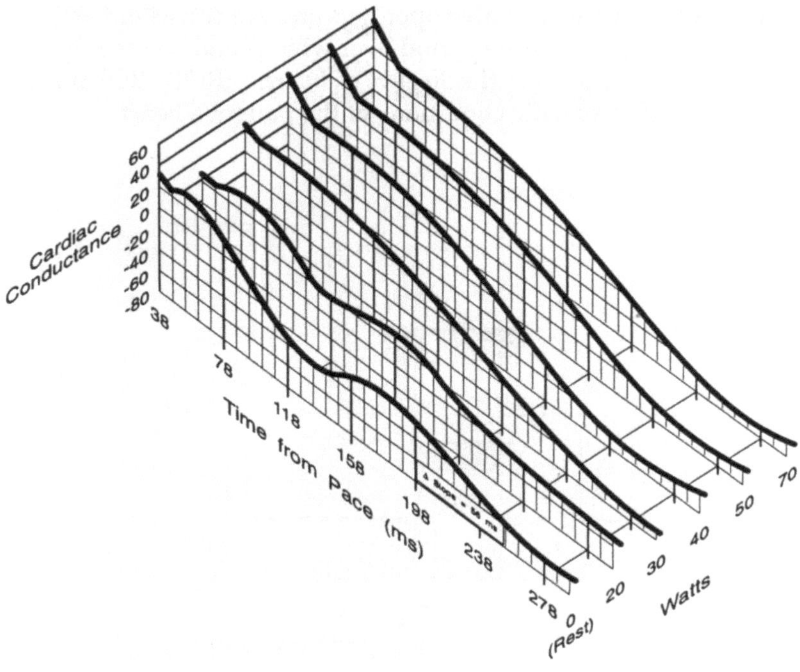

Figure 77. Cardiac conductance waveforms (changes under 50 W load).

The RQ method, however, allows a narrower detection window (region of interest - ROI) compared to the PEP* principle, resulting in a significant power reduction. ROI is defined as the time window that exhibits the highest significance change in RQ in response to exercise challenges. Figure 77 exhibits the gradual shift in the conductance curves in response to exercise. The left ventricular feature results in a shift of 56 ms during the transition from rest to 50 W. As an additional example, the systolic timing events are demonstrated in Figure 78. In a 62-year-old male under resting and exercise conditions, the conductance shows the expected shift. Three regions of interest (A, B, C) are marked and the valve status under rest are identified for timing reference. In all three regions, the detection of

ANS information is statistically significant. ROI A occurs during isovolumetric contraction and should not be selected for ANS control. ROI C takes place during left ventricular rapid ejection phase, serving as a third choice. The primary region (B) for extracting ANS-control information for this patient is associated with both valves opening and, therefore, most closely associated with systolic events reflecting the inotropic influence on the myocardium. For lead placement in the apex of the right ventricle, the timing of the cardiac cycle was obtained using echo cardiography. These were paced events and the timing of the right ventricle value events preceded the corresponding events in the left ventricle.

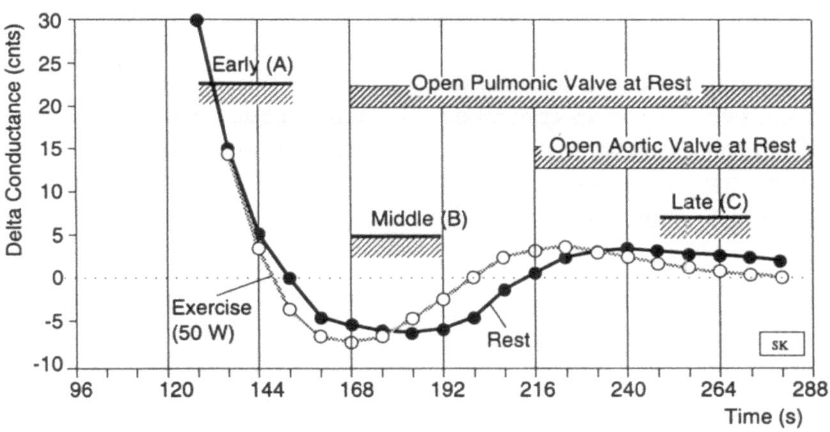

Figure 78. Cardiac conductance waveforms with three regions of interest (A), (B), (C).

RQ is defined (see Figure 75) as the conductance difference, measured between two time points (ROI), where the exercise response of the difference in slope of the corresponding waveform is most significant. This can be explained in a simplified manner as a result of a data-processing method optimizing extraction of sympathetic information while suppressing all disturbing influences such as respiration, beat-to-beat variability of atrial contributions, preload, afterload, stroke volume, etc. This quantity, VIP, enables the pacemaker to reliably control the rate, reestablishing the chronotropy in a closed loop. VIP defines the complex factors influencing the inotropic state of the heart.

Using intracardiac conductance Y, VIP is defined as:

$$VIP = \frac{(RQ - RQ_{rest}) \cdot Slope^{-1}}{(RQ_{exercise} - RQ_{rest}) \cdot Slope^{-1}}$$

RQ definition:

$$RQ = Y_{monitored} - Y_{reference}$$

As shown in Figure 79, VIP independent of PEP and *stroke volume* (SV) represents the desired ANS information, providing the advantage of being detectable by unipolar, intraventricular measurements. The ANS-controlled pacemaker measures the RQ value of alternate cardiac cycles, and calculates VIP that is finally used to adjust the pacing rate between the base adaptive rate (BAR) and the maximum adaptive rate (MAR).

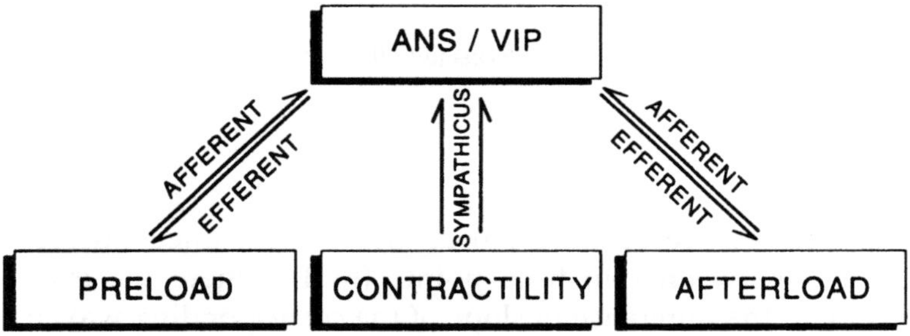

Figure 79. *VIP monitors contractility via the sympathetic tone without being affected by preload or afterload.*

Unipolar Right Ventricular Conductance Measurement

As mentioned above, for the ANS closed-loop concept, a unipolar electrode in the right ventricle using the pacemaker housing as the counter electrode allows the acquisition of the ANS information from the inotropic state of the heart.

The unipolar electrode arrangement for making an intracardiac conductance measurement is shown in Figure 80. The conductance measurement is made by injecting continuous square-wave excitation current through the pacing tip and monitoring the resulting voltage, requiring about 2 μA average current drawn from the battery. The current passes through the electrode, the blood, and myocardial tissue, across the thoracic cavity and intervening pulmonary tissue, back to the pacemaker housing. Since the conductive area of the pacing lead is much smaller than the conductive area of the pacemaker housing, the current density is highest near the electrode tip. Most of the induced voltage occurs within the 0.5 cm^3 of blood and tissue surrounding the electrode primarily reflecting conductance phenomena close to the electrode surface. Therefore, the term intracardiac conductance will be used to emphasize the fact that the signal measured and processed, as described here, is dominated by contraction phenomena local to the ventricular apex.

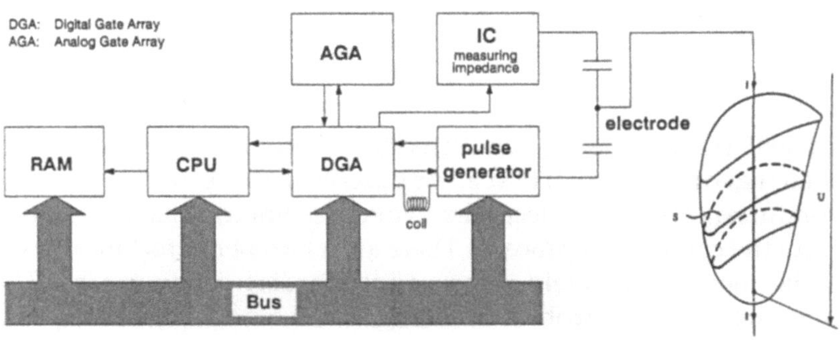

Figure 80. *Simplified block diagram of the ANS-controlled pacemaker with unipolar conductance electrode in the right ventricle (Biotronik Neos-PEP).*

Due to the thin palmate shape of the right ventricular chamber, the geometric changes that take place in the septal area are a result of the left ventricle contraction. Bulk volume changes in the ventricle do not contribute to the unipolar intracardiac conductance signal. Most of the volume of the right ventricle is more than five times farther from the detecting electrode than can be sensed by the current field distribution. From the modeling, as well as from the clinical observations, it can be concluded that there is no place in the right ventricle where the unipolar electrode can be placed that will permit measurement of chamber volume for stroke volume.

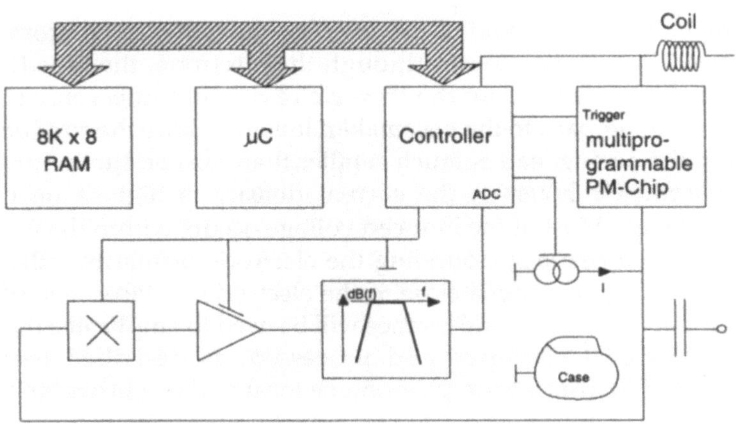

Figure 81. *Block diagram of the hardware in the ANS-controlled pacemaker (Biotronik Neos-PEP).*

At 4096 Hz, the specific conductivity of blood is three times higher than that of myocardial tissue. As described above, the intracardiac conductance signal reflects the evoked geometric changes arising from the contraction process. There are a number of technical tasks to be met to accurately measure intracardiac conductance. The measured voltage attributed to intracardiac conductance changes is small, typically 0.2 to 1.2 mV. It exists in the presence of the much larger induced amplitudes attributed to the electrode resistance, which is in the range of 20 mV for 500 ohm electrode impedance. To further complicate the detection of intracardiac conductance, the voltage is superimposed on much larger and rapidly changing pacing after-potentials and the IEG. Intracardiac conductance has frequency components ranging from about 0.5 to 35 Hz. Figure 81 shows the pacemaker block diagram with the elements used to extract the sympathetic information from the intracardiac conductance from these interfering signals. The excitation is a 4096 Hz square wave, the induced voltage is synchronously demodulated, and the resulting voltage passed through a Bessel bandpass filter with corners at 0.3 and 40 Hz. This filter rejects most of the low-frequency respiratory components, all of the basic lead impedance, and all of the high-frequency demodulation products, which include the IEG and the pacing repolarization components. The Bessel design minimizes signal distortion [78].

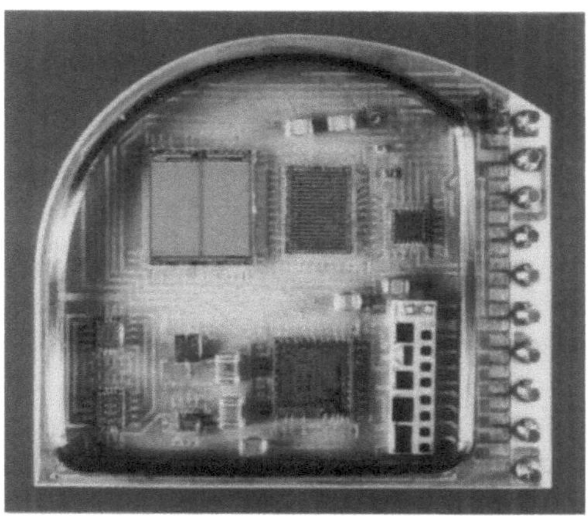

***Figure 82.** Hybrid circuit of the ANS-controlled pacemaker (Biotronik Neos-PEP).*

The signal-to-noise ratio of the filtered information is 36 dB, based on a 10 Ω intracardiac impedance. The signal is digitized using an 11-bit analog-to-digital (A/D) converter. The gain of the filter stages is programmable to adjust the impedance amplitude to exceed a minimum of 6-bit resolution during the A/D conversion. The generation of the excitation current, demodulation, and the filtering of the demodulated signal is performed by a custom low current IC. The A/D conversion is performed 128 times per second by an additional custom IC. Thus, intracardiac impedance is extracted from potential sources of interference with sufficient precision to assure that measurement noise is not the major source of signal variance. The complexity of the microelectronic design is demonstrated in Figure 82. The hybrid circuit contains 11 integrated circuits including an 8-bit CMOS microprocessor and 8K byte static RAM.

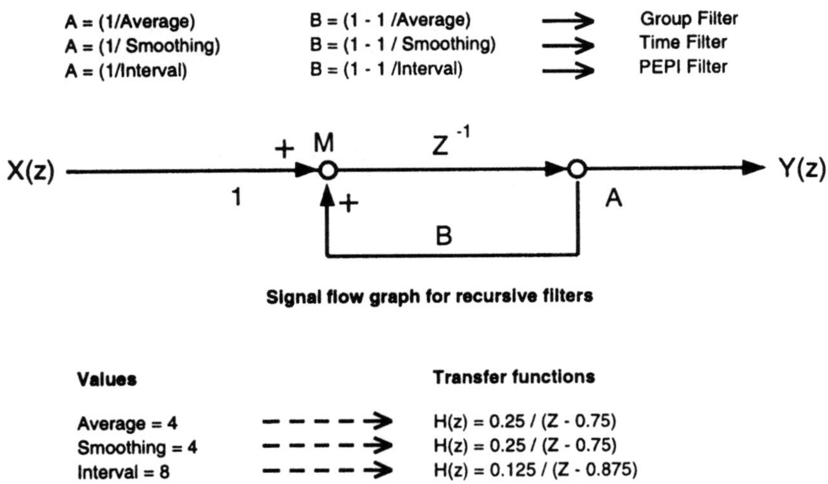

Figure 83. *Principle of recursive filter transfer functions.*

It should be emphasized that the major determinant of intracardiac conductance measured by a unipolar electrode is the local specific conductivity of the blood and myocardial tissue near the electrode tip. Generally, measurements with alternating current are of a complex value. However, in blood and tissue, the admittance at the measurement frequency is almost exclusively real. Thus, the extracted signal is the Intracardiac Conductance (IC). The shape, sign, and magnitude of the conductance changes depend on the lead position. A major complication associated with using the IC is the beat-to-beat variability of the conductance waveforms attributed to changes in AV synchrony when some atrial function is present. There are also shape and timing differences based on how the ventricular tissue was stimulated (*e.g.*, intrinsic, PVC, or paced events). Finally, there is a residual contribution of the respiratory cycle. Figure 83 shows the recursive digital filters used to constrain the variability. The time or smoothing filter was optional and generally not required. Waveform averaging was important and is indicated as a Group filter. These filters were time-discrete recursive filters with a single feedback path of 3/4 or 7/8. To achieve a smooth transition between pacing rates, the final heart rate was also filtered using a recursive filter with a single feedback path of 7/8. The combination of these filters allows the pacemaker to achieve 50% of a set change in heart rate over approximately 12 cardiac cycles.

Table X shows the Technical Specifications for the Biotronik Neos-PEP pacemaker with ANS-Control.

Table X
Biotronik Neos-PEP Pacing with ANS-Controlled Pacemaker

Configuration:	Unipolar
Pulse Form:	Biphasic, asymmetrical, cathodic
Programmer:	Neos-PEP Version 2.0 on an IBM compatible PC
Mode:	VVI, VOO, VVI/R, VOO/R
Fixed Rates	30 ... 110
Rate-Adaptive	
Base Rate (BAR):	50 ... 110
Upper Rate (MAR):	60 ... 150
Pulse Amplitude:	2.4, 3.1, 3.6, 4.8, 6.2, 7.2 V
Pulse Width:	0.25, 0.5, 0.74, 1.0 ms
Sensitivity (V):	0.8, 1.6, 2.4, 3.2, 4.0, 4.8, 5.6 mV, Off
Refractory Period:	VVI/R: 350, 400 msec; VVI: 250, 300, 350, 400 msec
Magnet Rate:	90 ppm VOO, BOL, 80 ppm VOO, Elective Replacement Period
Safety Rate[1]:	BAR - 5 ppm
Runaway Protection[2]:	160 ppm
Energy Source:	Li/I, 2.55 Ah
Circuitry:	LSI-CMOS Hybrid Electronics
Housing:	Titanium
Lead Connection :	PIN-Lock PE
Dimensions:	66 x 48 x 10 mm

1) Back-up pacing rate in event of failure of rate-adaptive circuitry.
2) Patient protection in the event of oscillator failure.

The technical realization of the ANS-controlled pacemaker took advantage of much experience in the design of rate-adaptive systems based on motion, central venous temperature (CVT), and end-diastolic volume [59, 61, 70, 99]. The use of microcomputer technology and digital signal processing was previously and successfully implemented in the Biotronik Thermos, a CVT-controlled implantable VVI,R pulse generator [69]. The Biotronik Neos-PEP system is based on the same microprocessor architecture with the replacement of the temperature sensing by intracardiac conductance sensing.

In addition to the standard VVI multiprogrammable features of the Biotronik Neos, Mikros, and Nanos families, the range of automatic ANS-controlled rate adaptation can be preselected by the lower- and upper-rate limits. The implantable device functions like a normal multiprogrammable pacemaker. When the ANS control is on, the conductance-related functions are initiated through the programming unit. The CPU is activated by interrupts of the VIP time marks to reprogram its timer registers, thus triggering a pacing impulse. In case of a failure, a hardware pacemaker circuit automatically takes over the function of the software-controlled pacemaker. A hardware-based, high-rate protection provides an additional safety guard.

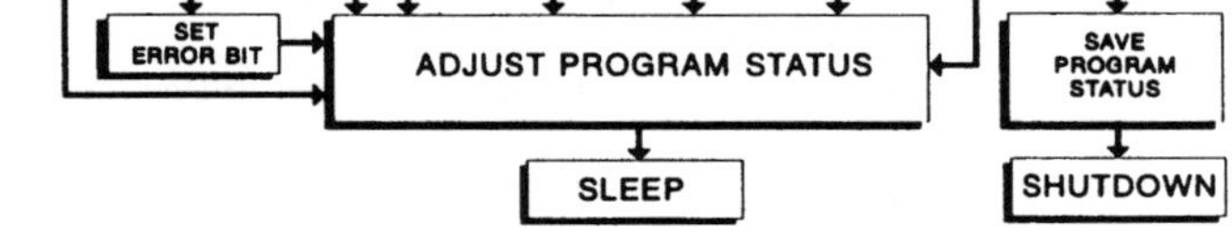

Figure 84. Simplified flow chart of software for ANS-controlled pacemaker.

To establish a certain flexibility, three *Region of Interest* (ROI) time windows are provided. Adjustments can be made in the rate response to VIP for optimal performance. The individual adjustment into the ANS-controlled, closed-loop system is activated by a simple calibration procedure. Over a short period of rest followed by exercise, the system monitors the intracardiac conductance and automatically determines the ROI and establishes the VIP to rate-transfer function.

The software consists of the following modules (Figure 84):

- Initialization after master reset
- Sleep/wake-up module (switching the processor on/off under supervision)
- Interrupt handling (selection of activity depending on HW- and SW-status)
- Communications (transmission and reception of data)
- Impedance measurements (time-sequence control, data acquisition)
- Pacing (digital filters, determination of the pacing rate)
- Recording (storage of impedance data in the RAM)
- System monitoring (self-checking the hardware and software)
- Error handling (error recovery)
- Switch-off (shutting down the ANS section)

Clinical Results

In the first phase of clinical evaluation, the ANS-controlled, closed-loop information was obtained using the PEP* algorithm. In a 68-year-old female patient, a PEP*-controlled system was implanted as replacement to an existing lead (9 years). As shown in Figure 85, the rate response to workload following a standard protocol demonstrated adequate closed-loop control, reestablishing chronotropy, taking the sinus rate (SR) as the indicator for optimal control.

In subsequent phases, ANS control was implemented in the device using the advantage of the RQ principle. A typical autonomic response to the exercise protocol is shown in Figure 86. Of particular interest is the physiologically desirable temporal rate response

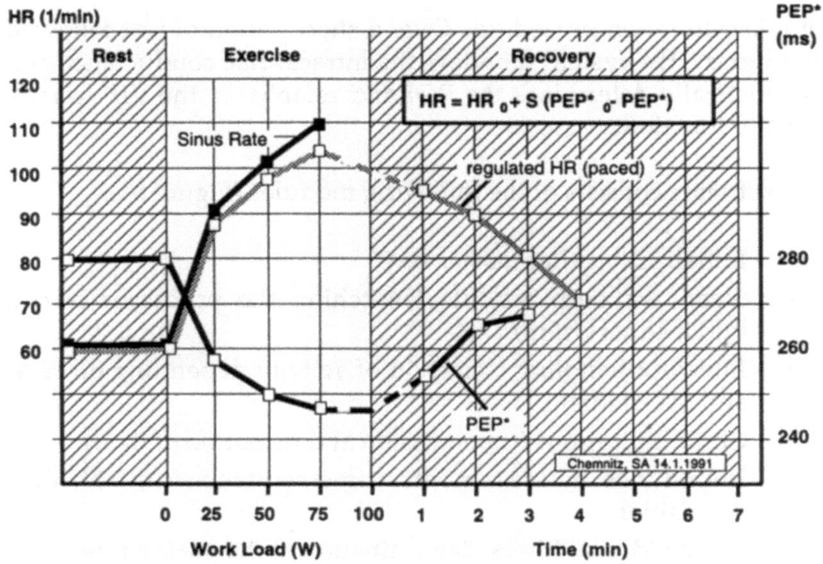

Figure 85. Comparison of PEP-controlled pacing rate to atrial rate in a patient with 3rd AV-blockage and intermittent sick sinus syndrome.*

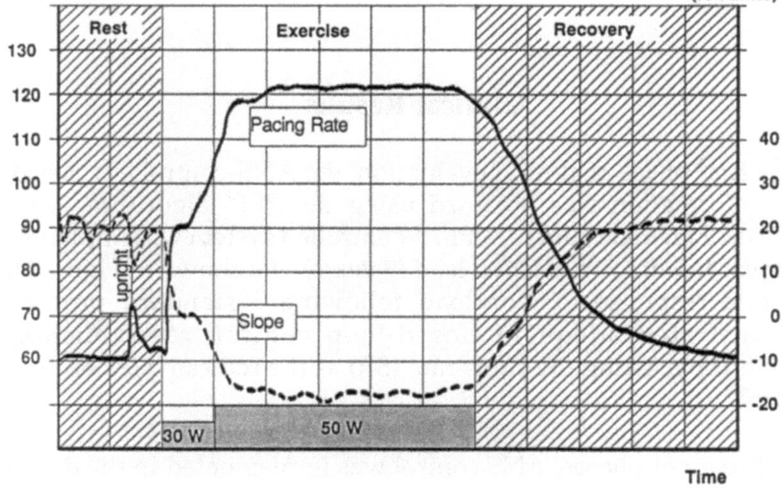

Figure 86. RQ control signal extracted from right ventricular conductance and pacing rate.

during exercise and recovery. Orthostatic changes during the transition from supine to upright position are supported with increase of heart rate.

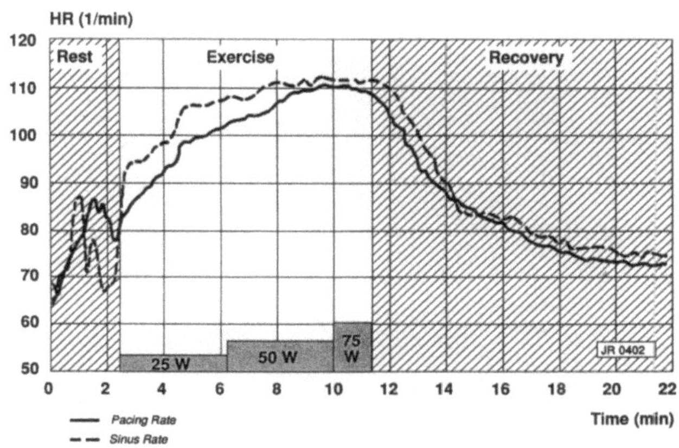

Figure 87. ANS-controlled pacing rate in a patient with AV-blockage during different phases of the evaluation protocol compared with the spontaneous sinus rate. Measurements from another patient (67-year-old male) make clear that with correct ANS-controlled frequency adaptation the mean atrial blood pressure rises only negligibly, even under a heavy workload (up to 100 W).

Using the RQ-time information, the cardiac pacing rate was modulated. In accordance with a standardized protocol that included changes in posture and an incremental increase in physical workload by bicycle ergometry, the pacing rate responded to sympathetic activity. Figure 87 shows an example of such an exercise protocol obtained from a 69-year-old male patient suffering complete AV block. The spontaneous sinus rate (broken line) allows comparison with the pacing rate (solid line). It is obvious that both curves agree during the entire exercise protocol, although there are minor deviations during the orthostatic challenge. Therefore, the excellent correlation of the curves demonstrates the quality of closed-loop controlled pacing.

Figure 88 illustrates a difference between the sinus rate and the pacing rate due to an incorrect parameter selection during another exercise test for the same patient. Under these conditions, the patient could only tolerate 56 W for about 1 minute instead of 75 W for 1 minute, as shown in Figure 87. Furthermore, during exercise, the sinus rate increases to nearly 140 beats per minute as compared to

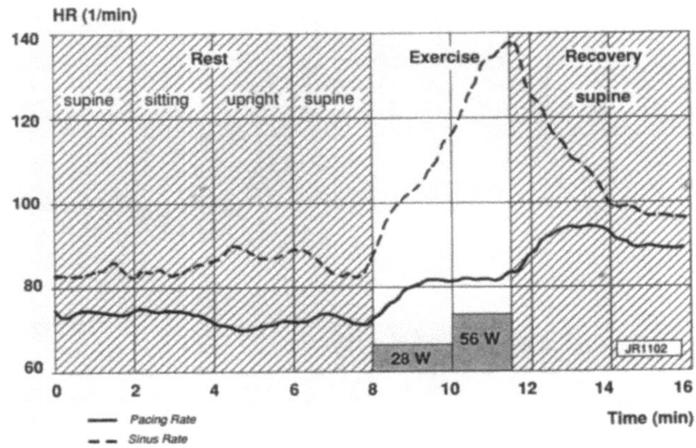

Figure 88. Same as previous patient, with ineffective pacing control due to insufficient detection of the ANS information.

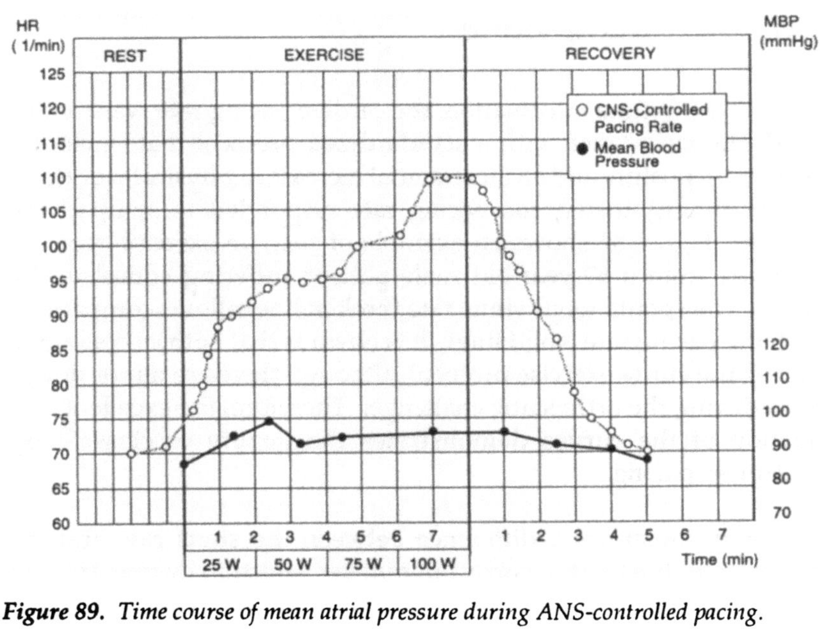

Figure 89. Time course of mean atrial pressure during ANS-controlled pacing.

about 110 beats per minute during proper ANS-controlled, rate-adaptive pacing (Figure 88). Another clinical example serves to demonstrate the validity of the principle: A 70-year-old male with sick sinus syndrome performed the standard exercise protocol (Figure 89). The ANS-controlled pacing followed the exercise in an expected way. The pacing rate during recovery follows in an appropriate profile down to the resting rate within 4 minutes. The excellent control of MABP, in this case, illustrates the value of VIP control. As expected, MABP remains constant within the tolerances available by the measuring methodology. The blood pressure remains stable during the 100 W physical workload.

The dynamic performance under normal daily exercise conditions is demonstrated in Figure 90. The patient's heart rate was adjusted in response to different types of exercise. The time course and magnitude of the rate adjustment were physiological. The ANS-controlled, closed-loop, rate-adaptive pacing allowed achievement of former exercise capability.

For an artificial pacemaker to achieve physiological rate adaptation, it must operate within a closed-loop control system that behaves very much like the physiological control system. With the ANS concept, this requirement can be met. The control circuit makes indirect but effective use of all the physiological elements that remain functional in a patient with chronotropic insufficiency:

- the baroreceptors with their static and dynamic characteristics,
- both afferent and efferent pathways of the ANS, and
- information processing in the circulatory centers of the medulla.

As a parameter for the representation of ANS activity, the ventricular inotropy parameter, VIP, fulfills all expectations and, in particular, is easily adjusted to the individual situation. The determination of VIP, by impedance measurement using the implanted unipolar electrode, is equivalent to a multimodal determination of all the disturbing factors that can affect the hemodynamic control system under everyday conditions. The results obtained thus far confirm the correctness of the ANS concept and the practicability of applying it.

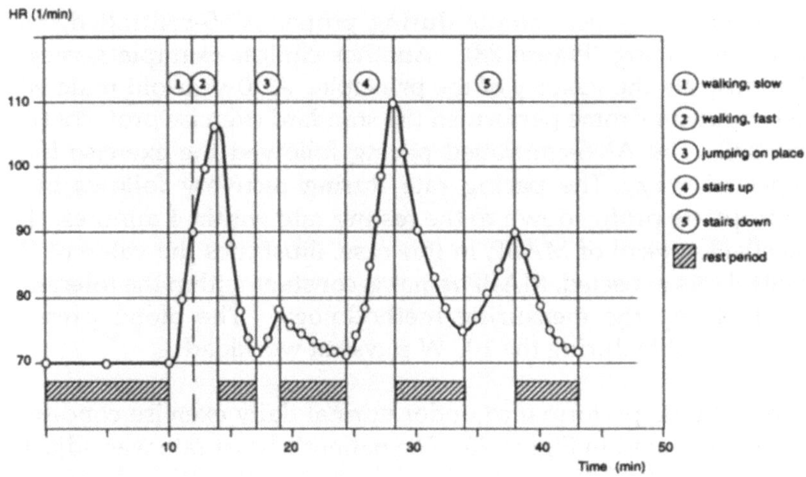

***Figure 90a.** ANS-controlled pacing rate during daily rest and exercise conditions.*

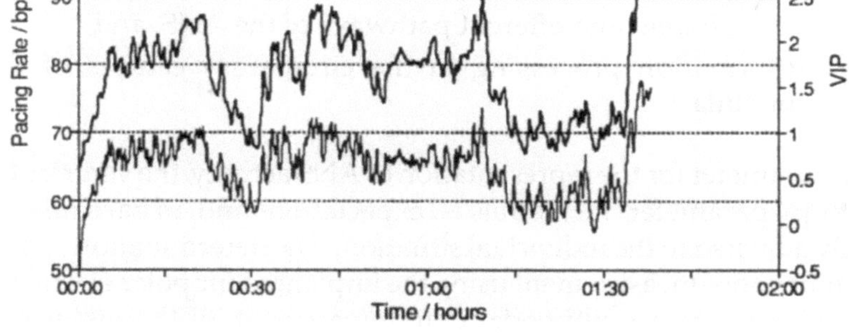

***Figure 90b.** Long-term monitoring of heart rate. Readout of the pacemaker's own data storage (24-hour).*

In addition to the measurement advantages of intracardiac impedance, the sensor system for the ANS-controlled pacemaker is accommodated by standard pacing electrodes. Therefore, these pacemaker systems can also be connected to existing implanted electrodes at the time of pacemaker replacement. The intracardiac impedance measurement provides the opportunity of including additional physiological information in the control loop, *e.g.*, to extract the posture-based preload information (orthostasis), as well as metabolic and ergonometric information (motion energy). The various items of physiological information can be extracted from the impedance signal by separating time-specific and rate-specific information. The sensor principle permits the control and supervision of electro-stimulation. Without additional technical effort, it is possible to implement a regulated pacing system using several physiological parameters that comprehensively contains the therapy for chronotropic insufficiency (Figure 97, page 142). For the first time, it is now possible to reestablish the physiological contraction sequence in a manner responsive to circulatory demand considering the inotropic condition of the heart.

Employment of Volume Information for Rate Adaptation

To increase cardiac output, the natural circulatory regulation has three mechanisms at its disposal:

1) augmentation of the end-diastolic volume as the result of an increased filling pressure,
2) reduction of the end-systolic volume caused by an increased contractility, and
3) increase in the heart rate.

In healthy individuals, only the last two of these mechanisms can be observed. The first mechanism, however, is an important factor in augmenting the cardiac output if the heart rate is low compared to the workload level. Thus, within certain limits, the natural circulatory regulation can adapt the cardiac output to match the actual requirements, even if the heart rate is constant. The same is true for artificially introduced rate changes at a constant workload. Again, the stroke volume will adjust in the opposite direction to the rate changes to keep cardiac output at or close to the level necessary to satisfy the metabolic needs.

This negative feedback between the heart rate and the stroke volume may be used to adapt the rate of an artificial pacemaker [70]. The pacemaker's task associated with a changing cardiac output consists of adapting the pacing rate so that the stroke volume remains within physiological limits, thereby preserving a range of variation for the stroke volume. This range, in turn, allows the natural circulatory regulation to adjust the cardiac output according to need. The heart is kept at an operating point that permits adaptation of the cardiac output through variation of the stroke volume within the framework of the existing natural circulatory regulation and myocardial reserve. Using this principle, the heart rate is controlled by a closed-loop system (Figure 91). Where the end-diastolic volume or the stroke volume represents the controlled variable, the pacing rate is the controlling variable, and the cardiac output the disturbance of the regulator.

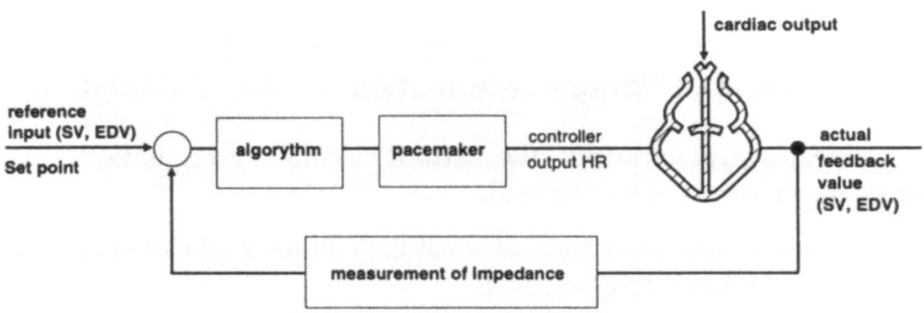

Figure 91. Control circuit for adaptation of pacing rate during exercise.

As previously discussed, the heart's ventricles have a very complicated geometrical shape. This makes it relatively difficult to determine the exact volume of the beating heart. Therefore, a model of the ventricle is generally used. Such models usually consist of relatively simple structures (*e.g.*, ellipsoids) or a summation of such structures. The available information is then entered in the model and approximating assumptions are made for the missing dimensions. If, as is the case here, only relative changes in volume are needed and the assumption of geometrically similar changes in shape is made, then the result is independent of the model used. This is because the volume of any model is proportional to the third power of its linear dimensions. Similarly, the information about a cross-sectional area can be used. In this case, the volume is proportional to the third power of the square root of the cross-sectional area [69].

Acquisition of a cross-sectional area (A) of the ventricular chamber is based on intracardiac impedance plethysmography. For this purpose, the impedance signal is recorded by means of a four-electrode arrangement (Figure 92).

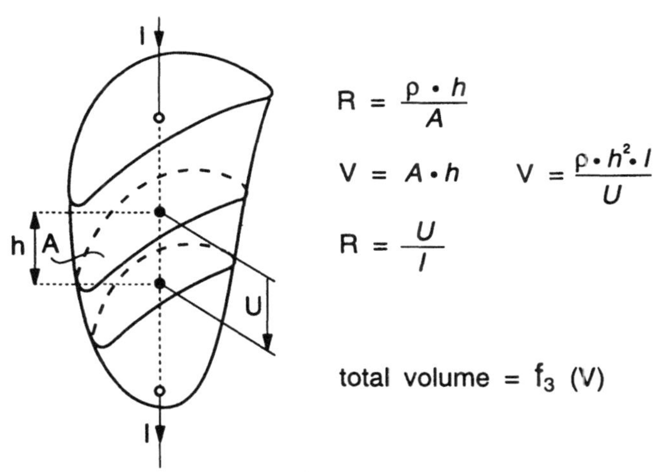

Figure 92. *Quadrupolar electrode design for intracardiac impedance plethysmography.*

Similar to the above-mentioned VIP principle, a current, I, is injected through the outer pair of electrodes and the resulting voltage, U, between the inner electrode pair is measured. At some distance from the feed-in electrodes, the applied current has spread out uniformly over the available ventricular segment. Consequently, a constant current density can be expected at the site of the voltage-measuring electrodes. Using a simple cylinder model, the cross-sectional area A is calculated as follows:

$$A = (h \cdot I)/(k \cdot U) \qquad (1)$$

(h = spacing of the voltage electrodes, k = specific electrical conductivity)

This representation of the measurement process, however, still contains two important simplifications, which are discussed below.

First, an exact cylindrical shape cannot be presupposed for the component volume measured. If one defines an average cross-sectional surface A corresponding to

$$A = (\Delta V/h) \qquad (2)$$

(V = component volume defined by the voltage electrodes),

every deviation from exact cylindrical shape causes a smaller average area A to be calculated from equation (1) for the cross-sectional surface A than is specified by equation (2). However, the resulting measurement error is negligible when the voltage electrodes are arranged at a small distance from one another.

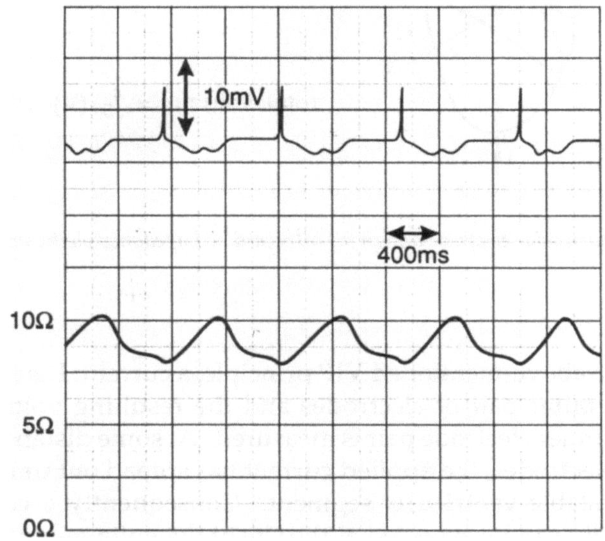

Figure 93. *Registration of the intracardiac electrogram and the impedance signal with measurement frequency 4 kHz and constant current ±10 μA*

Second, the intracardiac blood volume is not surrounded by insulating material as is assumed in the derivation of equation (1). The specific electrical conductivity of the surrounding tissue is clearly less, but nevertheless non-negligible. Part of the imposed current flows outside the intracavitary volume. If the measured voltage is inserted into equation (1), too large a value will, therefore, result for the cross-sectional area A.

This fact can be taken into account by taking the measured admittance Y^+ as a parallel circuit consisting of the admittance of the intracavitary blood volume and the admittance Y_p of the parallel-connected tissue:

$$Y^+ = (k \cdot A/h) + Y_P \qquad (3)$$

With equation 3:

$$A = (Y^+ - Y_P) \cdot h/k = (1/V - Y_P) \cdot h/k$$

Measurements have shown that Y_p, to a good approximation, can be assumed independent of the cardiac cycle.

The optimal electrode position is identical for pacing and for the impedance measurement; thus, the same catheter can be used for pacing and for determination of the geometrical intracardiac segment. Theoretical considerations and clinical studies with the described system confirm the necessity of the four-electrode arrangement for determining the volume. The right ventricular impedance signal is shown in Figure 93.

The technical implementation of the impedance-controlled pacemaker takes into account safety perspectives in connection with the use of microcomputers in implants. This objective was met by designing the digital circuit of the pacemaker to comprise three independent and different component systems:

- the microcomputer-controlled pacing system,
- the monitoring unit, and
- the redundant pacing system, which maintains simple security pacing in the event of failure.

Figure 94a. *Monolithic design of a universal rate-responsive dual-chamber pacemaker (Biotronik)*

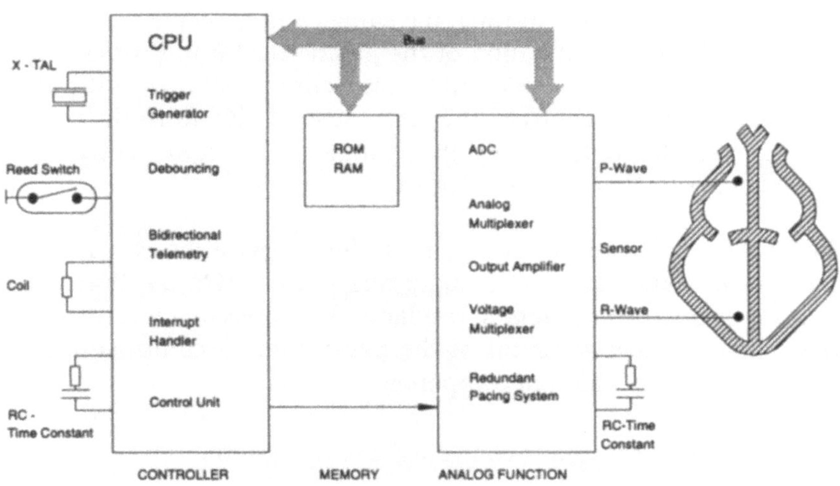

Figure 94b. *Block diagram of a universal rate-adaptive dual-chamber pacemaker controlled by multiple sensors including volume changes monitored in the right ventricle.*

The task of the monitoring unit is to detect faulty behavior of the microcomputer circuit. For this purpose, the monitoring unit is in a permanent dialogue with the microcomputer and also checks the pacing intervals with an independent time base. If erroneous signals from the microcomputer or runaway pacing is detected, the microcomputer is switched off, while the redundant pacing system is activated simultaneously.

The redundant pacing system involves a simple VVI pacemaker that operates from a third time base. It is activated only after a fault detection. During normal operation, the redundant pacemaker does not represent a load on the power source.

The safety concept of the implant offers the advantage that individual failure of one of the three component systems does not cause danger to the patient.

Technical progress, especially in the mastery of smaller Field Effect Transistor (FET) dimensions, and the introduction of low-voltage Complementary Metal Oxide Semiconductor (CMOS) processes, fulfills the preconditions for the technological realization of this regulated pacemaker concept. The application of microcomputers offers decisive advantages compared to so-called hard-wired logic because the application program can be stored in read-write memories and, therefore, can be replaced at any time even after implantation.

The pacing system electronic circuit has the following components:

On a single integrated chip (Figure 94a) there is a static 8-bit CPU comprised of

- a clock generator
- the monitoring unit
- a debouncing circuit for the magnetic reed switch
- a hardware interrupt handler
- 12 eight-bit timers (cascadeable, in part)
- bidirectional telemetry circuitry
- ECG detection and pacing circuitry for the atrial and ventricular channels

- an impedance measurement circuit
- an analog-to-digital converter (ADC) for the acquisition of physiological and operational data
- the redundant pacing system

Another 8k bytes of RAM is added on the hybrid substrate.

The block diagram in Figure 94b shows the functional units of the dual-chamber pacemaker with volume-controlled rate adaptation.

Figure 95 shows the left block, designated as "controller," and Figure 96 shows the right block, designated as "analog functions," with finer resolution.

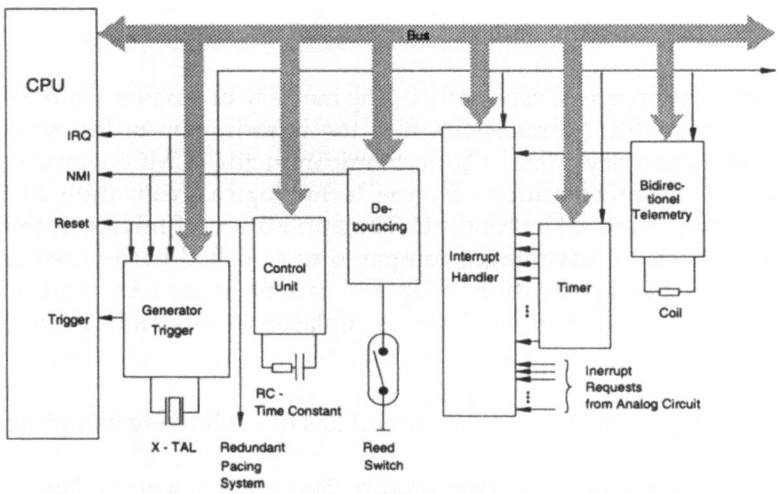

Figure 95. Block diagram of controller.

The microcomputer operates in real time. To achieve the lowest possible power consumption, the CPU switches its clock off during waiting times by a write access to the clock pulse generator. A service request from a peripheral unit causes the clock to be switched "on" again. Possible events that may cause the clock to be switched "on" are the following:

- The monitoring unit detects a fault (reset).
- The programming head is put in place, indicated by the closing of the reed switch (NMI).
- A service request occurs, which is forwarded by the priority switching mechanism (interrupt handler) to the CPU.

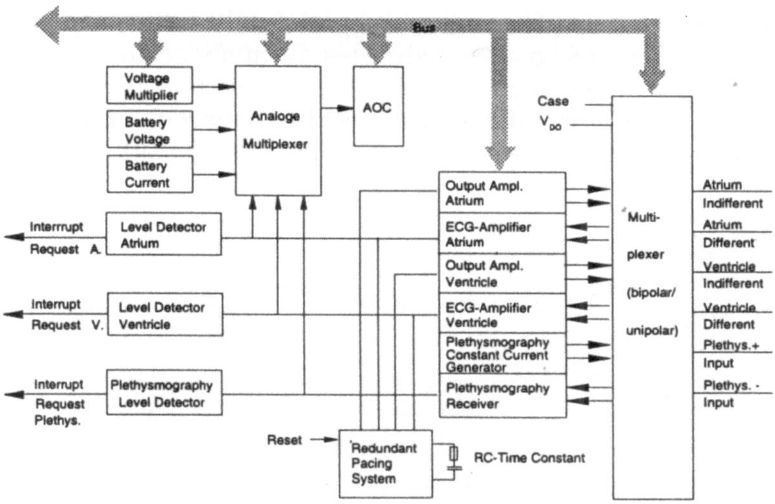

Figure 96. *Block diagram of analog circuit with redundant pacing system.*

The priority switching mechanism permits individual service requests to be disabled per program. If several enabled service requests are pending, a logic circuit in the hardware determines the priority of processing; *e.g.*, the CPU is activated by the crossing of a threshold in the level detector for the atrial electrogram (EGM). It then programs a timer with the atrio-ventricular delay and again switches itself off. The next event that activates the CPU is either the detection of a ventricular EGM—if the conduction system of the heart has transmitted the excitation to the ventricle—or the lapsing of the timer, if an AV block is present. In the latter case, the CPU triggers a ventricular pacing pulse by means of a write instruction to the pacing output stage. Possible subsystems that can request servicing are the telemetry circuit, the timers, the level detectors of the analog section, and the ADC. The signal from the reed switch and monitoring unit cannot be disabled.

Conclusions

The pacemaker's electronic resource is able to handle more challenging tasks, *e.g.*, to respond appropriately to acute pathophysiological conditions. By means of an adaptive learning algorithm, it seems possible to automatically select the preferred pacing mode and pacing rate to be determined from the metabolic, contractile, and hemodynamic data. The engineering task for the future is to make the pacemaker intelligent enough to handle minor acute pathological episodes. However, this biomedical objective presupposes detailed knowledge of the relationships between sympathetic activity, type and degree of stress, and individual factors.

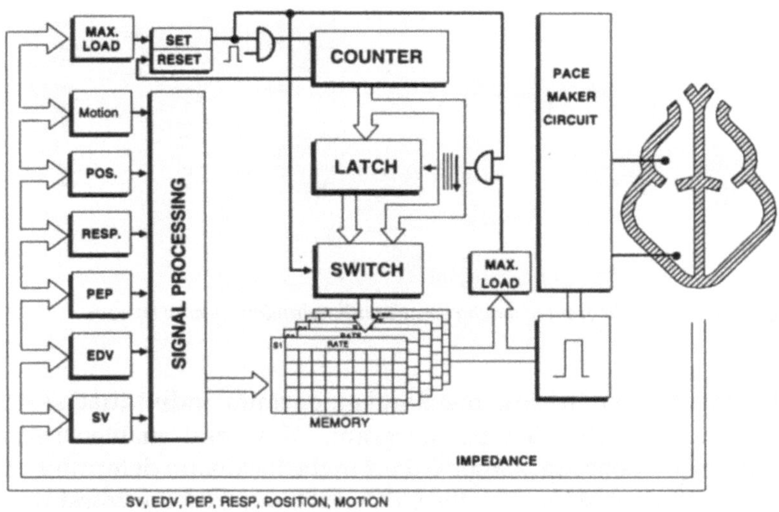

Figure 97. Universal concept of a multiparameter rate-responsive pacemaker based on intracardiac impedance plethysmography.

To assist with the future design of rate-adaptive and intelligent pacemakers, a valid regulatory model of the circulation needs to be created. Such a model would allow the biomedical engineer to test the algorithm to make certain that the exercise-related increase of cardiac output among various patient groups is comparable to that of a control group. In the design of the model, system regulation dynamics must receive close attention. The model should include the hydrostatic and hydrodynamic conditions of the cardiovascular system and must incorporate physiological control pathways [118]. In addition to simulation of the hemodynamics, the model must also represent the cardiac pumping function, the neural and humoral control of the heart and the peripheral circulation, workload, ambient temperature, and fluid exchange. The physiological modeling task remains multi-disciplinary in nature.

THE STIMULATING ELECTRODE

The principal method of artificial cardiac pacing involves the transvenous endocardiac placement of electrodes, usually unipolar, which transmit cathodal stimulus pulses to the myocardium as introduced by Furman in 1958 [119]. The response of an excitable cardiac muscle cell to an external stimulus depends on the electrical field strength—that is, on whether the field depolarizes the diastolic membrane potential sufficiently to generate spontaneous depolarization of the cell—an all-or-none AP which propagates excitation of the entire myocardium. Physiological experiments in which intracellular electrodes are used to excite the heart have shown that charge-carrier injection of 10^{-11} As suffices to trigger the AP [120]. Artificial pacing with the electrodes currently in common use requires a greater charge, about 10^{-6} As.

This difference in the amount of energy needed to excite the heart suggests the following technical goals of artificial pacing:

- The physical nature of the stimulating current must be such as to avoid irreversible redistribution of the intra- and extra-cellular ions resulting in tissue changes. This is a crucial prerequisite if the electrode is to function for the rest of the patient's life.
- Current consumption should be minimized so that the unit functions as long as possible with the energy stored in the battery.

Electrode/Myocardium Interface

The surface of the electrode tip and the immediately surrounding tissue have the characteristics of an electrochemical phase boundary. That is, on one side the electrical current is carried by electrons, and on the other side, by positively and negatively charged ions. The concentration and electrical conductance of the electrons differ by orders of magnitude from those of the various ions. Furthermore, it is characteristic of such a phase boundary that at the transition between metal and electrolyte, the electrons must overcome a reaction inhibition.

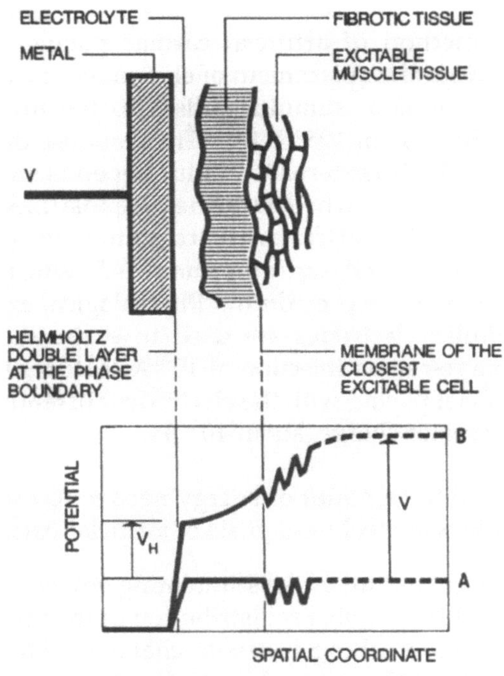

Figure 98. *Schematic illustration of the phase boundary between stimulating electrode and myocardium showing the local potential distribution* ***without*** *an applied stimulus voltage (A) and* ***with*** *an applied stimulus voltage (B).*

The geometric arrangement of the phase boundary and the potential across it are shown in Figure 98. The excitable cardiac muscle tissue is not in direct contact with the metal surface, but is separated from it by a layer of inexcitable fibrous tissue, a pathological change induced by the electrode. This tissue reaction to the presence of a foreign body takes the form of a chronic inflammation, limited to the site at which the stimulus acts, which extends only to the immediately adjacent myocardium. As a result, the electrode gradually becomes encapsulated in connective tissue. A delicate connective-tissue sheath encloses the electrode tip after only four weeks.

Because of the diversity of the electrical charge carriers in the metal, in the cell membranes and in the intra- and extracellular space, even without current flow, there is a complicated potential distribution in this region, illustrated schematically by curve A in Figure 98. It is characterized by abrupt changes at the so-called Helmholtz doublelayer and at the cell membranes. When an external stimulus voltage (V) is applied, the entire potential curve is shifted by this amount (curve B). A pacing stimulus is effective if, at a favorably located cell membrane in the excitable tissue, the membrane potential is lowered sufficiently to trigger a myocardial action potential (AP).

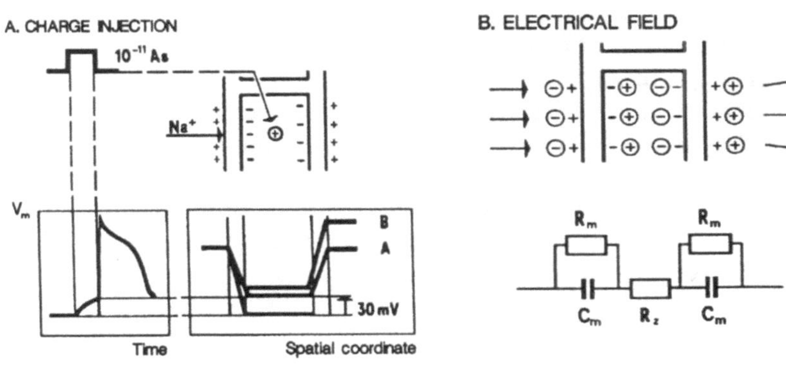

Figure 99. *Artificial stimulation of a myocardial cell (upper portion) by injection of charge carrier through a microelectrode and (lower portion) by an external field.*

Experiments have shown that the stimulation voltage required to exceed the threshold for myocardial contraction rises as the duration of the pulse decreases. It is clear *a priori*, then, that the excitable myocardium cannot be approximated by a simple ohmic resistance;

however, the conductivity of the tissue may be approximated by an RC network.

A precise description of its electrical properties is made more difficult by the fact that the tissue is not isotropic; it is a complex structure in which the cells and fibers are interlaced in three dimensions. The basic electrical organization of a single excitable cell is shown in Figure 99. The crucial elements are the cell membrane, the interior of the cell, and the extracellular space.

The active charge transport processes take place across the living membrane and can be described by a nonlinear component to the electrical properties of the myocardium. Nonlinear relationships analogous to those in membranes also characterize the metal surface. The interface between metal and electrolyte behaves like the cell membrane in a certain range of potentials. The reason lies in the "impermeability" of the phase boundary to the electrolyte ions and, in the obstacle it presents, to the electrons of the metal.

A direct electrical current toward the cathode can arise only by passage of electrons through the phase boundary and their reactions with suitable components of the solution. Current transport toward the anode is brought about when solution components, under the influence of the electrical field, enter into reactions at the phase boundary that result in the transition of electrons into the metal. All electrochemical reactions of this kind are subject to specific forms of inhibition, associated on the one hand with the limited diffusion of solution components to and from the boundary and, on the other, with the movement of electrons across it.

Helmholtz Doublelayer

The phase boundary between metal and electrolyte, according to the above description, acts primarily like a capacitor. As shown in Figure 100, this capacitor consists of the metal surface and those ions closest to the metal surface. Most of the lattice sites on the electrode surface are occupied by water molecules. Hence, there is a first layer of water molecules, adsorbed to the metal surface, and a second layer consisting of part of the hydration envelope of the ions that have accumulated at the phase boundary under the influence of the electric field. This arrangement is known as the Helmholtz doublelayer. It corresponds to a plate capacitor, the capacitance C_H which is given by the distance separating the opposite charges and by the mean dielectric constant of the water-adsorption layer. The capacitance of the Helmholtz layer at the smooth metal surface is of the order of 5 to 50μF/cm². As shown in Figure 100, the current of a metal electrode consists of three basic components:

A = The transport of electrons to the surface and of positive ions into the vicinity of the surface, with no crossing of the phase boundary by any charge carrier, is a purely capacitive current not associated with any electrochemical reaction.

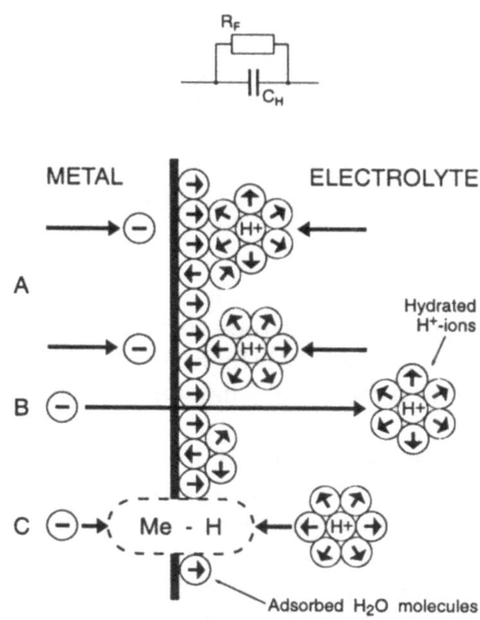

Figure 100. *Parallel plate capacitor model of the metal/electrolyte phase boundary during cathodal polarization. A - capacitive charge redistribution; B - electrochemical reaction; C - surface redox reaction.*

B = As soon as electrons pass through the phase boundary and react with components of the solution, a so-called Faraday current flows; it is represented in the equivalent circuit diagram of the electrode by an ohmic resistance R_F. It is important to point out that this faradic resistance R_F is not constant but depends greatly on the electrode potential. A typical example of a cathodal reaction is the reduction of dissolved oxygen

$$O_2 + 2\,H^+ + 4e^- \rightarrow 2\,OH^-$$

Other possible reactions include the electrolysis of water with the production of hydrogen and the reduction of biological redox systems.

C = In addition, there can be surface redox reactions associated with the adsorption or desorption of solution components and changes of the oxidation state of metal atoms at the electrode surface. A typical example of such redox reactions is the formation of a surface hybrid, which can occur only with a platinum or platinum/iridium electrode:

$$Me + H^+ + e^- \rightarrow Me\text{-}H$$

Faradic currents of Type B are negligible with modern stimulating electrodes. Only when the electrode surface is very small (less than 1 mm^2) and the current density is high, do electrochemical reactions of this kind occur and these are usually irreversible.

Figure 101 summarizes the most important properties of the metal/tissue phase boundary in an equivalent circuit diagram. Both the cell membrane and the electrode surface are represented by parallel capacitance and resistance. The ohmic resistance is composed of the resistance R_F which arises in the electrolytes, and the fibrous layer between electrode and excitable tissue, the resistances R_m, R_2 and R_i including the extracellular fluid and the resistance R_2 of the remainder of the tissue between the stimulating and the indifferent electrode. The diagram is a unidimensional approximation; in reality, especially in the case of the excitable cell, the components form a complex three-dimensional network.

Application of a constant battery voltage U_b from a pulse generator produces voltage changes and current flow, the time course of which is crucially determined by the capacitive components in the equivalent circuit diagram. During the pulse, the capacitors C_O, C_H and C_m are charged through the ohmic resistances, and when the voltage U_b is turned off, the capacitors discharge with current flow in the opposite direction.

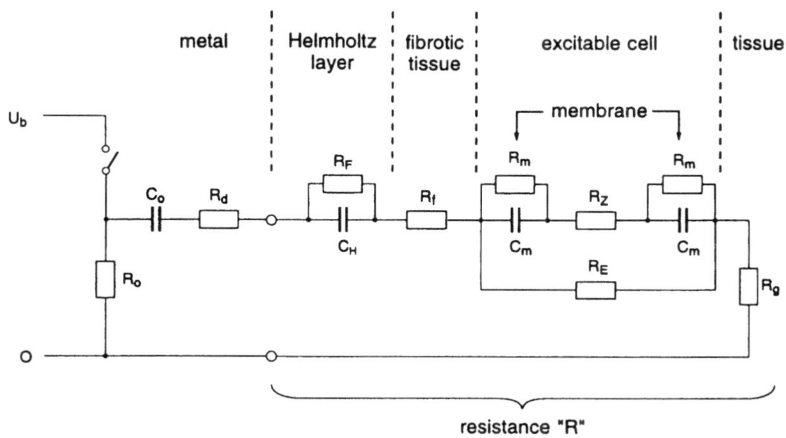

Figure 101. Properties of the metal/tissue phase boundary (R_E = extracellular resistance).

Electrical Characteristics

Improvements in the electrical properties of the stimulating electrode would be desirable on three fronts [121]:

1. Minimizing the energy required for stimulation.

2. Maximizing the detected signal amplitudes.

3. Minimizing the afterpotential as a prerequisite for reliable assessment of efficacy.

For all three of these requirements, the stimulating electrode—in particular its phase boundary with the surrounding electrolyte or tissue—is a crucial element.

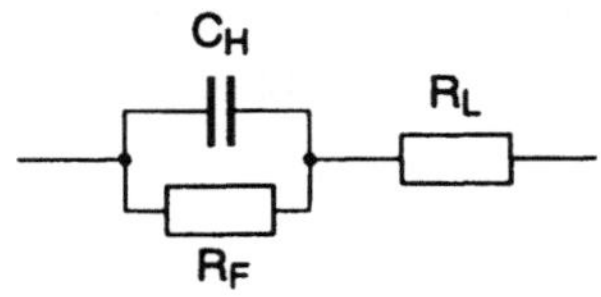

$$Z_{DL} = \frac{1}{\frac{1}{R_F} + i\omega C_H} + R_L$$

Figure 102. Simplified equivalent circuit of an interface.

In terms of electrical engineering, the phase boundary between a stimulation electrode and an electrolyte can be represented in simplified form by a circuit consisting of a Helmholtz capacitance C_H and a faradic resistance R_F in parallel, connected in series to a lead resistance R_L(see Figure 102). Therefore, the impedance of the electrode system Z_{DL} depends on the frequency ω of the applied voltage, as given by

$$|Z_{DL}| = \left(\frac{R_F^2 + 2\, R_F\, R_L}{1 + R_F^2 \omega^2 C_H^2} + R_L^2 \right)^{1/2}$$

According to the "fundamental law of electrostimulation" [122], to excite cardiac muscle, a certain charge is required, which can be calculated as the integral of the stimulating current I(t) over the pulse width T. If the time-dependent current is taken to be periodic, it can be expressed as a Fourier series, giving to a first approximation

$$Q(T) = \int_0^T I(t)dt = \int_0^T \left(c + \sum_{\omega=1}^{\infty} \frac{U\omega}{Z_{DL}(\omega)} \cos(\omega t) \right) dt$$

The stimulus pulse has a width of about 0.5 to 1 ms; that is, the pulse has a frequency spectrum with a bandwidth of a few kHz as shown in Figure 103. Therefore, to meet the first of the above requirements, the interface impedance should be made low for this bandwidth, which, in practical terms, means that the Helmholtz capacitance should be maximized. Large surface area, equivalent to a high Helmholtz capacitance, results in low cut-off frequencies and low

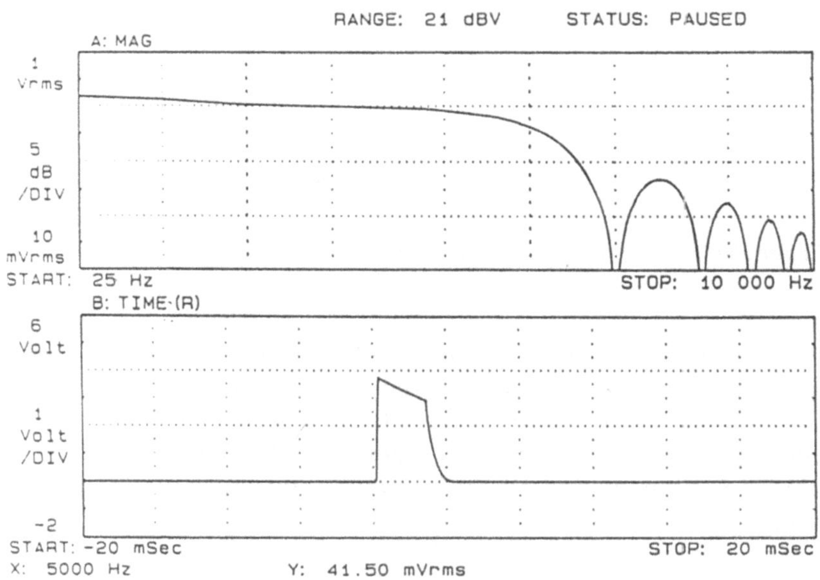

Figure 103. *Frequency analysis of a stimulation pulse.*

impedances over a large frequency range (see Figure 104). Similar requirements apply to the sensing performance. As shown in Figure 105, the frequency analysis of an intracardiac electrogram contains components up to 100 kHz. Therefore, in the application of smooth as well as sintered electrodes, attenuation of significant signal components occurs, making the signal detection task for the pacemaker circuit difficult. A high Helmholtz capacitance fulfills both requirements of low threshold and good sensing performance.

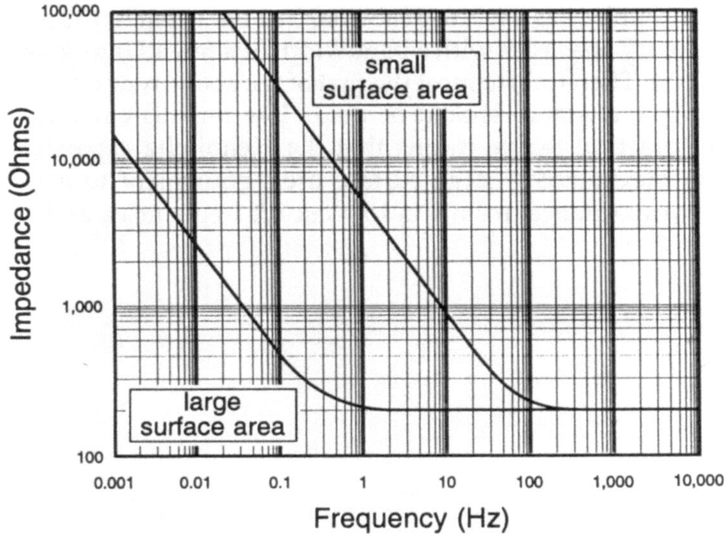

Figure 104. *Effect of electrode surface area on impedance.*

An entirely analogous conclusion is derived with respect to the second point listed above, since the voltage signal detected by the electrode U_{det}, a proportion of the actual intracardiac voltage U_{ECG}, is determined over all frequency ranges by the total impedance of the electrode system and the pacemaker input Z_s and the phase-boundary impedance, according to the equation

$$U_{det} = U_{ECG} \left(\frac{Z_S - Z_{DL}}{Z_S} \right)^{-1}$$

The third and last requirement, that the pacing electrode should exhibit little afterpotential, can also be described mathematically. Neglecting the faradic resistance and taking the charge voltage U_r across the Helmholtz capacitor to be constant during the stimulus pulse, the voltage drop at the capacitor U_C during the pulse is given by

$$U_C \, (t < T) = U_r \left(1 - exp\left(\frac{-t}{R_L C_H} \right) \right)$$

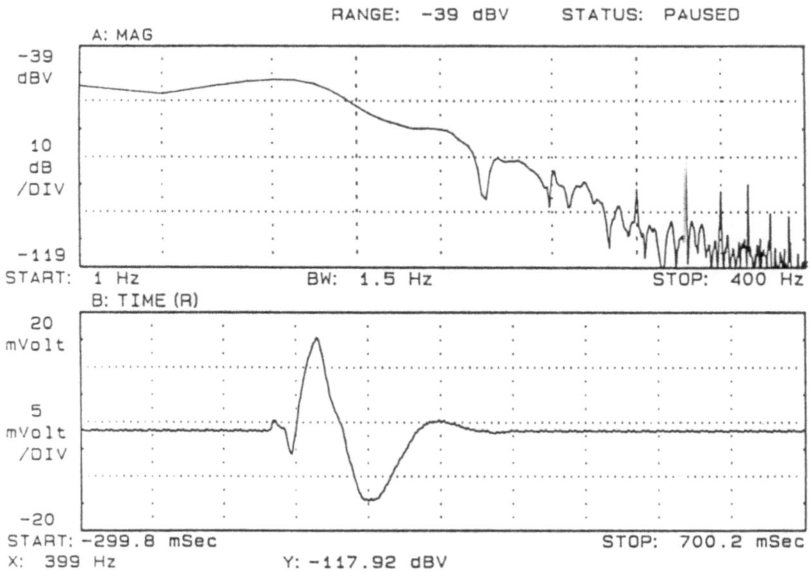

Figure 105a. *Frequency analysis of intracardiac ECG signals.*

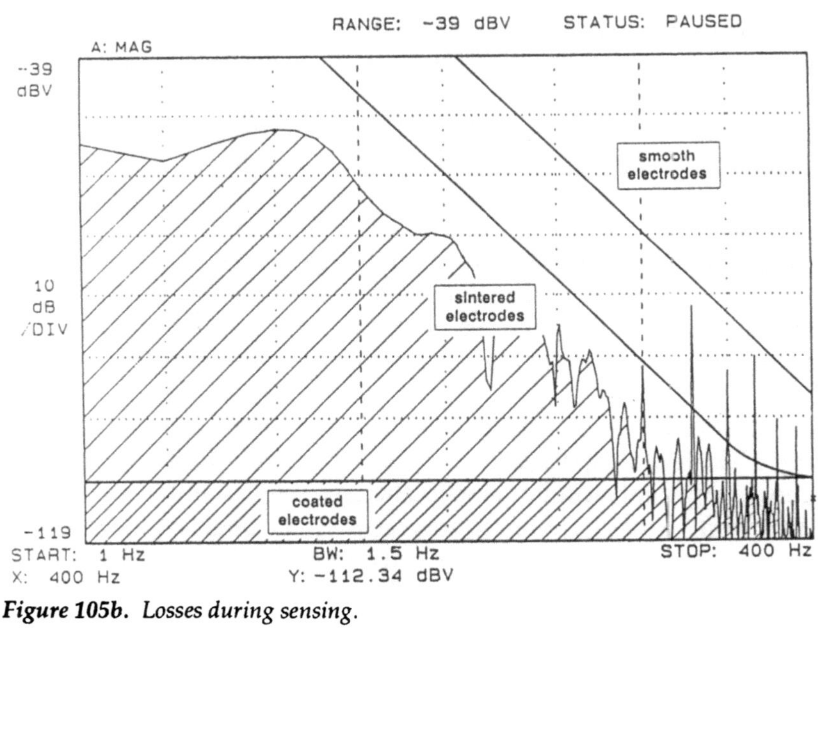

Figure 105b. *Losses during sensing.*

and the afterpotential by

$$U_C\,(t>T) = U_r\left(1 - exp\left(\frac{-T}{R_L C_H}\right)\right) \exp\left(\frac{-(t-T)}{R_L C_H}\right)$$

Here, again, an increase in the Helmholtz capacitance produces the desired result: a smaller afterpotential and, hence, an optimization of the pacing system. An increase in R_L would have the same effect with respect to polarization but is unacceptable because it would increase energy consumption and impair the sensing function of the electrode.

All three properties expected of stimulating electrodes can be combined in a single criterion that the Helmholtz capacitance C_H be maximized, with

$$C_H = \varepsilon \cdot \varepsilon_0 \frac{A}{d}$$

That is, capacitance is proportional to the dielectric constant ε of the adjacent water dipole and to the active surface area A, and inversely proportional to the thickness d of the Helmholtz layer. Whereas ε and d are hardly influenced by varying the material or structure of the solid body, the active surface area can easily be changed simply by enlarging the electrode. However, measurements of the chronaxie-rheobase curves for electrodes of different sizes but made of the same material have shown that enlargement of the electrode increases the charge required for stimulation approximately in proportion to the surface area and, hence, offers no solution. This result can be explained by the change in size of the region influenced by the electrode; in a refinement of the "fundamental law of electrostimulation," constant charge density should be specified.

It follows from these considerations that it will be useful to increase the Helmholtz capacitance only if the current-limiting electrode geometry is not simultaneously changed. As an initial step in this direction, considerable improvement has been observed with sintered pacing electrodes constructed of platinum-iridium or titanium alloys [123, 124] and with platinized platinum electrodes [125]. By use of a porous network of small spheres, the active surface area and the interface capacitance has been more than doubled (see Figure 106). The crucial breakthrough, however, was not achieved until porous superficial layers of suitable materials were applied to traditional electrodes by modern vacuum-coating procedures such as sputtering or ion-plating. The fractal geometry (see Figure 107)

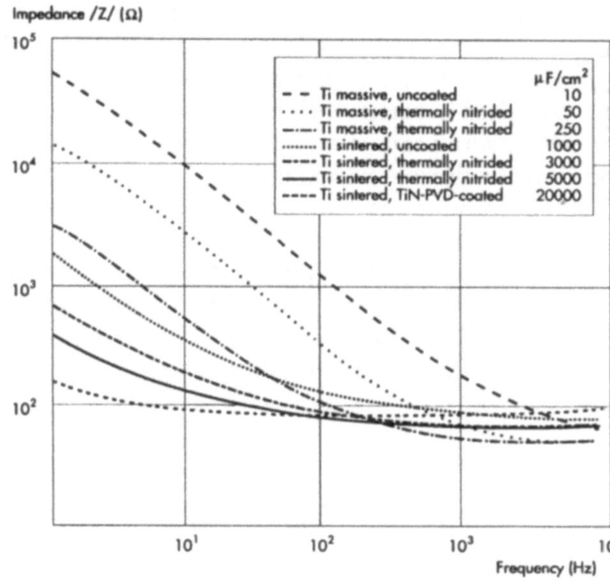

Figure 106. *Frequency dependence of Ti-electrodes with different coatings.*

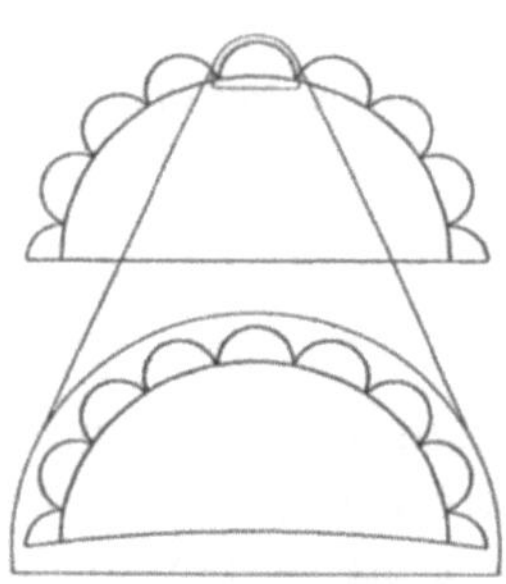

Figure 107. *Fractal folding of the surface to demonstrate the increase in Helmholtz capacitance.*

of the layers ensures a surface enlargement by a factor of 1000 or more [126]. Clinical experience with an electrode coated with titanium nitride (TiN) has confirmed that this principle does, indeed, provide a workable solution [121].

One problem, however, remains. Because of the tendency of these materials (especially titanium) to oxidize, a thin insulating or semi-conducting oxide layer forms at the electrode surface. This layer amounts to a capacitor in series with the Helmholtz capacitor and, hence, reduces the overall capacitance (see Figure 108). Under equilibrium conditions or during cathodal polarization, its influence is negligible [127], but under anodal polarization, OH^- ions drift into the solid material and increase the thickness of the oxide layer. The result is a further reduction of the interface capacitance and hence an increased electrode impedance. As a consequence, for instance, it restricts the application counterpulses to TiN electrodes such as are used to evaluate efficacy of stimulation.

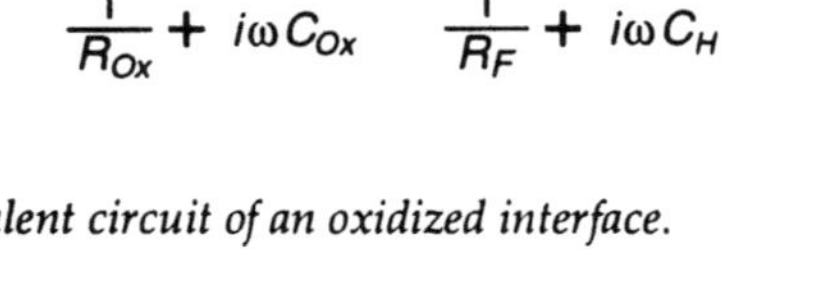

$$Z = \frac{1}{\frac{1}{R_{Ox}} + i\omega C_{Ox}} + \frac{1}{\frac{1}{R_F} + i\omega C_H} + R_L$$

Figure 108. *Equivalent circuit of an oxidized interface.*

Inert materials such as platinum and iridium offer a solution to this problem. When these are vacuum deposited, they form a fractal superficial layer with a large active surface while still ensuring long-term stability. Furthermore, a slight boron doping additionally increases the conductance of the surface.

Electrode Materials

Materials suitable for pacing electrodes must be well tolerated by the body and resistant to corrosion. The materials chosen according to these criteria are either platinum and its alloys or cobalt-based alloys. Cobalt-based alloys prove especially advantageous for the lead spiral. However, as a material for the electrode tip, both materials still resulted in significantly restricted sensing performance. The disadvantage was the high-pass behavior of the system formed by the electrode and the signal amplifier, resulting from the mainly capacitive properties of the electrode, and the relatively low input resistance of the signal amplifier. Significant improvements were made by optimization of the metal surface.

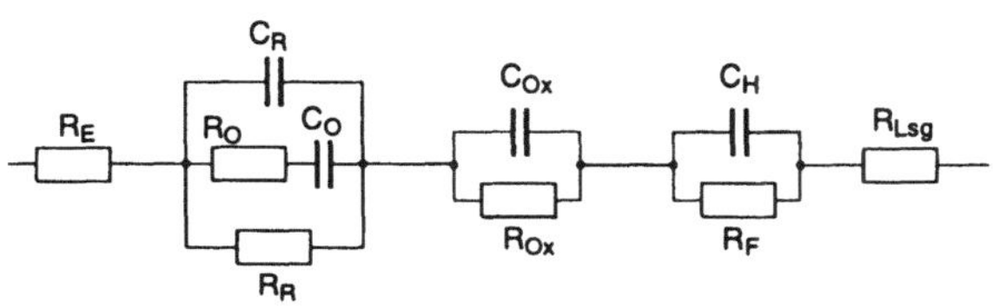

$$Z = R_E + \frac{1}{\frac{1}{R_R} + i\omega C_R + \frac{1}{R_O + \frac{1}{i\omega C_O}}} + \frac{1}{\frac{1}{R_{Ox}} + i\omega C_{Ox}} + \frac{1}{\frac{1}{R_F} + i\omega C_H} + R_{Lsg}$$

Figure 109. *Equivalent circuit diagram of the entire stimulation system.*

In a first attempt to improve the electrode, its surface was coated with a low-conductivity layer by anodal oxidation (Ta_2O_5 and TiO_2), which produced purely capacitive coupling [128, 129]. With these electrodes, chemical reactions that changed the tissue irreversibly—with the consequence, for instance, of an increased threshold—were suppressed and the electrodes were stable over long periods. However, the amplitude of the sensed signal was so severely attenuated that the electrodes were unusable. This experience demonstrated that the behavior of the phase boundary can be ascribed to the structure of the doublelayer [130-133]. The simplified equivalent circuit diagram (Figure 109) of the electrode/myocardium contact shows that the potential distribution is basically determined by the

Helmholtz capacitance and the Faradic impedance. The greater the doublelayer capacitance, the smaller the losses in stimulation and the larger the amplitudes of the sensed signals.

Qualitatively, then, porous electrodes with fractal surface structure are preferable to smooth metal electrodes as described [128, 134]. It is clear that technical improvements of the pacing system are achievable by the provision of surface layers. Regarding biocompatibility, compounds of titanium, tantalum, niobium, and zirconium are candidates, in the form of oxides, carbides, and nitrides [135-137]. The conductivity of the layer can be adequately adjusted by stoichiometric alterations; thus, titanium dioxide and tantalum pentoxide have been successfully used to coat electrodes or to produce sintered elements [129, 138]. The same applies to electrodes made of sintered carbon, which are known to be biocompatible and have a microporous structure [139-143]. In addition, fractal coating of platinum and iridium improves the aging performance by two orders of magnitude, as shown below.

The necessary characteristics of electrode materials can be summarized as follows:

(a) the material must be tolerated by the body so connective-tissue reactions are suppressed;

(b) the doublelayer capacitance must be as large as possible in order to keep the polarization losses during the pacing pulse small and, for sensing purposes, to ensure low impedance down to the lowest possible frequencies.

The large doublelayer capacitance in itself, by lowering the threshold amplitude of the pacing pulse, favors suppression of electrochemical tissue reactions. Furthermore, because less current flows during the sensing of intracardiac signals, the signal-to-noise ratio is improved.

Porous Electrode

Initial approaches toward a technical implementation involve making the electrode itself of a porous material such as Ta_2O_5, activated glassy carbon, etc. As a further development in this direction, the electrode is coated with TiN, a material well tolerated by the tissues [144, 145]. In principle, a variety of similar electrodes can be manufactured by coating metallic substrates with the carbides and nitrides of Ti, Zr, and Ta by *physical vapor deposition* (PVD) or *chemical vapor deposition* (CVD) processes [146, 147]; however, the

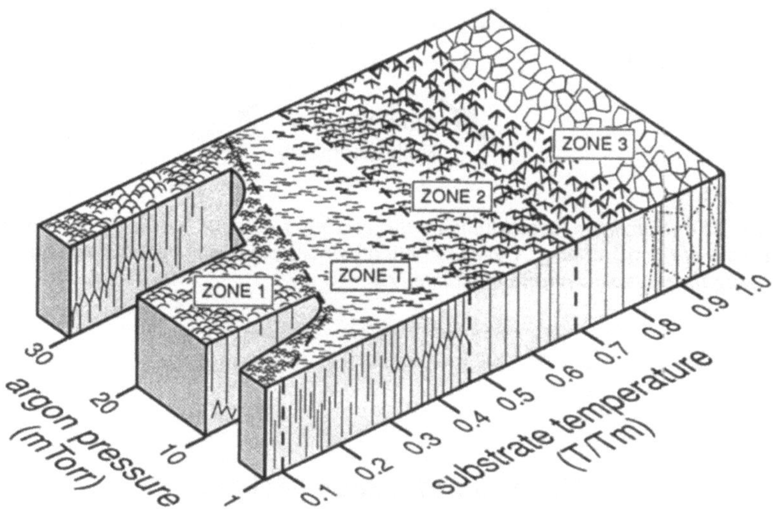

Figure 110. *Dependence of surface structure on deposition parameters [127].*

most advantageous has proved to be the PVD coating of Ti electrodes, which creates an outer layer with the desired characteristics. When the angle of incidence of the material condensing on the substrate is restricted and the substrate is kept at a low temperature (as compared with the melting point of the coating material), the surface layer tends to grow in the form of columns separated by gaps. The superposition of this columnar growth, as illustrated in Figure 110, can be used to describe the dependence of deposition structures such as TiN and IrN [148].

The anisotropic character of the layer affects both its electrical properties and its action as a barrier. Processes involved in growth of the layer and interactions are diagrammed in Figure 111.

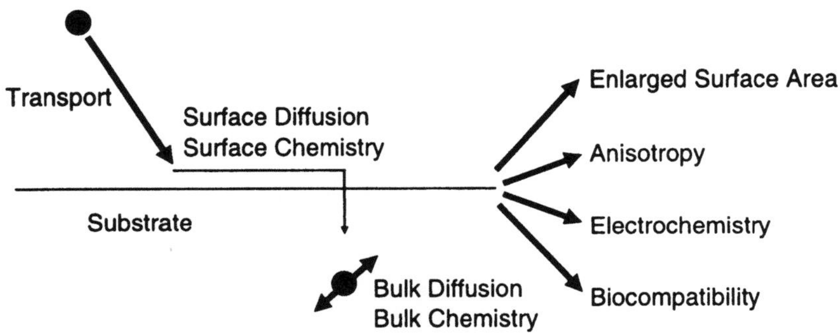

Figure 111. *Simplified description of the coating technique resulting in different properties of the materials. A wide variety of surface parameters can be generated by changing the physical deposition parameters.*

TiN as an Electrode Material

Other major applications of the PVD procedure—whether in microelectronics or to provide increased hardness and wear resistance—require that the surface coating be dense and nonporous. For stimulating electrodes the aim is to produce an open, porous structure that increases the surface area and endows the electrode with a large doublelayer capacitance [149, 150]. The generally undesirable flaws formed in the layer as it grows are advantageous here. To this end, one important factor is the low mobility of the atoms being deposited at a low substrate temperature, which promotes the formation of gaps in the shadows cast by the crystals growing from the condensation nuclei; another is the suppression of high-energy ion bombardment with the consequent low rate of back-scattering. The substrate topography is also significant. For example, a topography that provides for oblique angles of incidence enhances shadowing.

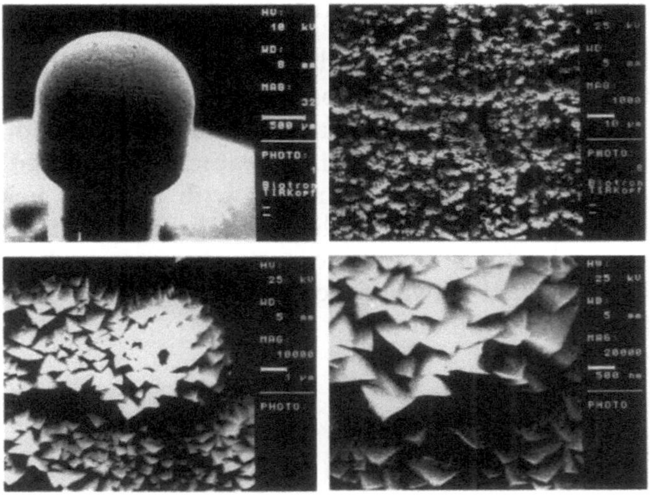

Figure 112. *Microstructure of the TiN layer on pacemaker electrodes.*

The scanning electronmicrographs of Figure 112 show the structure of the TiN layer at various magnifications. The layer is a few micrometers thick. The high porosity, clearly visible here, gives a doublelayer capacitance of about 10 µF/cm^2 under implantation conditions. The physical properties of TiN are summarized in Table XII.

Table XII
TiN Physical Properties

Characteristic	*Value*
Molecular weight	61.9
Density	5.21 g/cm^3
Lattice	cubic face centered
Lattice constant	42.3 nm
Hardness	1800...2500 HV100
Thermal expansion	9.35×10^{-6} / K^{-1}
Conductivity	4×10^4 (Ohm cm)$^{-1}$

Comparative Studies

Data from potentiostatic impedance measurements (Figure 113) show that *iridium nitride* (IrN)-coated electrodes have considerably better transfer characteristics than electrodes coated with TiN, Ti (Boron-doped) as well as platinized platinum or sintered carbon. Clinical findings confirm this result: with unipolar ventricular electrodes, both the pacing threshold and the intracardiac potential live up to expectations. The intraoperatively measured R-potentials significantly exceed the values for comparable electrodes made of activated, sintered carbon.

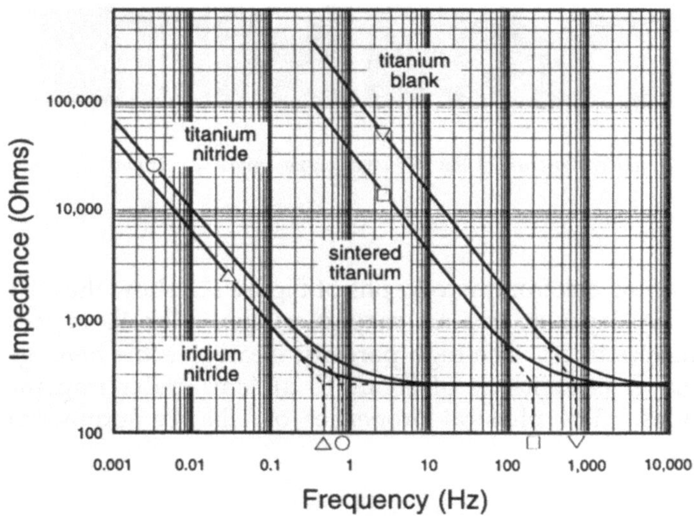

Figure 113. *Damping characteristics of various electrode materials under potentiostatic conditions.*

Figure 114 shows the distributions of intracardiac signal amplitudes and pacing thresholds. The measured thresholds of IrN (Figure 115) demonstrate that the doublelayer at equilibrium determines the behavior of pacemaker electrodes. The pacing threshold and the sensing behavior are directly affected by biocompatibility and doublelayer capacitance. This example of the effect of TiN and IrN layers documents the extent to which electrotherapy can be improved by choosing suitable materials and a suitable electrode

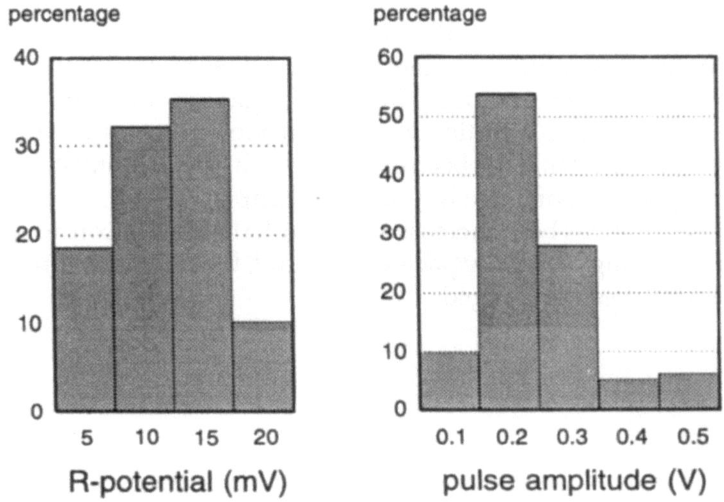

Figure 114. *Intracardiac R-potentials and pacing thresholds obtained with TiN-coated pacemaker electrodes.*

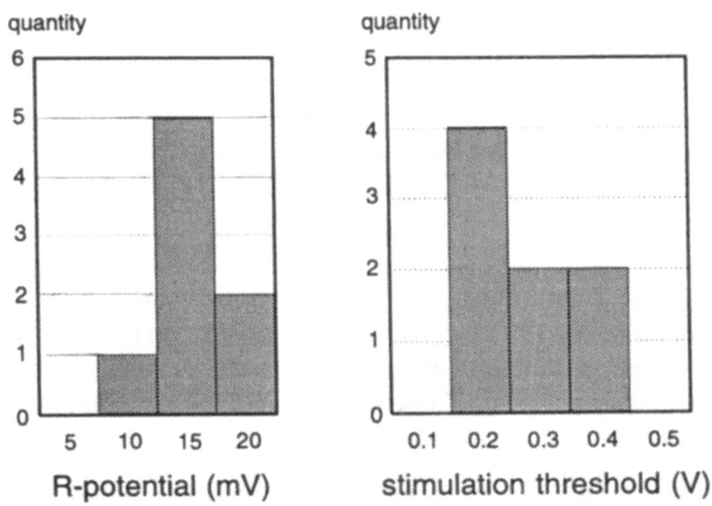

Figure 115. *Intracardiac R-potentials and pacing thresholds obtained with IrN-coated pacemaker electrodes.*

geometry. To demonstrate the reliability of the surface-coated electrode system, the long-term performance of electrodes in the right-ventricular position shows an insignificant increase in threshold. This, in addition, can be considered as an indication of biocompatibility that has been observed with titanium and its chemical compound in orthopedic and dental implants [144]. The signal-to-noise ratio has been increased and, with the energy consumption reduced owing to the low pacing threshold, the functional life of the system has been prolonged.

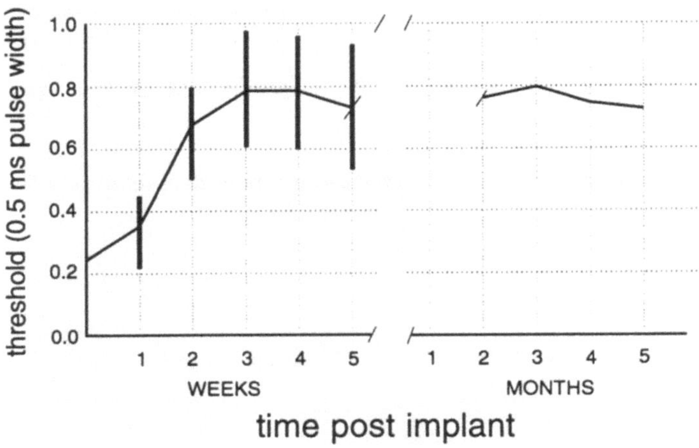

Figure 116. Long-term stability of stimulation threshold of TiN-coated electrodes (0.5 ms pulse width).

Clinical long-term results in regard to the threshold (see Figure 116) demonstrate for TiN-coated electrodes permanent thresholds below 0.9 V monitored on more than 500 implanted systems over a period of more than one year. These results convincingly demonstrate the importance of the electrochemical interface impedance and prove that the assumption of an ohmic behavior of the stimulation electrode interface is not valid. The cut-off frequency under pacing conditions in the TiN system can be reduced from about 1 kHz for smooth surfaces to 0.3 kHz for PVD-coated electrodes. With regard to the energy losses, Figure 117 demonstrates the influence of the

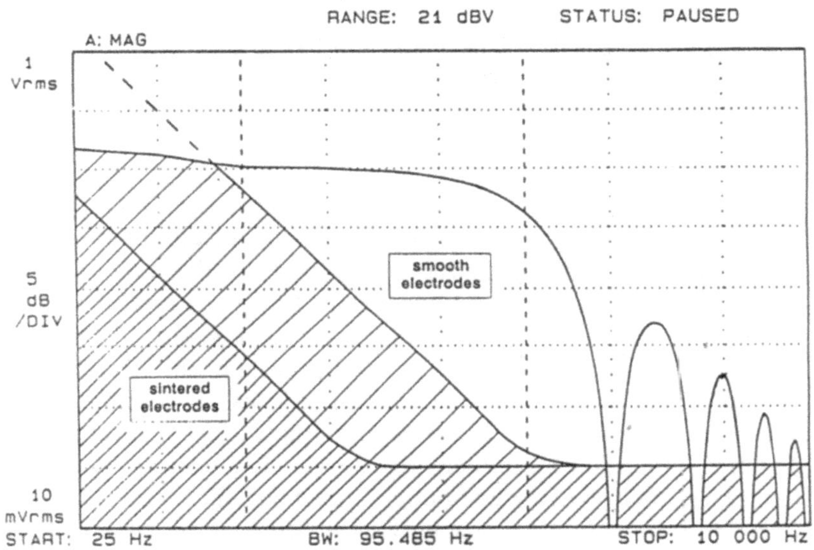

Figure 117. Energy losses during stimulation.

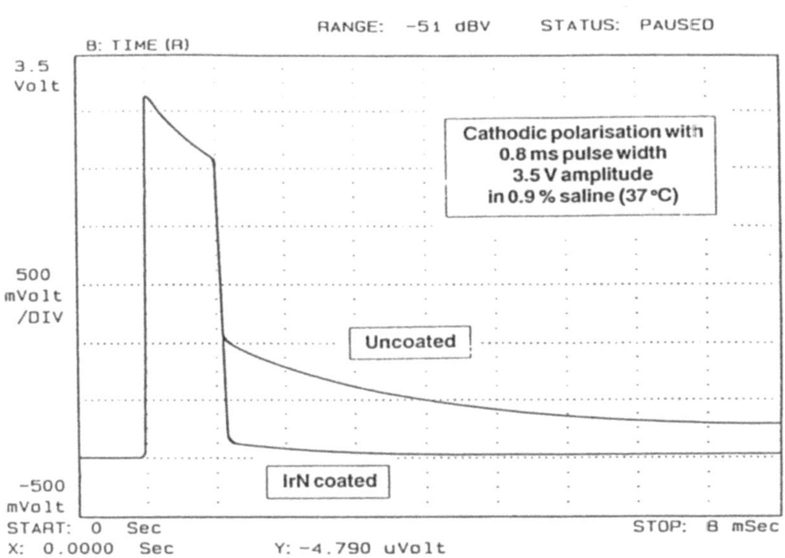

Figure 118. Polarization behavior of different electrodes.

surface structure on energy losses during stimulation based on *in vitro* experiments. The polarization behavior shown in Figure 118 for IrN PVD coating, in addition to the advantages of long-term stability, prove capability of the strategic requirement for capture recognition in low afterpotentials.

In conclusion, the importance of the surface morphology in the sensing and pacing performance of pacemaker electrodes can be demonstrated. Besides the biocompatibility of the materials (TiN and IrN), the ratio of the active surface to the geometric area has proven to be of paramount importance. Instead of the well-established sintering technique, fractal coating of the electrode surface by PVD and reactive sputtering allows an increase of the geometrical surface by a factor of more than 1,000, resulting in the desired reduction of interface impedance. The clinical results confirm the reduction of the permanent threshold, the increase of sensitivity, and the reduction of polarization. By the introduction of IrN as a coating material, an additional improvement of long-term stability and lower polarization was clinically demonstrated. As a result of their biocompatibility and stability, all systems utilizing TiN and IrN supersede the steroid electrodes with their limited reservoir of eluting drugs [148].

MATERIALS IN PACEMAKER TECHNOLOGY

In the present state of the art, the usefulness of the implantable pacemaker is defined, not only by design constraints in achieving a shape appropriate to the physiological and biomedical requirements of the desired function, but also by the properties of the materials of which the device are made. The use of alloplastic materials in replacement surgery has a long history, and has mainly involved implants in the skeletomuscular system [151-153]. The implantation of artificial parts in the cardiovascular system have only recently been developed [154-157]. The major role played by implantable devices in modern medicine can be illustrated most simply by some statistics. Every year more than 1,500,000 people worldwide are provided with vascular prostheses; artificial heart valves are implanted in 100,000 patients, and about 220,000 receive an implantable cardiac pacemaker [158]. In addition to the electronic requirements discussed previously, the biocompatibility of the implantable pacemaker is of importance in the long-term success of arrhythmia treatment.

Definition of a Biomaterial

Considering the engineering aspects, there are usually a number of materials that can be used to develop and/or manufacture an implantable device. Medical requirements are more stringent, because the device, while restoring impaired function, must also be guaranteed not to do any long-term damage to the human body. Hence, the choice of material is critical. It is an interdisciplinary task to achieve biocompatibility, defined as follows:

A material of optimal biocompatibility is one that does not lead to an acute or chronic inflammatory response and that does not prevent a proper differentiation of implant-surrounding tissues [159].

Interactions between the material and the surrounding tissue can also produce complications involving the implant itself; leaching, degradation, corrosion, and mechanical abrasion as well as foreign-body reactions can affect the stability of the implant (see Figure 119). These factors considerably restrict the range of materials available.

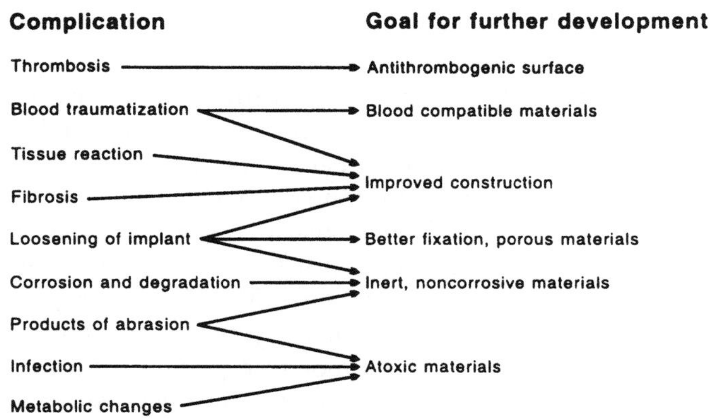

Figure 119. *Phenomena involved in interactions with implantable materials.*

Summary of Biomaterials in Common Use

The biomaterials needed for the implantable pacemaker are alloplastic, that is, not biological in origin. They include metals, ceramics or glasses, and polymers. From a physical point of view, the main difference between these groups of materials is the type of chemical bond which holds the materials together. Metals have typical metallic bonds, characterized by free electrons in a lattice of positive atomic cores; it is this structure that gives metals their high electrical conductivity. Apart from orthopedic implantable devices, metals are employed almost exclusively in pacemaker housings and stimulation electrodes. The transfer of experience from orthopedics to pacemaker technology has contributed significantly.

The bonds in ceramics and glasses, as a rule, have a large ionic component. Those most commonly used as biomaterials are metal oxides, which are used as feedthrough material in the design of the implantable pacemaker.

Polymers are distinguished by an elongated molecular structure with a large proportion of carbon and predominantly covalent bonds. The length of the chains and hence their molecular weight crucially affect the parameters of the materials. The mechanical properties of polymers are temperature-dependent; as a rule, there are two characteristic transition temperatures. Below the so-called glass temperature, polymers are solid bodies. Their structure is either completely amorphous or partially crystalline, with intermingled amorphous domains, and can be modified by the conditions of manufacture. When the glass temperature is exceeded, the polymer is a viscous fluid. Once the melting temperature is reached, it becomes liquid.

Table XIII
Biomedical Applications of Polymers

Plastic	*Processing*	*Characteristics*	*Applications*
HDPE	Extrusion, die casting; injection molding	Small dielectric constant; low resistance	Water-permeable coating; packing; yarns
LLDPE	Extrusion, die casting, injection molding	Straight chains	Foils; electrode insulation
Polytetrafluorethylene (PTFE)	Sintering	90–95% crystallinity; sensitive to gamma radiation	Impregnation of fabrics; reinforcement of pacemaker leads
Polyamides	Extrusion	Hydrophilic (water uptake 10%); resistant to abrasion and chemicals; dimensionally stable	Yarns; fabrics; packing; parts subject to abrasion; pacemaker lead introducers
Polydimethyl siloxane, polyurethane	Molding; extrusion	Abrasion resistant; dimensionally stable; flexible	Insulation; pacemaker technology; catheters; stimulation electrodes
Epoxy resins	Molding	Chemically resistant; low shrinkage; good adhesion	Laminates; encapsulation; pacemaker connector block

Polymers are ordinarily available as granules, films, or other semiprepared forms. Processing is material- and product-dependent; customary methods include extrusion, injection molding, or vacuum casting; mechanical processing methods are also possible in some cases. A survey of the various procedures, their limitations, and their possibilities can be found in many review articles [160-165]. Table XIII summarizes the polymers most commonly employed at present in pacemaker technology.

General Characteristics of Biomaterials

In developing implantable devices, the tasks are to improve and invent biomaterials and to construct the device in a manner suited to the material as well as to the intended function. Furthermore, the device should be designed to remain functional for the lifetime of the patient. The extent to which these conditions can be met is still a matter for conjecture since only long-term experience can bring a definitive answer.

Given the complexity of the problems, the need to select or develop a material with precisely those characteristics that meet the special requirements imposed by the nature and site of the implant solutions will have to be approached step by step. In terms of the physiological conditions to which implantable devices are exposed *in vivo*, three groups of materials can be distinguished, as follows:

Materials for tissue implants. Examples include the housings and leads of cardiac pacemakers, neurostimulators, defibrillators, systems for the storage and timed release of drugs, tissue adhesives, and materials for plastic and reconstructive surgery [166, 167].

Blood-compatible materials. These comprise all the substances used in vascular prostheses, oxygenators, artificial kidneys, heart valves, systems to assist circulation and substitute for the heart, catheters, and stimulation electrodes in temporary or permanent contact with the blood [168-169].

Materials for orthopedic implants and dentistry. This group includes metals in pure form and as alloys, polymers, ceramics, and composite materials. In short- and long-term implants, these serve to adapt, reinforce, and stabilize bone fragments or to replace destroyed and malfunctioning joints in the upper and lower extremities [170].

The various criteria by which all these materials must be evaluated also fall into three categories, depending on the particular application:

Physical and mechanical characteristics, *i.e.,* density, gas permeability, hardness, bending strength, shear modulus, electrical and thermal conductivity, thermal coefficient of expansion, specific heat, surface roughness;

Chemical and electrochemical parameters, *i.e.,* corrosion resistance, chemical stability, resistance to solvents, sterilizability, water absorption, surface tension;

Biological and physiological aspects, *i.e.,* toxicity, carcinogenicity, thrombogenicity, immune reactions, allergic and inflammatory reactions.

Especially in regard to pacemaker design, these material properties and the desired functional properties are combined in different ways. These are the central problems in the evaluation and development of biomaterials, and can be solved only by close collaboration between physicians, scientists, and engineers. As with most implants made of alloplastic materials, function, construction, and material must always be considered in relation to one another. The materials for skeletomuscular replacement parts as well as for pacemakers must be highly resistant to corrosion and degradation, *e.g.,* the transvenous electrode. Blood compatibility clearly has priority for implants in the vascular system. In the current state of materials development, much effort is being devoted to making materials antithrombogenic by special surface treatments to suppress interactions at the interface between material and blood. The development of materials for artificial (alloplastic) heart valves and intracardiac stimulation electrodes is progressing in a different direction, with the exploration of ways to coat the surface with electronic structures that will optimize stimulation performance and at the same time prevent activation of the proteins involved in the clotting mechanism. In each case, the emphasis is on the reaction- and tissue-specific boundary conditions, which, on historical grounds, are often formulated in terms of the difficulties yet to be overcome and the complications that still arise. Among these are thrombosis, blood traumatization, tissue reactions, fibrosis, infection, metabolic disturbances, loosening of the implant, corrosion, and degradation.

So far, these problems and potential solutions are very generally considered. Biocompatibility, the interpretation in terms of specific physical parameters, applies only in special cases. Thus far, the development of materials has been largely empirical. Where a solution to a particular problem has been found, it will not necessarily be valid for other applications.

Biomaterials in Soft-Tissue Implants

Although some characteristics of a biomaterial are constrained by the nature and purpose of the implant for which it is to be used, all materials must fulfill certain fundamental requirements. Ideally, there should be no change in the chemical and physical properties of the material as a result of exposure to the intracorporeal milieu. In particular, pacemakers as long-term implants should, therefore, be made of materials in a stable, passive state. The requirement for biocompatibility is also a general one. In the ideal case, neither the material itself nor components that may be released, such as products of corrosion or degradation, leached additives or abraded particles, would do even the slightest damage to the surrounding tissue or interfere with it in any way. In reality, the tissue has a certain level of tolerance; for example, noxious elements may be continually removed. However, the tissue tolerance limits must not be exceeded, or the useful lifetime of the implant may be reduced by granulation and necrosis [171].

At present, implantable pacemakers are constructed mainly of metal, polymers being used only to cover the surfaces of the implantable device and to form the connector block.

Among the metals, titanium and its alloys have proved to be the biomaterials of first choice. The need for corrosion resistance, biocompatibility, favorable mechanical characteristics, with a suitable modulus of elasticity and durability appropriate to the application, and availability at reasonable cost, rules out a large number of metallic materials at the outset. Those known to meet these criteria for biocompatibility include:

- Stainless steels (ISO 5832/1 and AISI 316L);
- CoCrMo alloys (vitallium), such as cast CoCr30Mo6 (ISO 5832/4); CoNiCr alloys, such as CoNi35 and forged Cr20Mo10 (ISO 5832/6);
- MP 35 N
- Titanium materials, including technically pure titanium (ISO 5832/2) and titanium alloys such as TiAl6V4, TiAl5Fe2,5 and TiAl6Nb4 (ISO 5832/3);
- Niobium, technically pure; and
- Tantalum, technically pure (ASTM 560-78).

In the following sections, the various groups of metallic biomaterials are compared with respect to the criteria listed above [172].

Resistance to Corrosion

The pH of the electrolytes in the body is 7.4 under normal conditions. However, after surgery there may be an initial transient increase to 7.8, followed by a fall to below 7.0, and then a gradual return to the original value over a period of several days. Not only are these transients a problem, even during the postoperative phase, this creates an extremely aggressive condition for the material surroundings.

Titanium and its alloys, niobium and tantalum, have proved resistant to corrosion. Somewhat less resistant are rolled CoNiCr, cast CoCr, and stainless steel [173, 174]. All these materials are distinguished by a passive state, so that relatively little current flows and only a few micrograms of the metal enter into solution.

Nevertheless, friction and processing, *e.g.*, welding, can increase the rate of corrosion by orders of magnitude. The same applies to crevice corrosion, galvanic corrosion or pitting, stress and fatigue corrosion, which determine the behavior of the metal in contact with the body fluid. In the crevice below a metal plate that has been screwed into place, the medium can become very acidic, down to pH 1. *In vitro* studies with the redox system $Fe(CN)_6 4-/Fe(CN)_6 3-$

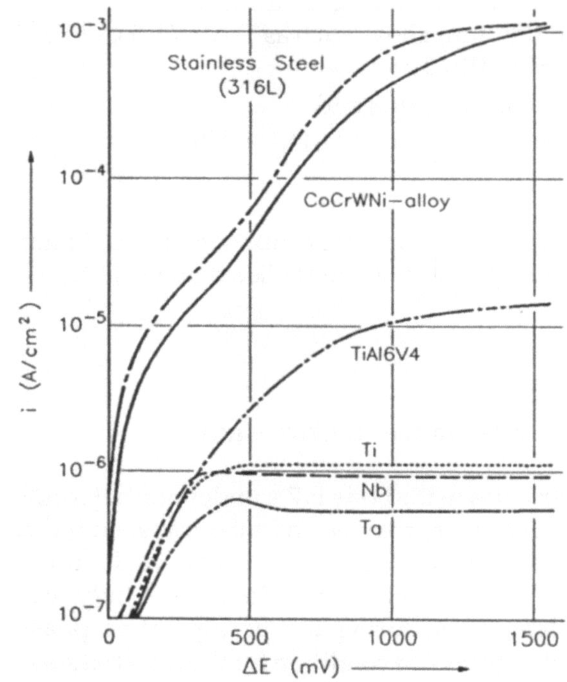

Figure 120. *Curves of current density vs. potential for various materials in 0.9% NaCl solution with a stable redox system* $[Fe(CN)_6^{4-}/Fe(CN)_6^{3-}]$

in 0.9% NaCl solution have shown that titanium (and its alloys), tantalum, and niobium (see Figure 120) are less reactive than either the stainless steel AISI 316L or rolled CoNiCr alloys. Measurements of the polarization resistance of the various materials have led to the same conclusions (Table XIV) [175].

Table XIV
Polarization resistance (R_p) of metallic biomaterials in physiological saline solution with a redox system [$Fe(CN)_6 4-$/$Fe(CN)_6 3-$]

Material	R_p *in* $k\Omega\ cm^2$
Au	0.28
FeCrNiMo (316L)	4.38
CoNiCr (forged)	3.32
cp-Ti	714
TiAI6V4	455
cp-Nb	455
cp-Ta	1430

The various metallic implant materials also differ in resistance to corrosion, in the same order, when tested for breakdown voltage in Hank's solution. Whereas technically pure titanium and the alloy TiAl6V4 have high breakdown voltages, 2.4 and 2.0 V respectively, stainless steel and the CoCr and CoNiCr alloys, whether cast or rolled, had values of only 0.2 and 0.42 V (see Table XV) [176]. Furthermore, titanium and its alloys, as well as niobium and tantalum, are known to belong to the group of metallic materials that show no pitting when implanted [177].

Table XV
Breakdown voltage and repassivation time of metallic materials under simulated conditions (Hank's solution and 0.9% NaCl) [174]

Material	*Breakdown voltage in V (calomel electrode)*	*Repassivation time in ms*			
		t_e		$t_{0.05}$	
		−0.5V	*+0.5V*	*−0.5V*	*+0.5V*
FeCrNiMo (316L)	+0.2 ... 0.3	>72,000	35	>72,000	>6,000
CoCr (cast)	+0.42	44.4	36	>6,000	>6,000
CoNiCr (forged)	+0.42	35.5	41	>6,000	5,300
TiAI6V4	+2.0	37	41	43.3	45.8
cp-Ti	+2.4	43	44.4	47.4	49
cp-Ta	+2.25				
cp-Nb		47.6	43.1	47	85

Because of the vulnerability of the passive layer to mechanical disruption, the repassivation behavior of metallic implant materials is especially important. In comparative studies, the repassivation time has been shown to be a useful criterion, after the electrode surface has been activated by grinding. As Table XV shows, the passive layer grows more rapidly on titanium and its alloys than on the other materials. Titanium is also readily coated with hard surface layers, which protect the passive layer from mechanical damage and confer favorable tribological properties. Optimal results are obtained by coating with TiN, applied by ion implantation and reactive sputtering. It has been shown that when TiN layers are applied to CoNiCr alloys, both the friction behavior and the resistance to corrosion are improved. The breakdown voltage is increased from 0.83 V to 1.16 V [178]. Of course, permanent surface protection is obtained only if the layers are homogeneous with no fissures.

Biocompatibility

As metallic prosthesic parts implanted in the body react, the flow of electrons in the implant causes a corresponding flow of ions in the living tissue so that the physiological ion concentrations are not preserved. Other effects, including an inorganic reaction of the implant and the appearance of primary products of implant corrosion as metal ions become dissolved in the body electrolytes, are not restricted to the vicinity of the implant; they may result in an accumulation of foreign substances, in various organs, sufficient to reach the specific toxicity limits for those substances. A direct organic reaction between the implant or its primary corrosion products and the tissue proteins is also possible and can lead to inflammation or to other problems. Inflammation, in turn, can cause the formation of H_2O_2 and, hence, of hydroxyl radicals, which upset the physiological equilibrium.

The probability of such interactions depends fundamentally on the physical and chemical properties of the implant materials. Titanium, tantalum, and niobium are especially biocompatible since they spontaneously form protective surface layers of nonconducting oxides that prevent the exchange of charge carriers across the phase boundary and the resulting loss of ionic equilibrium in the tissue.

On the basis of this shielding action, correlated with the dielectric constants of the various oxides (see Table XVI), the oxides can be subdivided into three biocompatibility groups. TiO_2 (rutile), Fe_2O_3 and Nb_2O_5 have dielectric constants larger than that of water; Al_2O_3, Cr_2O_3 and Ta_2O_5 shield less effectively because their conductivity is higher; and nickel and vanadium oxides conduct so well that they have no measurable dielectric constants.

Table XVI
Dielectric constant ε, enthalpy of formation ΔH and solubility p_k of the primary corrosion products

Primary corrosion product	ε	ΔH°_{298} kJ/mol	p_k
Ai_2O_3	5 ... 10	1675	+14.6
$Ai(OH)_3$		916	
CoO		239	−12.6
Cr_2O_3	12	1141	+18.6
CrO_3		595	
$Cr(OH)_3$		988	−1.8
FeO		267	−13.3
Fe_2O_3	100	822	−14
$Fe(OH)_2$	30 ... 38	568	+2.3
MoO_3		712	+3.7
NiO		240	−12.2
$Ni(OH)_2$		538	
NbO		486	
Nb_2O_5	280	1905	>20
Ta_2O_5	12	2090	>20
TiO		518	
TiO_2 Anatase	48	935	
Brookite	78		+18
Rutile	110	943	
VO		410	
V_2O_5		1560	+10.3
H_2O	78	273	+14

The high biocompatibility of Ti and its alloys is, therefore, the most ideal material for pacemaker encapsulation, which is derived from the insulating action of titanium oxides, and the fact that the dielectric behavior resembles that of water. Given that the inorganic and organic interactions between metal and tissue are due to the formation of primary corrosion products, the differences in biocompatibility of metallic materials must also be due to differences in their thermodynamic stability. For example, the oxides and hydroxides of aluminum, chromium, niobium, tantalum, titanium and vanadium are stable because of their high enthalpy of formation, whereas the oxides and hydroxides of cobalt and nickel have low enthalpy (see Table XVI) [179]. The lower the enthalpy of formation of the oxides, the more likely an interaction between these oxides or hydroxides and the electrolytes in the body. The solubility (or solubility product p_k) of the primary corrosion products provides an additional indicator of biocompatibility. The oxides of titanium, tantalum, noibium, and chromium have p_k values 14 so that no hydrolysis occurs; cobalt, iron, and nickel oxides actually have negative p_k values, which explains their great toxicity [180].

It is partly on the basis of these considerations that the vanadium-containing alloy, TiAl6V4, although once widely used, has been replaced by TiAl5Fe2,5 and TiAl6Nb7. These now represent the state of the art [181-183].

In addition to the inorganic reactions of metal ions, reactions with proteins can also occur. Thermodynamically stable, primary corrosion products with low solubility in the body electrolytes have only a very slight tendency to react with proteins. When inert and biocompatible materials are implanted, the immediately adjacent cells are always found to have a good vascular supply, whereas in the vicinity of toxic materials, inflammation and dead cells are found. Polarization resistance is also a measure of biocompatibility, as shown by the classification of the various materials in Figure 121. The toxic effect of some elements (Co, Cr, Ni) is clearly correlated with a low polarization resistance, whereas the biologically inert behavior of titanium and its alloys, and of niobium and tantalum, are associated with a high polarization resistance.

In summary, the biocompatibility of an implant material is closely related to its corrosion behavior, the solubility of the corrosion products, and other physicochemical parameters of the material [184].

The mechanical properties of implant materials are summarized in Table XVII. The modulus of elasticity of titanium and its alloys, like that of niobium, is in the range 100,000 to 120,000 N/mm^2, so that

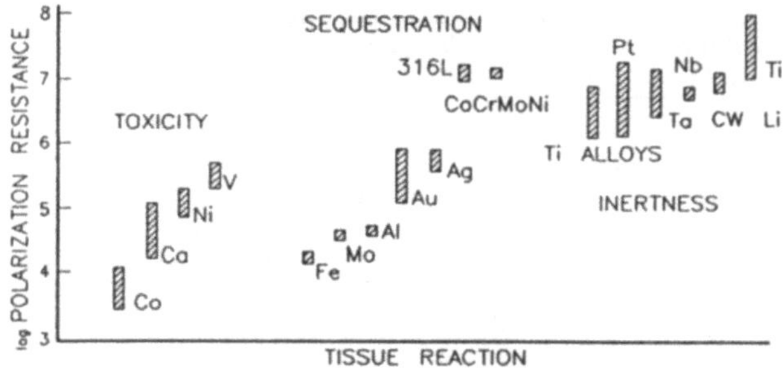

Figure 121. *Polarization resistances grouped according to tissue reaction of metals.*

with respect to elasticity, this group of materials is closer to bone (10,000 N/mm^2) than the other metallic materials. With respect to durability, titanium and its alloys are equivalent to the other alloys or, in some cases, superior; they can be adapted to biomechanical conditions over a wide range and, hence, are the material of choice in pacemaker technology. The main properties of the metals and metal alloys in common use as implant materials are summarized in Table XVIII.

Table XVII
Summary of the mechanical properties of metallic implant materials (E = Young's modulus, $R_{p0.2}$ = 0.2% yield stress, σ_b = bending strength, A5% = breaking elongation)

	2CrNiMo 18 12 AISI 316L	CoCr cast	CoNiCr forged	TiAl6V4	TiAl5 Fe2.5	cp-Ti	cp-Nb	cp-Ta
$E \bullet 10^3$ N/mm^2	210	200	220	105	105	100	120	200
$R_{p0.2}$ N/mm^2	450	500	850	900	900	300	250	300
σ_b N/mm^2	250	300	500	550	550	200	150	200
A5%	40	8	20	13	15	30	70	40

Table XVIII
Summary of material characteristics relevant to implant applications

	2CrNiMo 18 12	CoCr cast	CoNiCr forged	TiAl6V4	TiAl5 Fe2.5	cp-Ti	cp-Nb	cp-Ta
Corrosion resistance	–	–	+	+	+	+	+	+
Biocom-patibility	–	–	–	(+)(V)	+	+	+	+
Bioad-hesion	–	–	–	+	+	+	(+)	(+)
$\sigma_b/E^a \cdot 10^3$	1.2	1.5	2.3	5.2	5.2	1.8	1.3	1.3
Process methods[b]	CDWP	CWP	CDWP	CDWP	CDWP	CDWP	CDWP	CDWP

[a] Biofunctionality
[b] C = casting, D = die stamping, W = welding, P = powder metallurgy possible

Quality Testing and Reliability

In addition to the questions of biocompatibility and resistance to corrosion, for long-term function of an implantable device to be guaranteed, it is essential that the construction meet certain quality criteria and that there be thorough quality checks during the manufacturing process. Conformity to national and international standards for chemical composition, structural properties, and mechanical characteristics of the alloys and other materials used for clinical implants is essential to the extent that such standards exist. For orthopedic implants, as for many other technologically exacting industrial products, the processing steps (melting, casting, stamping, final processing, and sterilization) should occur in a documented sequence so that the final product is accompanied by the appropriate documentation and any problems can be traced back to the source. Before an implantable device is released for clinical trials, all mechanical, physical, and chemical characteristics of the material, of the manufacturing process, and of the finished product must be tested with respect to suitability for the specific implant. The methods include those customarily employed in the processing of polymers and metals: spectroscopic and texture analyses, determination of elastic limit, tensile strength and breaking tension, as well as stability under long-term vibration. To these are added analyses of corrosion, tests of biocompatibility, measurements of resistance to abrasion, and studies of fatigue fracturing in which the construction and the material are tested to the point of failure under *in vitro* implantation conditions. When the implant has qualified for release by all these criteria, additional quality-control procedures are necessary to ensure that standards are maintained during the subsequent manufacture by testing and documentation of the technical data during the actual production process.

Physiological Characteristics of the Blood

Before the possible interactions of the blood with artificial surfaces such as the pacemaker electrode can be estimated, its chemical and cellular composition must be known.

Blood is an inhomogeneous fluid consisting of a light yellow plasma in which blood cells (chiefly erythrocytes, leukocytes, and thrombocytes) are suspended. The cells account for about 44% of the blood volume (ml cells/dl blood). The plasma is composed of 90% water and 6.5-8% protein. The remainder comprises substances of low molecular weight. The blood pH is kept at about 7.4 by the electrolytes listed in Table XIX.

Table XIX
Average concentrations of dissolved substances in human blood plasma

	Cations in mg/dl		*Anions in mg/dl*	
Electrolytes	Sodium	328	Chloride	365
	Potassium	18	Bicarbonate	61
	Calcium	10	Phosphate	4
	Magnesium	2	Sulphate	2
			Proteins	7000
Non-electrolytes	Glucose	90 ... 100		
	Urea	40		

One of the most important components of human blood is the group called plasma proteins. Their main function is trophic since they constitute a rapidly available reservoir of protein for cell nutrition.

The most abundant of these proteins, at 4000 mg/dl, is albumin. The globulins have a predominantly vehicular function; that is, they are responsible for transporting lipids, iron and so on. Cationic blood components such as calcium are readily bound to proteins and transported by them. Some of the globulins, known as gamma globulins, are a major element in the immune system since they can be activated as antibodies against foreign proteins or antigens.

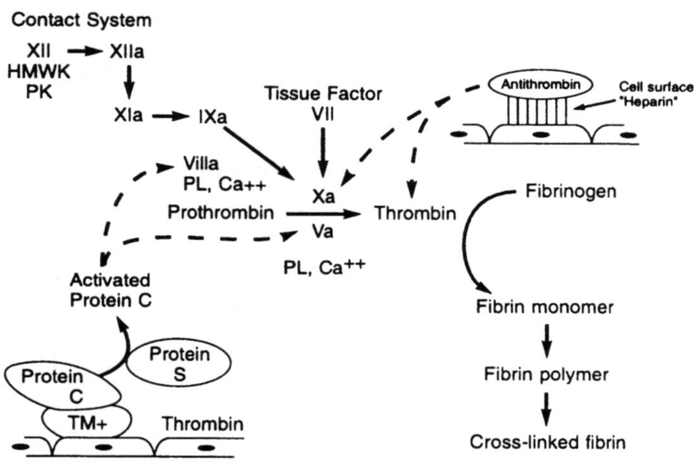

Figure 122. Schematic three-step mechanism for blood clotting.

With respect to biomaterials technology, however, probably the most important function of the blood proteins is hemostasis—the closure of wounds in blood vessels to stop blood loss. This clotting process can also be activated by artificial surfaces. For comparison, first we consider the natural course of events. Figure 122 is a diagram of the final stages in secondary hemostasis, which can be activated both by contact (intrinsic system) and by tissue injury (extrinsic system). In both cases the activation phase ends with the formation of thrombin. This peptidase splits off four fibrinopeptides from the fibrinogen molecule. The fibrin monomers that initially remain group themselves into electrostatically linked chains; later these are stabilized by covalent bonds induced by the fibrin-stabilizing factor (FSF, or Factor XIII). In this way, fibrin threads are formed that can trap blood cells and connect to one another to form a three-dimensional network—the thrombus. Activation of the clotting system by artificial surfaces mainly occurs at the final irreversible step—the splitting of fibrinogen. This process will be considered in more detail in the following discussion of microscopic explanations.

Interactions Between Blood and Artificial Surfaces

In principle, the interactions between blood and material are reciprocal, not unilateral. Figure 123 illustrates some of the most important processes that can take place at the phase boundary. Ways in which the blood can alter the properties of artificial materials are shown at the top of the figure. Metals and metal alloys can be corroded especially when they are under mechanical stress (frictional corrosion) or at vulnerable geometric features such as sharp edges or gaps. Special problems appear where the same or even different metals are joined since welding generally changes the microstructure of the metal, creating local elements in which one of the partners has a much increased corrosion rate. Plastics are also altered by contact with blood, a phenomenon that has been called "biodegradation" [185, 186]. A familiar example is the hydrolysis of the urea or urethane bonds in polyurethane, which can actually be catalyzed by

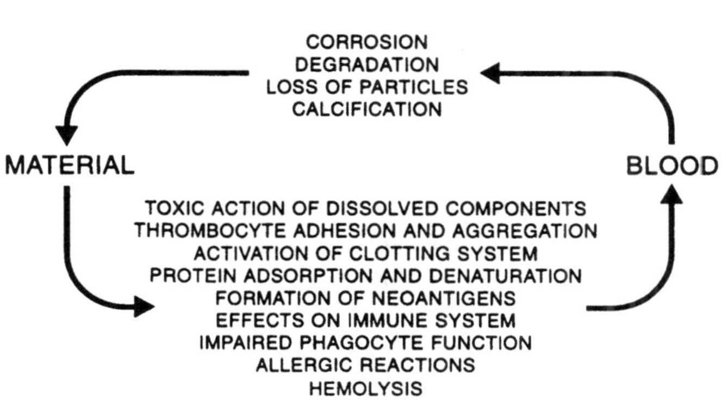

Figure 123. *Possible interactions between blood and material. The arrows show the directions in which influence is exerted.*

phagocytes or enzymes such as papain [98]. Another group of materials in common use, the ceramics and glasses, usually have low rates of corrosion, but the superficial regions can be dissolved or leached by contact with blood.

The substances that enter into solution during all these processes can, in turn, affect the blood and trigger toxic reactions. However, the corrosion products need not always be chemically dissolved; for example, it has been shown that small particles of the oxide layer are released from titanium implants.

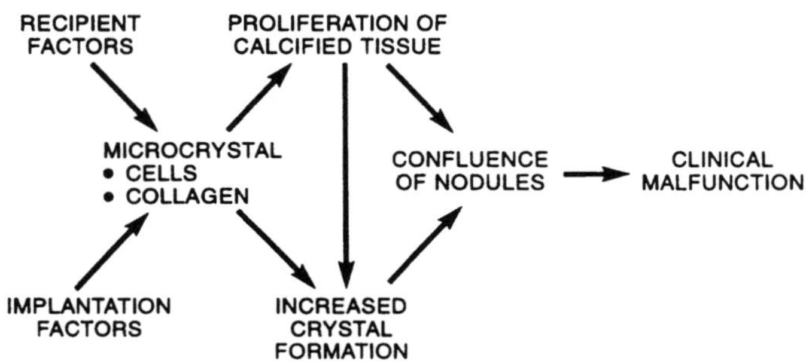

Figure 124. *Diagram of the calcification of polymers.*

Other effects on materials can be attributed to individual components of the blood. For example, Figure 124 illustrates the calcification of polymers, showing the mechanisms by which, once microfractures have formed, calcium deposits cause the material to become more brittle and to eventually lose mechanical and electrical performance of the implanted device.

Of the effects of the material on the blood, listed in the bottom portion of Figure 123, activation of the clotting system is one of the most severe since it promotes the development of thromboses and embolisms. The schematic drawing of an artificial surface in Figure 125 shows an influence of morphology independent of the material.

If the depth of surface roughening is too great, corpuscular elements in the blood may collect at the surface; one result, for example, is that blood platelets release an activating factor in the intrinsic clotting mechanism [187].

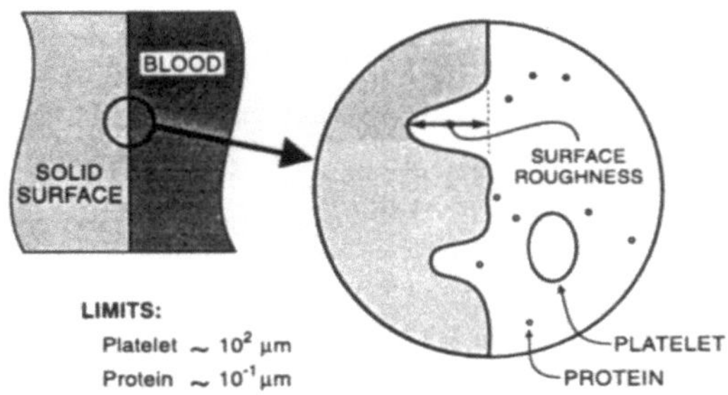

Figure 125. Definition of the parameters to describe a real surface.

Smooth surfaces also interact with the blood. The first step is adsorption of blood proteins at the interface. As far as is known at present, all artificial surfaces adsorb proteins in differing amounts and combinations [188]. An example of the time course of adsorption of proteins by polysiloxanes is shown in Figure 126. The resulting protein film, in turn, induces the adsorption of blood corpuscles, especially platelets, erythrocytes, and leukocytes. The consequences are similar to those of accumulation of these materials at a rough surface, though the nature and intensity of the subsequent reactions depend on the composition of the adsorbed protein film, which in turn depend on the material [189, 190]. For example, gamma immunoglobulin appears to promote the adhesion of platelets. Because it seems likely that material-specific protein adsorption is systematically related to the properties of the various materials, this area remains the subject of extensive research [191].

Another point at which a material can interfere in the natural clotting process is the transition from fibrinogen to fibrin which, as mentioned previously, can be directly induced by artificial surfaces [192].

Both phenomena direct fibrin production and the enhanced adsorption of blood cells leads to the formation of a thrombus although its corpuscular composition will depend on the flow conditions. "White" thrombi consist predominantly of blood platelets formed where the shear forces are strong; "red" thrombi, which collect many red corpuscles, form in regions of low-shear force.

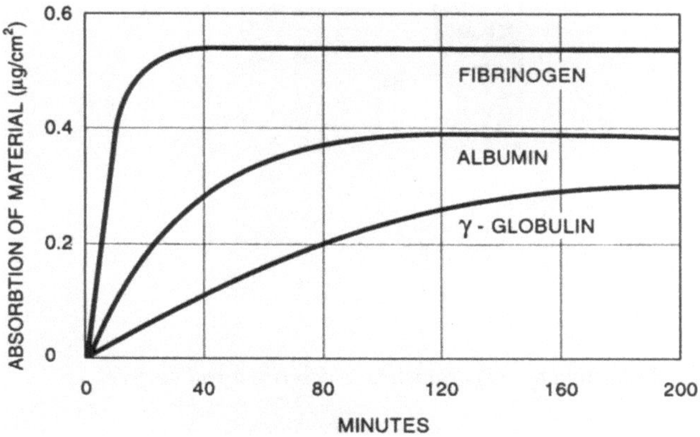

Figure 126. *Time courses of adsorption of blood proteins onto silicone rubber.*

Some of the adsorbed proteins return to solution. If the secondary or tertiary structure has changed as a result of the adsorption process, they can become enzymatically active in the blood [193] or have immunological effects by forming neoantigens and, for example, produce inflammation resulting in pannus formation. Another kind of protein impairment has been observed in studies of dialysis patients — a change in phagocyte function that resulted in an increased susceptibility to infection [194]. Implants or their dissolution products have also been found to trigger allergic reactions, especially when they contain cobalt, nickel, or chromium or have been sterilized with ethylene oxide [194].

The list of possible effects on the blood must also include hemolysis, although this problem is associated mainly with the design of the implantable device and its surface [195]. For instance, implants restricting blood flow resulting in high flow velocity can damage the blood cells.

Determination of Blood Compatibility

The qualitative interaction phenomena considered thus far must now be quantified since, at present, this is the only way to determine whether a material is suitable for use in implantable devices such as cardiac pacemakers.

Relatively good methods to evaluate the degree to which a material is affected by contact with the blood are to measure corrosion electrochemically and to carry out long-term exposure tests. Scanning electron microscopy has special possibilities since it clearly demonstrates structural changes down to the microscopic level. Corrosion layers, sometimes extremely thin, can also be analyzed by surface-sensitive procedures such as SIMS [196], XPS [197], or Auger-electron spectroscopy.

The most reliable methods are *in vivo* experiments in which a sample of the material to be tested (*i.e.*, in the shape of a small tube) is introduced into the circulatory system of an animal. A number of such "ring tests" have been developed [198], the best-known of which is the vena cava ring test of V. L. Gott [199].

Finally, it should be pointed out that neither *in vitro* nor *ex vivo* experiments can predict the long-term *in vivo* behavior of materials. At best, these can assist in the preliminary selection of a material [200]. Long-term observations are the most reliable measures for blood compatibility.

The results of investigations into microscopic causes are, of course, also applied to the development of new materials. For instance, a major area in present-day medical technology is concerned with the vacuum-coating of devices, which has already proven useful in combining the advantages of two different materials. This added processing is worthwhile because the superficial layers that form naturally, such as the oxide layer on titanium, are too fragile under mechanical stress and are not homogeneous enough. One approach, therefore, is to coat parts with antithrombogenic materials such as α SiC:H, TiN, and TiC. Artificial heart valves exemplify a kind of surface treatment; here the electronic characteristics of the surface are altered with the specific aim of preventing direct fibrin activation [201, 202]. In the case of polymers, the trend is toward coating with proteins or anticoagulants such as heparin or prostaglandin [203]. On the other hand, the technique of making polymer surfaces more hydrophilic, by means of an oxygen plasma, has already been shown to be of value [204]. Ion implantation may, in addition, contribute to the improvement of polymers as biomaterials. The efficiency of PVD or CVD as a coating for the active surface of the intracardiac stimulation electrode has been demonstrated in the previous chapter.

The successful development of materials in pacemaker technologies, together with the applications in biomedical technology, make clear that the paramount task in the development of biomaterials is to investigate the phase boundary between implant and body. It is essential to clarify the fundamental physical phenomena that underlie the complications involving organic substances in the body, in order to progress from purely empirical science to the organized development of materials with specified characteristics. For this reason, much effort is being expended on basic research whether in the study of protein adsorption and reaction kinetics or of the potential distribution at the phase boundary [205, 206].

From the approaches tried thus far, it is already evident that implantable devices must be made from combinations of materials because simple systems cannot meet all the demands of the complex situation. Whatever group of materials is considered, whether ceramic-coated metals or polymers with active, organic surface layers, surface coatings or treatments consistently offer the possibility of marked improvement. New ways of treating materials and processing methods, such as sintering, also widen the scope of opportunity. Clearly, in the field of implantable pacemakers, as elsewhere, the mastery of technologies is the central prerequisite for further development.

PACEMAKER POWER SOURCES

Semiconductor microelectronics has played the same important role in the development of the implantable pacemaker as in the technology of electrochemical power sources. Well over five million patients have benefitted from the use of battery-powered implantable devices. The lithium battery has evolved into the principal power source since its introduction in 1972 [207]. The lithium-based primary cell has dominated pacemaker technology since the early 1980s, requiring power sources capable of delivering currents in the 10 to 100 microampere range. Compared to other electronic applications, the amount of energy required to operate a cardiac pacemaker is very small. In general, a stimulus pulse of 5-19 mA is delivered at 1-10 V for 0.25-1.0 ms with a rate from 30-150 bpm. For a typical application, a stimulus pulse of 10mA is delivered at 5 V for 0.5 msec at 70 bpm. Assuming a pacing rate of 70 pulses per minute, the average continuous energy drain is 30 μW.

These lithium iodine (Li/I_2) systems operate in a nonaqueous electrolyte, such as propylene carbonate, containing a mobile ionic charge carrying species such as lithium perchlorate. In some cases, such as iodine cathode cells, the electrolyte is a solid ionic conductor (LiI). The cathode, separator, and materials used determine many of the system properties. A summary of lithium battery chemistries that have been used in human implants is shown in Table XX [208].

Many other energy sources have been experimentally used, including piezoelectric, biologic, nuclear, and biogalvanic, as well as other electrochemical cells containing solid electrolytes and nonaqueous electrolytes. In addition to the mercury/zinc and lithium anode battery systems, two other battery types have been used in pacemakers implanted in humans. These are nuclear isotope batteries and nickel/cadmium batteries. The latter is inductively recharged by transmission of energy to an implanted receiver. Although these systems were rendered obsolete by the development of lithium cells for pacemakers, they are still useful for other implantable devices [209].

Table XX
Properties of Implantable Lithium Anode Cells

Cathode	Theroretical Limits[a]				Practical Limits[b]		Electrolyte Types
	Wh/cm^3	Wh/g	Ah/cm^3	E_{oc} V	E_L, V	Ah/cm^3	
PbI_2, PbS	1.4	0.31	0.73	1.9	1.8	0.28	Solid
CuS	1.8	0.58	0.86	2.1	2.1	0.29	Liquid
I_2	1.9	0.57	0.61	2.8	2.7	0.28	Solid
MnO_2	2.5	0.86	0.70	3.0	3.0	0.26	Liquid
Ag_2CrO_4	2.1	0.5	0.64	3.3	3.1	0.19	Liquid
Br_2	2.5	1.1	0.71	3.5	3.0	<0.2	Liquid
$SOCl_2$	2.3	1.7	0.47	3.6	3.6	0.25	Liquid
V_2O_5	1.36	0.46	0.40	3.4	2.8	0.08	Liquid

[a] Volumetric energy density, weight energy density, volumetric specific capacity, and open-circuit voltage, respectively.
[b] Load voltage and practical volumetric specific capacity.
[c] Based on a high-rate design.

Battery Development

Since 1958, the pacemaker battery of choice was the zinc/mercuric oxide cell. Although this cell demonstrated problems of self-discharge and gas evolution, it made pacemaking possible. In 1972, the first lithium-powered pacemaker was implanted. Over the next several years a variety of different lithium systems were used in pacemakers. Among the chemistries used in pacemakers were Li/$SOCl_2$, Li/Ag_2CrO_4, Li/CuS, Li/I_2-Polyvinylpyridine (PVP), and, in more limited use, Li/LiI(Al_2)$_3$)/PbI_2,PbS, Pb. Of these, the Li/I_2-PVP battery is in principal use in today's pacemakers.

The design and construction of implantable batteries has presented a unique set of challenges. The obvious requirements of safety and reliability dictate that materials compatibility be carefully considered. For example, the use of corrosion-resistant glasses in liquid organic elecrolyte systems is required because of the longevity requirements of implantable devices. Electrochemical redundancies are engineered into the cell designs. A comprehensive system of quality control and detailed documentation of construction and inspection steps is required to maintain high quality production. Test methods are used to predict and verify long-term battery performance on the basis of accelerated testing as part of an extensive life-test program.

The first implanted pacemaker designed by Elmquist in 1958 used a rechargeable power source. This method was later repeated by Pacesetter Systems Incorporated and the Johns Hopkins Applied Physics Laboratory, producing a pacemaker with a nickel cadmium rechargeable battery. The cell reaction is

$$2NiOOH + Cd + 2H_2O \leftrightarrow 2Ni(OH)_2 + Cd(OH)_2$$

The open circuit voltage is 1.25 volts. The cell was cylindrical with a nominal capacity 190 mAh. Several thousand of these units were produced and implanted. However, there were medical objections to the necessity for recharging by the patient, and the development of lithium systems quickly made the rechargeable pacemaker obsolete.

The mercury/zinc oxide alkaline aqueous cell was developed and commercialized by the P. R. Mallory Company; the so-called RM-1 cell powered hundreds of thousands of pacemakers in a three- to six-cell configuration stimulating the heart with an impulse voltage of four to eight volts. The basic cell reaction is

$$Zn + HgO \rightarrow ZnO + Hg$$

providing an open circuit voltage of 1.35 volts. The original electrolyte was an aqueous solution of potassium hydroxide, which was later replaced with sodium hydroxide. The cell design is of a cylindrical shape with a central zinc anode surrounded by layers of separator and a toroidal mercuric oxide cathode.

Although the Zn/HgO cell made pacing possible, it presented several problems. Hydrogen out-gassed during cell operation, making is impossible to hermetically seal the pacemaker. Self-discharge at 37 C was also a problem, and catastrophic failures could occur because of dendrite growth resulting in internal shortening. Many improvements were made to the cell, but longevity of pacemakers using the RM-1 was rarely longer than 48 months. Improvements in the Zn/HgO cells were made by Paul Ruetschi, who designed a cell with reduced self-discharge, and by H. Cataldi and co-workers at General Electric, who designed a cell with redundant features to prevent internal short circuiting and a vent that permitted hermetic sealing. This latter design showed improved longevity [210].

The problems associated with both the Zn/HgO and the rechargeable systems together with advances in the technology of lithium batteries led to the development of several lithium power sources for implantable devices. From 1972 to 1976, at least five lithium-anode batteries were designed and used in pacemakers.

The Li/I_2-PVP system was invented by Moser and Schneider [211] after Gutmann, Herman, and Rembaum demonstrated in 1967 that cathodes could be fabricated by reacting halogens with organic donors. Greatbatch and co-workers first proposed this system for use in cardiac pacemakers in 1971, and the first lithium/iodine-PVP battery was implanted in Italy in 1972 [212].

The cell consists of a lithium anode, a cathode material comprising iodine and polyvinylpyridine, and a self-forming lithium iodide electrolyte. The cathode material contains a large amount of free elemental iodine and becomes electrically conductive with the PVP. The electrolyte is formed *in situ* and increases in thickness, causing the internal impedance to increase as the cell is discharged, causing the cell voltage to drop in a gradual, predictable manner that allows adequate end-of-life (EOL) warning. The basic cell reaction is

$$2Li + I_2 \rightarrow 2LiI$$

with an open circuit voltage of 2.8 volts.

The first commercially significant lithium/iodine battery was large, with nominal cell dimensions 14 mm by 45 mm by 52 mm, an energy density of 0.3 Wh/cm^3 and a rated capacity of 3.5 Ah. Thousands of these cells were implanted. Materials development decreased cell impedance with a reduction of cell size and an increase of energy density up to 0.9 Wh/cm^3 as achieved today.

Lithium/silver chromate cells were developed by Gabano and co-workers at the SAFT Company in the 1970s [213]. The cell consisted of a lithium anode, an organic liquid electrolyte of lithium perchlorate in propylene carbonate, and a cathode material of highly purified silver chromate. The basic cell reaction is

$$4\ Li + Ag_2CrO_4 \rightarrow 2Ag + CrO_2 + 2Li_2O$$

with an open circuit voltage of 3.35 volts. Cells were produced in cylindrical shapes employing polypropylene crimped seals and in rectangular shapes with glass-to-metal seals. More than 470,000 lithium/silver chromate cells were produced prior to 1984.

Lithium/cupric sulfide cells have been used in pacemakers since the mid 1970s. This cell has a two-stage discharge curve in which the lower plateau is useful as an indicator of approaching cell depletion. The cell employed a lithium anode, an organic liquid electrolyte consisting of lithium perchlorate dissolved in a mixed organic solvent, and a high-purity cupric sulfide cathode. Two different chemistries were used:

$$2\ Li + 2\ CuS \rightarrow Li_2S + Cu_2S$$

with an open circuit voltage of 2.12 volts, and

$$2\ Li + Cu_2S \rightarrow Li_2S + 2Cu$$

with an open circuit voltage of 1.75 volts.

The lithium/thionyl chloride system was used in pacemakers from 1976 to 1980 [214]. The cell reaction

$$2Li + SOCl_2 \rightarrow 2LiCl + 1/2\ S + 1/2\ SO_2$$

provides an open circuit voltage of 3.65 volts. The self-discharge of this system is a function of its current drain and determines the actual longevity. The cells showed good performance under accelerated testing, but actual longevities turned out to be significantly less than originally projected. The beginning of cell depletion was rapid, with no significant EOL warning.

A solidstate battery was developed by the P. R. Mallory Company with a lithium anode, an electrolyte of lithium iodide doped with aluminum oxide to enhance ionic conductivity, and a cathode comprising a mixture of lead iodide, lead sulfide, and elemental lead. The cell reactions

$$2Li + PbI_2 \rightarrow 2LiI + Pb$$

$$2Li + PbS \rightarrow Li_2S + Pb$$

provided open circuit voltages of 1.9 and 1.8 volts, respectively. A few implantable cells using titanium disulfide as a cathode material were evaluated.

A lithium/manganese dioxide cell was developed by Lehmann and Gerbier in 1979 with a central cathode enclosed by two lithium anodes. A few cells were used in France and Italy.

State of the Art in Pacemaker Power Sources

Almost all pacemakers being produced today employ the lithium/iodine-PVP system, of which two distinct types are in current use. The first employs a cathode in which the iodine and polyvinylpyridine are thermally reacted to form a paste-like cathode material which is poured into the battery in a molten state and solidifies upon cooling. The anode is coated with pure polyvinylpyridine. The coating has a dramatic effect on the electrical properties of the cell and affects the morphology of the discharge product.

The second employs a cathode fabricated by mixing particles of iodine and polyvinylpyridine and pelletizing them into the appropriate shapes without thermal treatment. This material remains solid throughout the cell life. The anode is flat, and to increase current-delivery capability, a cell containing two anodes using this pelletized-cathode technology has been developed.

Lithium/iodine-PVP cells may be produced in a variety of shapes and sizes such as rectangular, half-round, and asymmetric. Cells as thin as 5 mm are in current use resulting in pacemakers of a nominal thickness of 6 mm. Since nearly sixteen years of real-time testing have been completed, questions regarding self-discharge, reliability, and the approach to elective replacement voltage are well understood, therefore permitting the design and production of pacemakers with predictable performance.

The dominant power source for the implantable pacemaker in the foreseeable future will continue to be the lithium/iodine-PVP battery. However, the lithium/organic electrolyte/silver vanadium oxide LiSVO chemistry is a new technology finding use, in both pacemaker and defibrillator applications. In the future, advances in the state of the art of battery technology will result in new designs with smaller cells and more versatile shapes and sizes for more custom applications. Improved low-current consumption of microelectronics, together with low internal self-discharge characteristics, will lead to reduced size and a longer operational life of the implantable pacemakers.

The Elements of a Primary Battery

The chemical reaction in a battery must be described by the half-cell reaction associated with the material that loses electrons or, respectively, gains electrons. These reactions occur at specific physical locations in the battery. The anode is referred to as the reservoir of material that will be oxidized during discharge. An anode current collector is in physical contact with the chemical anode, and carries electrons produced by the oxidation of the chemical anode to the outside of the battery forming a battery's negative terminal. A cathode current collector returns electrons to the battery after their energy has been dissipated in the load. Like anode, cathode is a term used to describe the chemical cathode with the depolarizer, or the positive side of a battery.

The electrochemical reactions occurring in the battery result in discharge products, and the electrolyte keeps the reactants from direct chemical reaction. Therefore, a separator is needed that can be made of a porous paper or polymeric material such as polypropylene. With a solid electrolyte, it serves as a separator itself. As demonstrated in Figure 127, the chemical reaction in the two heart cells produces electrical energy in which the compounds are reduced or oxidized.

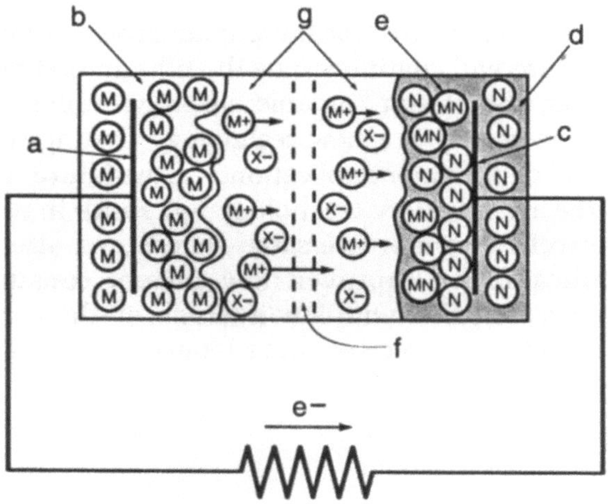

Figure 127. Electrochemical representation of a battery (a = anode current collector; b = chemical anode (M); c = cathode current collector; d = cathode depolarizer (N + carbon); e = discharge product (MN); f = separator; g = electrolyte containing M, X ions) [215].

The specific ratio of the amounts of chemical anode and chemical cathode sets the stoichiometry of the ratio and optimizes cell performance. Batteries are engineered with amounts of material near the stoichiometric value for the reaction and, therefore, matching the EOL performance of the battery. For example, depletion of the anode material metal may lead to a fast loss of voltage (see Figure 128a). Cathode materials lose activity much more slowly so that batteries with limited amounts of cathode material may exhibit a more gentle voltage decline as the battery depletes (see Figure 128c). A battery near depletion might show the voltage declining gradually near EOL or even to go through a voltage step (Figure 128b). At the EOL of all batteries, two specific voltages are important: the first, in which the EOL indicator voltage is set below the normal operating voltage, and the second, the voltage where the device no longer functions. The time between these two conditions is known as the elective replacement interval, demonstrated in Figure 128a. The selection of the elective replacement interval has proven to be about six months for pacemaker replacement.

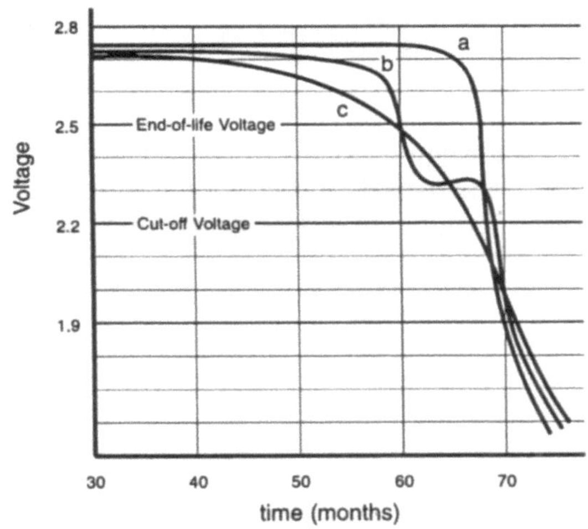

Figure 128. *Battery discharge curve showing critical operating points. a: early depletion of the anode. b: step-down performance. c: gradual decline due to cathode depletion.*

The nature of Li-cells requires hermetic encapsulation accomplished by a feedthrough consisting of a metal pin surrounded by a ferrule and isolated by a glass or ceramic insulator. The most common design of implantable medical batteries uses the case as a positive with respect to the feedthrough pin. An alternative uses the case as a neutral with both the anode and cathode isolated from the battery case. This requires two feedthroughs in addition to insulating material that will keep the anode and/or cathode separated from the case.

Battery Performance

The battery is defined by the open circuit voltage representing the characteristics of the electrochemical reaction described by the equation

$$\Delta G = -nFE$$

The free energy of the reaction, ΔG, the maximum electrochemical energy content, is related by the equation

$$\Delta G = \Delta H - T\Delta S$$

The enthalpy ΔH is the energy available in the form of heat from the chemical reaction. The term $T\Delta S$ corrects for entropy effects. This means that most of the chemical energy can be converted into

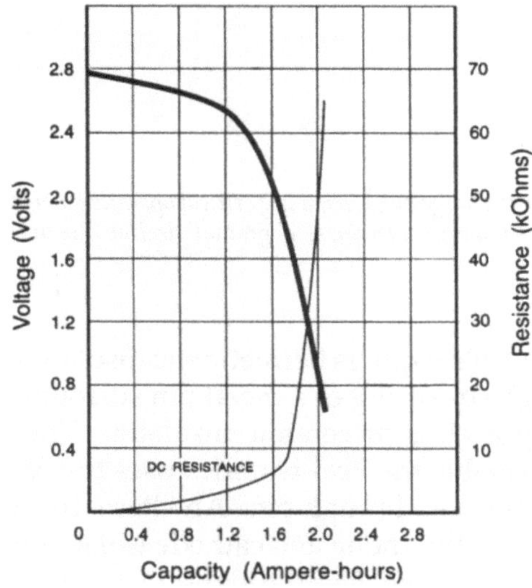

Figure 129. *Discharge curve for a single-anode lithium/iodide-PVP battery.*

electrical energy under theoretical conditions. The open circuit voltage of the implantable battery system (Li/I_2-PVP) is 2.792V.

The stoichiometric capacity is given by the amount of anode and cathode material required to react stoichiometrically, providing an estimate of electrochemical performance.

Battery performance is determined by the load, voltage versus time or versus charge at a particular discharge rate as given by the curves such as those seen in Figures 129 and 130. Figure 129 demonstrates, for a single-anode Li/I_2 battery, the increase of internal resistance as a function of remaining capacity rate; therefore, capacity decreases with lifetime. Figure 130 demonstrates a low internal resistance for a Li/I_2 double anode-battery, almost independent of the remaining battery capacity toward the end of life.

In long-term applications such as an implantable pacemaker with a life expectancy of 5-10 years, the influence of chemical parameters on discharge curves for the lithium/iodine system is shown in Figure 131 for discharge current densities ranging from 5 to 50 $\mu A/cm^2$. Special attention must be given to the self-discharge as a

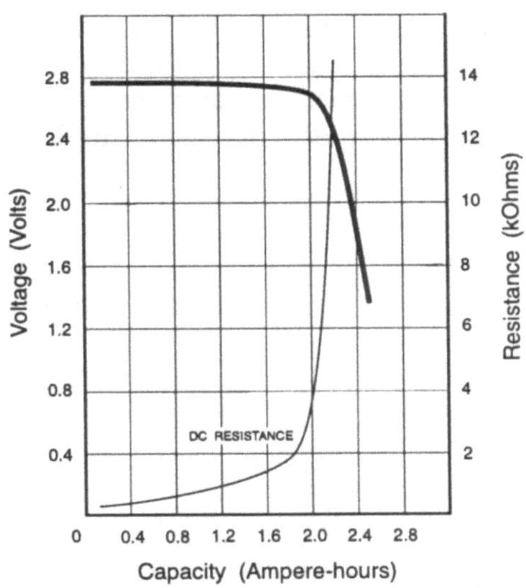

Figure 130. *Discharge curve for a double-anode lithium/iodide-PVP battery.*

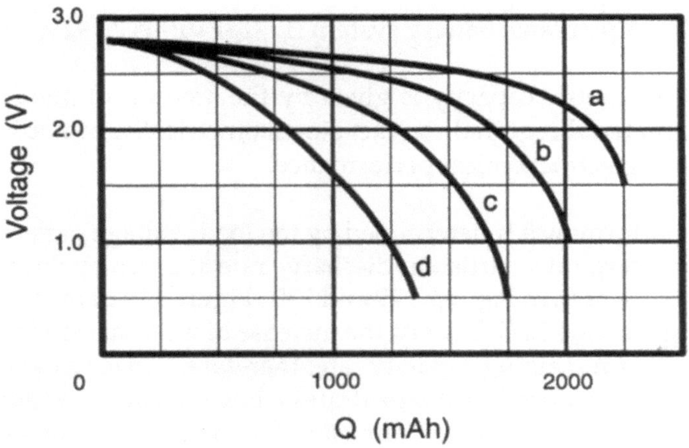

Figure 131. *Battery discharge curves as a function of current density (a through d: 5, 10, 15 and 50μA/cm²) [216].*

result of direct chemical reactions inside the battery, especially the reaction of iodine and lithium. The lithium iodide discharge product is not a perfect barrier to iodine diffusion. It may be formed with small cracks and grain boundaries, and the iodine may diffuse along these interfaces and react directly with the lithium resulting in a self-discharge reaction which decreases the useful capacity. Special performance is desired for medical batteries near EOL to ensure an appropriate warning and a sufficient elective replacement time (see Figure 132).

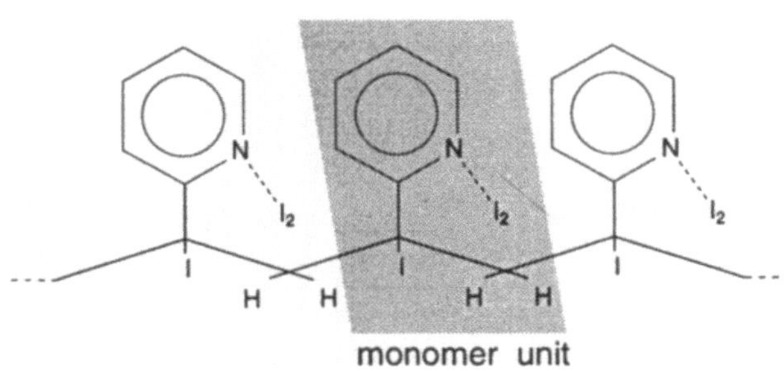

Figure 132. *Structure of iodine-PVP reaction product [217].*

Chemistry of the Lithium/Iodine-Polyvinylpyridine (Li/I_2-PVP) System

The basic cell reaction is

$$Li + \frac{1}{2}I_2 \rightarrow LiI$$

for the Li/I_2-PVP battery.

The Gibbs free energy of this reaction is -64.5 kcal/mol resulting in an open circuit voltage of 2.792V. The cathode material includes the compound poly-2-vinylpyridine, which reacts with a portion of the iodine to render the mixture conductive.

Lithium is an efficient anode material for high energy-density batteries. Its density is 0.534 g/cm^3 at room temperature. The first ionization potential of lithium is 5.39 eV, and the electrochemical equivalent is 3.86 Ah/g. It reacts readily with halogens at room temperature. Its physical properties are such that it is easily formed into sheets that can be cut to specific anode sizes, and it is easily pressed into specific anode shapes, such as the design of recent versions of the Li/I_2-PVP cell.

The melting point of lithium is 180.5 C. This fact is important because many lithium battery systems, including the LI/I_2-PVP system, may present hazards if heated to temperatures above the melting point of lithium. In many versions, the lithium anode is coated with a solution of PVP. The solvent is evaporated, leaving a contiguous film of pure PVP on the entire anode surface. Precoating of the anode has a profound effect on the nature of the discharge product and leads to lower cell impedance with enhanced performance. (See Figure 133.)

Iodine has an electron affinity of 3.24eV and an electrochemical equivalent of 0.211 Ah/g. Its melting point is 113.7 C and its density is 4.94 g/cm^3. The electronic conductivity of iodine is insignificant, rendering the pure material of no use in practical lithium/iodine battery systems.

Cathode materials are considered "charge-transfer complexes," defined as a system involving the partial transfer of one or more electrons from a donor aromatic hydrocarbon to an acceptor halogen or an organic compound of high electron affinity. One significant property of such materials is a relatively high electronic conductivity. The cathodes were made by mixing certain organic compounds with iodine to form these "complexes," One such organic compound may be polyvinylpyridine mixed with carbon black. The weight ratio of iodine to PVP is 50:1 [219].

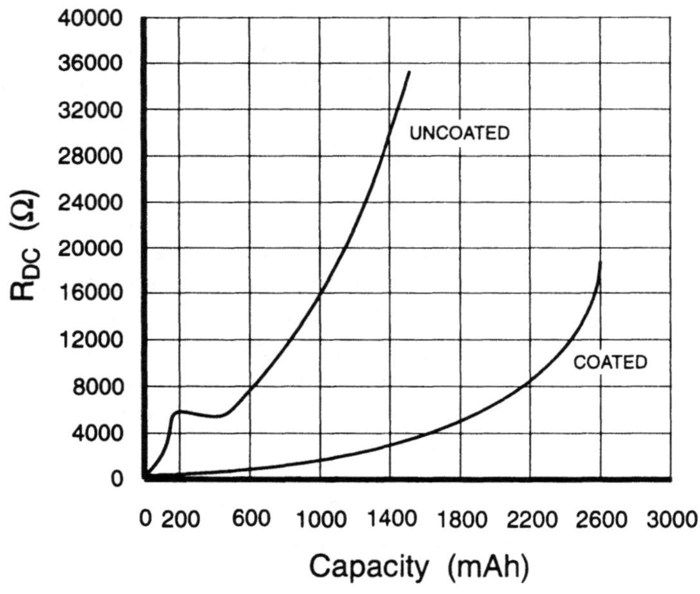

Figure 133. *Internal cell resistance as a function of capacity for coated and uncoated cells [218].*

Design of Lithium/Iodine-PVP Cells

The first commercially successful Li/I_2-PVP battery consisted of a central lithium anode surrounded by iodine-PVP cathode material. Because of the uncertainty regarding the reactivity of iodine with the metal case material, the battery was enclosed in a fluoropolymer support and then potted in a polyester casing. Internal connections were brought out via hermetic feedthroughs in the stainless steel lid, and welded into a case made of 304L stainless steel.

To provide more efficient encapsulation of cathode material, this material was completely encapsulated by the lithium anode. The anode material was then insulated from the case by either a fluorocarbon insulator or a polyester potting material. Anode and cathode were connected through glass-to-metal seals.

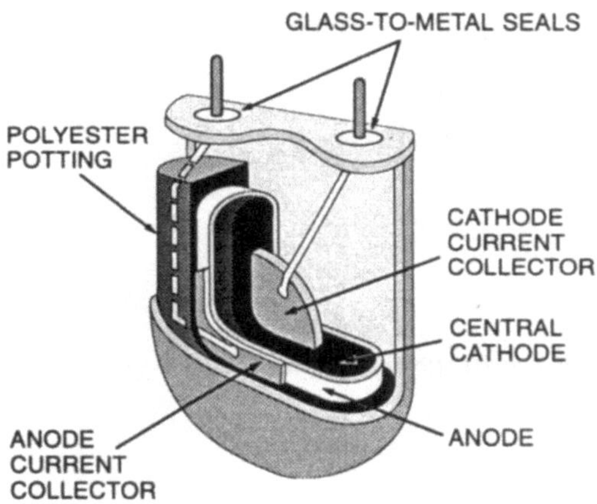

Figure 134. *Internal construction of central cathode cell encased in polyester.*

A cutaway view of one such design is shown in Figure 134. The design is a bipolar configuration in which two cup-shaped lithium anodes are pressed into a fluorocarbon mold holding the cathode material. The two subassemblies form a battery in which the cathode material is surrounded by lithium. The lithium anodes were pre-coated with polyvinylpyridine before addition of the cathode material. The battery is hermetically encapsulated into a stainless steel case.

The fact that the 304L stainless steel case may be affected by corrosion led to the development of the case-grounded battery. In this design, a precoated central lithium anode is surrounded by the cathode material. The case itself serves as the encapsulant of the cathode material and as the cathode current collector. Today's batteries employ an anode that is corrugated or "ribbed" during the pressing process to increase surface area, as shown in Figure 135 [220, 221].

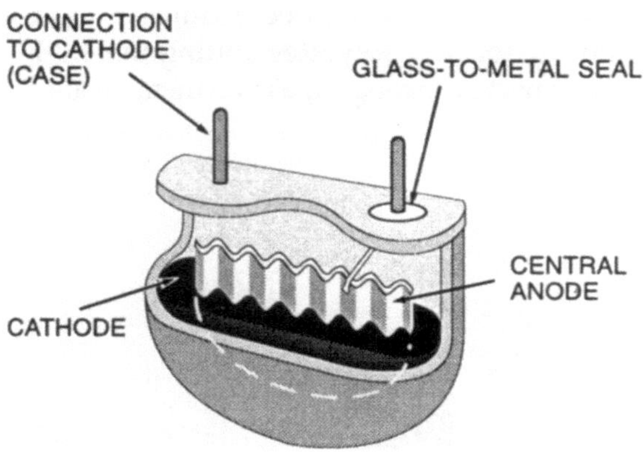

Figure 135. Internal construction of central anode/case-grounded cell with corrugated anode .

Discharge Characteristics of the Li/I_2-PVP Battery

The basic shape of the discharge curve results from thermodynamic, kinetic, and chemical considerations. Under a current drain of 10–100 μA the discharge voltage begins near the open circuit voltage of 2.8 V. As the layer of lithium iodide increases in thickness, the cell resistance increases and the voltage decreases. In the case of cells containing thermally reacted cathode material and uncoated anodes, this voltage decrease is linear. In coated-anode cells and cells with pelletized cathode material, the voltage decrease is less steep as discharge proceeds. As iodine is depleted from the cathode material the cathode conductivity follows the curve shown in Figure 133. This causes the discharge curve to become nonlinear and exhibit a higher rate of voltage drop near the EOL. The result is a discharge curve that provides a clinically desirable gradual approach to the EOL, normally 2.5 V.

Cell Design and Qualification

The reliability of implantable medical batteries begins in the design phase and concludes with long-term testing of the product. Qualification testing of new designs demonstrates acceptable performance under accelerated test conditions and resistance to environmental testing. In-process controls and testing ensure that cells are manufactured under stringent quality controls. Accelerated testing and simulation provide projections of cell performance. Finally, a life-testing program, on a sampled basis under simulated implant conditions, verifies the technical performance of the cells.

Each new cell model is subjected to a qualification testing program before an implantable-grade product is released for production. The final qualification program is performed on an early group of units made by production personnel using documented production processes. Figure 136 shows the qualification procedure used for lithium/iodine batteries. Among the tests performed are accelerated discharge, elevated temperature exposure, shock and vibration testing, thermal cycling, exposure to high temperatures and humidities, determination of the dependence of voltage on

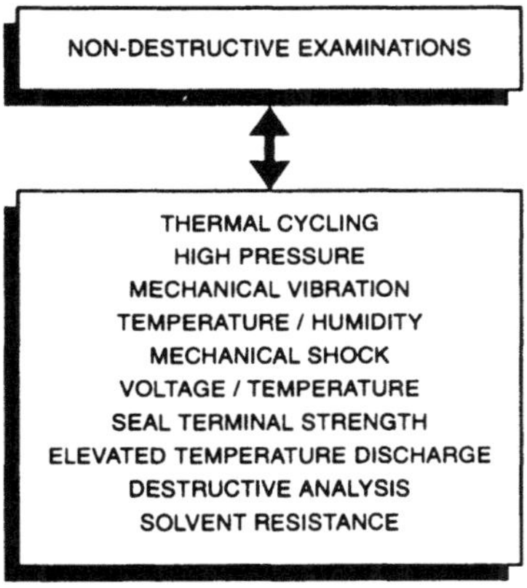

Figure 136. *Qualification flow chart for Li/I_2-PVP cells.*

temperature, and exposure to high pressures. After each test non-destructive examinations such as electrical testing, dimensional testing, X-ray analysis, and visual examination evaluate the response of the battery to the test. The completed qualification program forms the basis upon which device manufacturers can qualify the cell for use in implantable devices [222].

All pacemakers today use the Li/I_2-PVP system. This battery has achieved an outstanding record of reliability since its introduction. This record is the result of the implementation of a quality system that is based on the *Good Manufacturing Practices* (GMP) developed and issued by the United States *Food and Drug Administration* (FDA), coupled with a design and testing philosophy aimed at developing and evaluating batteries that are conservative in design and whose long-term performance is readily predicted from shorter-term accelerated tests.

Table XXI summarizes the 50% survival intervals for a series of Wilson Greatbatch lithium iodine cells that have reached EOL.

Table XXI
Cumulative Survival Testing

Model	*50 Percent Survival Interval (Months)*
7905	81-84
8031	87-90
8077	87-90
8205	81-84
8207	75-78
8304	72-75
8402	60-63

The life test sample for Wilson Greatbatch lithium iodine cells (manufactured from November 1981 until May 1990) totals 17,751 units and has accumulated 817,000 life test months. To date, no random failures have occurred, attesting to the stringent adherence to good design, manufacturing, and quality control practices.

REESTABLISHMENT OF PHYSIOLOGICAL REGULATION

A CHALLENGE TO TECHNOLOGY

The exemplary collaboration between engineering sciences and medicine has, in recent decades, developed electropacing of the heart into an extremely successful therapy. Due to high technological standards, cardiac pacemakers have become therapeutic agents that reliably reestablish cardiovascular function and improve the quality of life for a large number of patients. Their superiority over previous forms of treatment arises from the utilization of still existing physiological regulatory processes of the cardiovascular system. An understanding of natural impulse formation, impulse conduction, and replacement by artificial processes substantiates the success of the aforementioned interdisciplinary efforts.

Microelectronics became a key technology in computing and data processing in many areas of medical engineering, so the implantable cardiac pacemaker is taking advantage of the fast evolving technology of microelectronics. The complex analog and digital functions demonstrated in the previous chapters are now assembled using Very Large Scale Integration (VLSI) technology on integrated circuits. The advanced state of the electronic technologies has led to broad clinical applications supporting a wide range of pacing therapies. The classical form of physiological rate-adaptive pacing is dual-chamber pacing. DDD pacing employs dual sensors and pacing electronics. With an intact sino-atrial node, improved hemodynamic results are obtained when AV synchrony can be included in the pacing process. When natural sinus rhythm is absent, a series of sensors with complex microelectronic control circuits meets the need for rate-adaptive pacemaker systems while preserving AV synchrony.

In addition, the use of microprocessors allows complex calculations, and for the first time, the models of central neural cardiovascular circuits can be implemented in a more efficient way to establish chronotropy supplementing the lost parasympathetic control.

Pacemakers now in clinical evaluation achieve adaptive functionality through the direct control of the pacing rate in response to physiological signals. These devices function by cooperating with the intrinsic cardiovascular control processes. Optimal rate-adaptive pacemaker systems have been developed on the basis of intracardiac impedance measurements in which, the ANS information, as measured by VIP is used as the regulatory signal. VIP offers the advantage that the remaining regulatory signals can be included in the therapy, thereby restoring to the patient's homeostatic system the ability to control heart rate. For the first time, the restoration of closed-loop control was established.

New Concepts in Electrotherapy of the Heart

In spite of the technological advancements, the understanding of the physiological control mechanism remains a prerequisite for further improvement of the electrotherapy in the future. Based upon the experience in ANS-controlled pacing and the remarkable progress in neural cardiology [223-225], a general concept of electrical treatment of arrhythmia becomes possible if the *neural factors* in the *arrhythmogenesis* are considered. With the powerful tool of monitoring the sympathetic tone by intraventricular impedance measurements, the VIP that was introduced for the restoration of chronotropy will serve as a sensor of the increased neural activity of an impending arrhythmia, therefore making it possible to prevent tachycardia.

The evidence of the role of sympathetic nervous activity in the genesis of cardiac arrhythmia suggests the activation of the antagonist, the vagus nerve, thereby decreasing susceptibility to fibrillation in the ischemic heart. Corr, Gillis, Schwartz and others concluded that the presence of an intact n. vagi protects against ventricular fibrillation [226, 227]. In addition, effects of vagus nerve activation on the susceptibility to arrhythmia during ischemia and perfusion have been confirmed by Schwartz [228]. Vagus nerve stimulation reduces the incidence of ventricular fibrillation, thereby preserving jeopardized myocardial tissue from advancing ischemia. The antifibrillatory effect [229] was also demonstrated. Based on a number of experimental and clinical studies, it appears to be most promising for developing an improved electrophysiological intervention for the prophylaxis and treatment of malignant ventricular and supraventricular arrhythmias.

While Schwartz [230] describes the role of sympathectomy in cases of sympathetic imbalance, it seems to be feasible that the same effect can be accomplished with electrical parasympathetic stimulation. Preliminary results indicate that vagal stimulation, largely through its effect on heart rate, can prevent ventricular fibrillation induced by myocardial ischemia in dogs with healed myocardial infarction.

The ANS-controlled pacemaker will serve as the regulator of the autonomic tone establishing the neural equilibrium, thereby providing a more flexible tachyrhythmia prevention. The technology already developed for rate-adaptive cardiac pacemakers can be used to monitor the neural balance ***without*** the need of additional sensing electrodes. A standard pacing lead, at the same time monitoring the changes in heart rate and serving as an indicator of pending malignant tachyrhythmia, may also serve as a sensor for contractility and, therefore, sympathetic tone. Prior to the onset of tachyrhythmia, this neural stimulation system applies bursts of electrostimuli to the vagal afferent fibers of the vagus nerve endings at the endocardium. The block diagram of such an implantable system is shown in Figure 137.

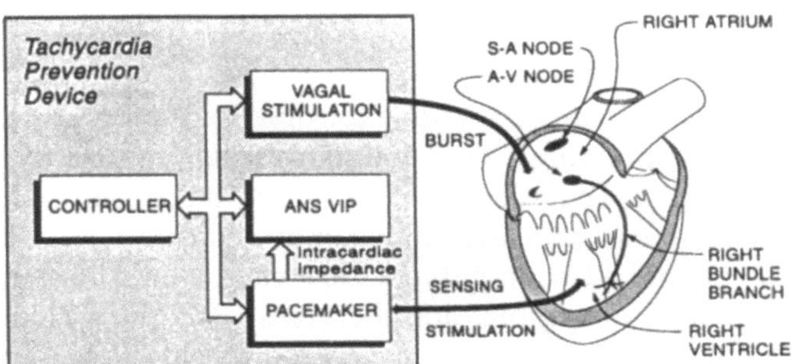

Figure 137. *Block diagram of the ANS-controlled vagal stimulator. ANS sensing in the right ventricle with endocardial pacing electrode; vagal stimulation in the atrium.*

Conclusions

Access to the flow of cardiac ANS information opens new avenues for cardiac electrotherapies [231]. Figure 138 depicts many of these new therapies. On the left are the rhythm, myocardial performance and perfusion disturbances. On the right are the therapies improved through the use of cardiac ANS information. At the upper right, the cardiac ANS signal is being used to control the chronotropic pacing therapy in response to physiologic demand. Physiological pacing avoids episodes of chronotropic insufficiency and, thus, prevents the

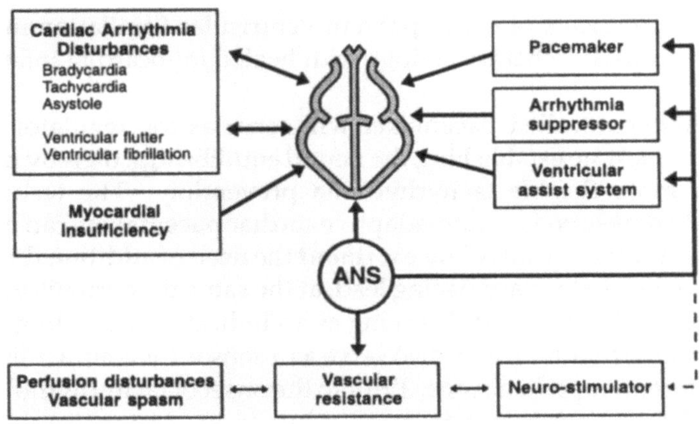

Figure 138. New concepts in the electrotherapy of the heart.

occurrence of excessively high inotropic excitation of the myocardium. Avoiding high inotropic states serves to directly prevent many ventricular arrhythmias. The second block shows that early detection of an incipient arrhythmia or an acute imbalance between sympathetic and parasympathetic tone can be used to trigger and control arrhythmia suppression therapies. The third block shows that the cardiac ANS signal can be used to modulate Ventricular Assist Systems including cardiac myoplasty preparation, which would further protect the myocardium from acute stress (see Figure

139). The fourth block covers circulatory disorders triggered by pain. In these cases, acute pain leads to anxiety and the systemic sympathetic response progresses to angina. The patient may be equipped with a neuro-stimulator to block the painful sensory stimulus, which may be invoked by a volitional act or be automatically triggered. The neuro-stimulator effects the corrective therapy by providing an epidural neuro-stimulation to block nociception and, thus, disrupts conditions leading to the emergence of angina. The effect of the therapy may be monitored by the analysis of the cardiac ANS signal and the intracardiac ECG (ST-segment).

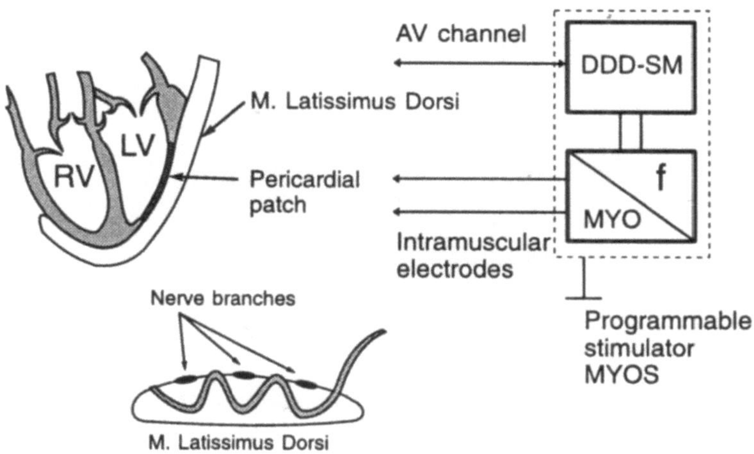

Figure 139. *Principle of ANS-controlled cardiomyoplasty.*

A nociceptive stimulus invokes a general "fight or flight" reflex in the autonomic nervous system that must be considered when evaluating and managing the cardiovascular performance of a patient (Figure 140). Afferent nociceptive signals stimulate the vasomotor centers and the ascending reticular activating system. This excitation leads to a suppression of vagal parasympathetic tone to the heart and a release of chronotropic inhibition. The excitation also leads to an increase in sympathetic tone and the associated increase in chronotropy, inotropy, peripheral vascular resistance, and a potential decrease in coronary vascular resistance. The combined increase in chronotropy, inotropy, and cardiac workload places an increased demand on the coronary circulation.

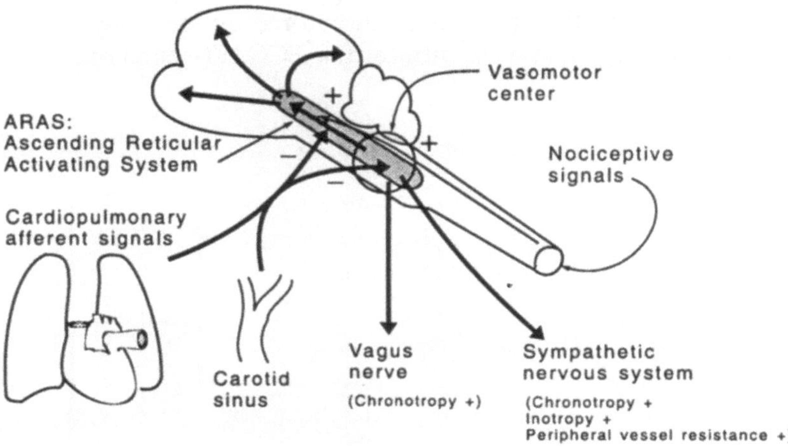

Figure 140. Effect of nociceptive stimuli on the cardioactive ANS activity.

In cardiology, and specifically in the field of electrotherapy [232], medical progress by continuous technological improvements in many disciplines has been made. These advances are leading to simpler, physiological, and more versatile pacing therapies. Presuppositions for practical solutions are being met since the state of microelectronics presented here offers the technical preconditions for the further development of pacemaker therapy.

BIBLIOGRAPHY

[1] Rosen, M. R., M. J. Janse, A. L. Wit, eds. *Cardiac Electrophysiology: A Textbook.* Mount Kisco, New York: Futura Publishing Co. (1990).

[2] Guyton, A. C. *Textbook of Medical Physiology.* 3rd edition. Philadelphia, London: W. B. Saunders Co. (1968).

[3] West, J. B., ed. *Best and Taylor's Physiological Basis of Medical Practice.* 12th edition. Baltimore: Williams & Wilkins (1985).

[4] Hurst, J. W. *The Heart, Arteries and Veins.* New York: McGraw-Hill (1986).

[5] Netter, F. H. *Heart*, 3rd edition. Summit, New Jersey: CIBA Pharmaceutical Co. (1974).

[6] Carvalho, A. P., B. F. Hoffman, M. Lieberman, eds. *Normal and Abnormal Conduction in the Heart: Biophysics, Physiology, Pharmacology, and Ultrastructure.* Mount Kisco, New York: Futura Publishing Co. (1982).

[7] Little, R. C. *Physiology of the Heart and Circulation.* 4th edition. Chicago: Year Book Medical Publishers (1989).

[8] Saksena, F. B. *Hemodynamics in Cardiology: Calculations and Interpretations.* New York: Praeger (1982).

[9] Randall, W. C., ed. *Neural Regulation of the Heart.* New York: Oxford University Press (1977).

[10] Cranefield, P. F. *The Conduction of the Cardiac Impulse: the Slow Response and Cardiac Arrhythmias.* Mount Kisco, New York: Futura Publishing Co. (1975).

[11] Mandel, W. J., ed. *Cardiac Arrhythmias: Their Mechanisms, Diagnosis, and Management*. 2nd edition. Philadelphia: Lippincott (1987).

[12] Platia, E. V. *Management of Cardiac Arrhythmias: the Nonpharmacologic Approach*. Philadelphia: Lippincott (1987).

[13] Conover, M. B. *Understanding Electrocardiography: Arrhythmias and the 12-lead ECG*. 4th edition. St. Louis: Mosby (1984).

[14] Chung, E. K. *Principles of Cardiac Arrhythmias*. 4th edition. Baltimore: Williams & Wilkins (1989).

[15] Zipes, D. P. "Specific arrhythmias: diagnosis and treatment." In *Heart Disease*. Vol. 1. 2d ed., edited by E. Braunwald. Philadelphia: W. B. Saunders Co. (1984), 667-725.

[16] Zipes, D. P. "Genesis of cardiac arrhythmias: electrophysiological considerations." In *Heart Disease*. Vol 1. 2nd edition, edited by E. Braunwald. Philadelphia: W. B. Saunders Co. (1984) 588-627.

[17] Bellet, S. *Essentials of Cardiac Arrhythmias: Diagnosis and Management*. Philadelphia: W. B. Saunders Co. (1972).

[18] Zipes, D. P. "Management of cardiac arrhythmias: pharmacological, electrical, and surgical techniques." In *Heart Disease*. Vol 1. 2nd edition, edited by E. Braunwald. Philadelphia: W. B. Saunders Co. (1984) 628-666.

[19] El-Sherif, N., P. Samet. *Cardiac Pacing and Electrophysiology*. 3rd edition. Philadelphia: W. B. Saunders Co. (1991).

[20] Moses, H. W., J. A. Schneider, B. D. Miller, G. J. Taylor. *A Practical Gude to Cardiac Pacing*. Boston, Toronto, London: Little, Brown and Co. (1991) 81-88.

[21] Furman, S., D. L. Hayes, D. R. Holmes, Jr. *A Practice of Cardiac Pacing*. Mount Kisco, New York: Futura Publishing Co. (1986).

[22] Barold, S. S., J. Mugica. *New Perspectives in Cardiac Pacing*. Mount Kisco, New York: Futura Publishing Co. (1988).

[23] Lüderitz, B. Herzschrittmacher. *Therapie und Diagnostik Kardialer Rhythmusstörungen.* Berlin, Heidelberg, New York, Tokyo: Springer-Verlag (1986) 430 S.

[24] Furman, S. The Present Status of Cardiac Pacing. *Herz* 16 (1991) 171-181.

[25] Sulke, N., J. Chambers, A. Dritsas, E. Sowton. A randomized double-blind crossover comparison of four rate-responsive pacing modes. *Jour. Am. Coll. Cardiol.* Vol. 17, No. 3 (1991) 696-706.

[26] Anderson K. M., A. A. Moore. Sensors in pacing. *PACE* 9 (1986) 954-959.

[27] Camilli, L., L. Alcidi, E. Shapland, S. Obino. Results, problems and perspectives with the autoregulating pacemaker. *PACE* 6 (1983) 488-493.

[28] Mehta, D., C. P. Lau, D. E. Ward, A. J. Camm. Comparative evaluation of chronotropic responses of QT-sensing and activity sensing rate-responsive pacemaker. *PACE* 11 (1988) 1405-1412.

[29] Nappholz, T. A., J. D. Maloney, T. Simmons, M. Masterson, S. Fischer, H. Valenta. A two-year research study of minute ventilation as an indicator for rate-responsive pacing. *PACE* 10 (1987) 1222.

[30] Schaldach, M. Present state and future trends in electrical heart stimulation. *Medical Progress Through Technology* 13 (1987) 85-102.

[31] Stangl, K., A. Wirtzfeld, R. Heinze, M. Laule, K. Seitz, G. Göbl. A new multisensor pacing system using stroke volume, respiratory rate, mixed venous oxygen saturation, and temperature, right atrial pressure, right ventricular pressure, and dP/dt. *PACE* 11 (1988) 712-24.

[32] Wirtzfeld, A., R. Heinze, K. Stangl, K. Hoekstein, E. Alt, H. D. Liess. Regulation of pacing rate by variations of mixed venous oxygen saturation. *PACE* 7 (1984) 1257-1262.

[33] Mindt, W., M. Schaldach. "Electrochemical Aspects of Pacing Electrodes." *Advances in Pacemaker Technology, Engineering in Medicine* 1, edited by M. Schaldach and S. Furman. Berlin, Heidelberg, New York: Springer-Verlag (1975) 297-305.

[34] Owens, B. B. *Batteries for Implantable Biomedical Devices*. New York, London: Plenum Press (1986) 358 S.

[35] Schaldach, M., M. Hubmann, R. Hardt, A. Weikl. Titannitrid-Herzschrittmacher-Elektrode. *Biomedizinische Technik* 34 (1989) 185-190.

[36] Dreifus, L. S., *et al.* Guidelines for Implantation of Cardiac Pacemakers and Antiarrhythmia Devices. A Report of the American College of Cardiology/American Heart Association Task Force on Assessment of Diagnostic and Therapeutic Cardiovascular Procedures. *Circulation,* 84:1 (1991) 455-464.

[37] Elmquist, R., A. Senning. Implantable pacemaker for the heart. *Proc. Sec. Int. Conf. on Medical Electronics,* Paris, June 1959. *Medical Electronics.* (1959).

[38] Schaldach, M. Reliability considerations in the design of implantable products. *Biomedizinische Technik* 32 (1987) 276-284.

[39] Friedberg, M. D., S. S. Barold. On hysteresis in pacing. *J. Electrocardiology* 6 (1973) 2.

[40] Castellanet, M. J., J. Garza, S. P. Shaner, J. C. Messenger. Telemetry of programmed and measured data in pacing system evaluation and follow-up. *Journal of Electrophysiology* 1 (1987) 360-375.

[41] Schaldach, M. Die Bedeutung der Mikroelektronik in der Implantattechnologie. *Biomedizinische Technik* 35 (Ergänzungsband 2) (1990) 3-13.

[42] Schaldach, M. Hybrid microelectronics in implant technology: A progress report. *Hybrid Circuit Technology* 4 (1987) 19-26.

[43] Schaldach, M. Stand und Entwicklungstendenzen der Mikroelektronik in der biomedizinischen Technik. *E und M Elektrotechnik und Maschinenbau,* 103 (1986) 127-138.

[44] Bernstein, A. D., A. J. Camm, R. D. Fletcher, R. D. Gold, A. F. Rickards, N. P. D. Smyth, S. R. Spielman, R. Sutton. The NASPE/BPEG generic pacemaker code for antibradyarrhythmia and adaptive-rate pacing and antitachyarrthmia devices. *PACE* 10 (1987) 794-799.

[45] Furman, S., Ph. Hurzeler, V. DeCaprio. The ventricular endocardial electrogram and pacemaker sensing. *The Journal of Thoracic and Cardiovascular Surgery* 73 (1977) 258-266.

[46] Schaldach, M. Entwicklungstendenzen in der Physiologischen Stimulation des Herzens. *Physiologische Stimulation des Herzens,* edited by A. Weil. Erlangen: perimed-Fachbuch-Verlagsgesellschaft (1984) 134-165.

[47] Castello, R., A. G. Grassi, S. Donati. A 500-nA sixth-order bandpass SC filter. *Journal of Solid-State Circuits* 25 (1990) 669-676.

[48] Shieh, S. P. *et al.* A scalable switched-capacitor filter implemented in 1.25 m technology. *IEEE J. of Solid State Circ.* 24 (1989) 174-177.

[49] Signell, S., K. Mossberg. Offset-compensation of two-phase switched-capacitor filters. *IEEE Trans. on Circuits and Systems,* 36 (1989) 31-41.

[50] Stangl, K., H. Heuer. Aktivität. In *Frequenzadaptive Herzschrittmacher,* edited by K. Stangl, H. Heuer, A. Wirtzfeld. Darmstadt: Steinkopff Verlag. (1990) 107-30.

[51] Grote, J. Gewebeatmung. Eds., R. F. Schmidt, G. Thews. *Physiologie des Menschen.* 23rd edition, Heidelberg, Berlin, New York: Springer-Verlag (1987) 633-48.

[52] Witzleb, E. Funktionen des Gefäßsystems. *Physiologie des Menschen*. 23rd edition, edited by R. F. Schmidt, G.Thews. Heidelberg, Berlin, New York: Springer-Verlag (1987) 505-72.

[53] Ulmer, H-V. Arbeitsphysiologie. In *Physiologie des Menschen*, 23rd edition, edited by R. F. Schmidt, G. Thews. Heidelberg, Berlin, New York: Springer-Verlag (1987) 683-704.

[54] Cowley, A. V., A. C. Guyton. Heart rate as a determinant of cardiac output in dogs with arteriovenous fistula. *Am. J. Cardiol.* (1971) 28:321-5.

[55] Baller, D., A. Hoeft, H. Korb, H. G. Wolpers, J. Zipfel, G. Hellige. Basic physiological studies on cardiac pacing with special reference to the optimal mode and rate after cardiac surgery. *Thorac. Cardiovasc. Surgeon* 29 (1981) 168.

[56] Bevegard, S., B. Johnsson, I. Karlöf, H. Lagergren, E. Sowton. Effect of changes in ventricular rate and central pressures at rest and during exercise in patients with artificial pacemakers. *Cardiovasc. Res.* 1 (1967) 21.

[57] Guyton, A. C., T. G. Coleman. Long-term Regulation of the Circulation: Interrelationships With Body Fluid Volumes, edited by E. B. Reeve, A. C. Guyton. *Physical bases of circulatory transport: Regulation and Exchange*. Philadelphia, London: W. B. Saunders Co. (1967) 179-201.

[58] Guyton, A. C., T. G. Coleman. Quantitative analysis of the pathophysiology of hypertension. *Circulation Research* 24 (1969) I-1-I-19.

[59] Hubmann, M., A. Weikl, R. Hardt, M. Schaldach. Bewegungsenergie als Steuergröße für die Anpassung der Stimulationsfrequenz. *Biomed Technik* 34 (1989) 191-6.

[60] Wirtzfeld, A., L. Goedel-Meinen, T. Bock, R. Heinze, H. D. Liess, J. Munteanu. Central venous oxygen saturation for the control of automatic rate-responsive pacing. *Circulation* 64 (IV) (1981) 299.

[61] Schaldach, M. Compensation of chronotropic incompetence with temperature-controlled rate-adaptive pacing. *Biomedizinische Technik* 33 (1988) 286-294.

[62] Schaldach, M. Herzschrittmacher mit temperaturgesteuerter Frequenzanpassung. *Herzschrittmacher* 9 (1989) 5-14.

[63] Jones, C. E. Control of Cardiac Output. *In Cardiovascular flow dynamics and measurements,* edited by N. H. C. Hwang, N. A. Normann. Baltimore, London, Tokyo: University Park Press (1977) 365-401.

[64] Schaldach, M. Systolic time intervals as a control of rate-adaptive pacing. *Proc. 11th Annual Int. Conf. IEEE Eng. in Med. & Biol.* (1989) 1407-1410.

[65] Schaldach, M. PEP-gesteuerter Herzschrittmacher. *Biomedizinische Technik* 34 (1989) 177-184.

[66] Baan, J., T. T. Aouw Jong, P. L. M. Kerkhof, F. A. Rodrigo, J. Koops. Continuous intraventricular measurement of stroke volume and cardiac output and estimation of regional wall motion. *Digest Comb Meet XII Int Conf Med Biol Eng V Int Conf Med Phys,* Jerusalem, Part 2: 29.3 (1979).

[67] Baan, J., E. T. van der Velde, H. G. de Bruin, G. J. Smeenk, J. Koops, A. D. van Dijk, D. Temmermann, J. Senden, B. Buis. Continuous measurement of left ventricular volume in animals and humans by conductance catheter. *Circulation* 70 (1984) 812-23.

[68] Baan, J., E. T. van der Velde, P. Stenndijk, J. Koops. Calibration and application of the conductance catheter for ventricular volume measurement. In *Intracardiac conductance volumetry,* edited by M. E. Valentinuzzi. *Automedica* 11. New York: Gordon and Breach Science Publisher (1989) 357-365.

[69] Boheim, G. Intrakardiale Impedanzmessung zur frequenzadaptiven Stimulation des Herzens. Ph.D. Dissertation. Erlangen: Zentralinstitut für Biomedizinische Technik (1988).

[70] Boheim, G., M. Schaldach. Frequenzadaption eines künstlichen Herzschrittmachers über einen Volumenregelkreis. *Biomedizinische Technik* 33 (1988) 100-105.

[71] Geddes, L. A., H. E. Hoff, A. Mellow, C. Palmer. Continuous measurement of ventricular stroke volume by electrical impedance. *Cardiac Research Center Bulletin* 4 (Houston, TX) (1966) 118-130.

[72] Herrera, M. C., O. E. Clavin, J. C. Spinelli, M. E. Valentinuzzi, E. I. Cabrera Fischer, R. H. Pichel. Multichannel tetrapolar admittance meter (MY) for intracardiac volume measurements in animals. *Med. Prog. Technol.* 11 (1986) 43-9.

[73] McKay, R. G., J. R. Spears, J. M. Aroesty, *et al.* Instantaneous measurement of left and right ventricular volume and pressure-volume relationships with an impedance catheter. *Circulation* 69 (1984) 703-10.

[74] Spinelli, J. C., M. E. Valentinuzzi. Conductivity and geometrical factors affecting volume measurements with an impedancimetric catheter. *Med. & Biol. Eng. & Comput.* (1986) 27:25-32.

[75] Chirife, R. The pre-ejection period: an ideal physiologic variable for closed-loop rate-responsive pacing. *PACE* 10 (1987) 425.

[76] Smith, J. J., *et al.* Physiological principles of a new method for rate-responsive pacing using the pre-ejection interval. *PACE* 11 (1988) 1545-1554.

[77] Niederlag, W., W. Rentsch, E. Wunderlich, P. K. H. Schmidt. Ist die Anspannungszeit ein brauchbarer Parameter für die frequenzadaptierte Herzstimulation. *Biomed Technik* 34 (Ergänzungsband) (1989) 62-5.

[78] Rentsch, W., W. Niedrelag, H. Foelske, E. Wunderlich, P. K. H. Schmidt. Zur physiologischen Frequenzanpassung von Herzschrittmachern mittels systolischer Zeitintervalle. Z. *Gesamte Inn Med* 42 (1987) 386-9.

[79] Schaldach, M., W. Rentsch, H. W. Rentsch. Advances in Intracardiac Impedance Plethysmography. *Proc. IEEE Eng. in Med. & Biol.* 12, No. 2 (1990) 711-713.

[80] Spodick, D. H., Y. L. Doi, R. L. Bishop, T. T. Hashimoto. Systolic time intervals reconsidered. Reevaluation of the pre-ejection period absence of relation to heart rate. *Amer. J. Cardiol.* 53 (1984) 1667-70.

[81] Schaldach, M. Automatic adjustment of pacing parameters based on intracardiac impedance measurements. *PACE* 13 (1990) 1702-1710.

[82] Ko, W. H. A review of implantable sensors. *PACE* 6 (1983) 482-487.

[83] Lau, C. P., J. R. R. Stott, W. D. Toff, M. B. Zetlein, D. E. Ward, A. J. Camm. Selective vibration sensing: A new concept for activity-sensing rate-responsive pacing. *PACE* 11 (1988) 1299-1309.

[84] Reswick, J. *et al.* Preliminary evaluation of the vertical acceleration gait analyzer (VAGA). *Proc. 6th Ann. Symp. External Control Human Extremities.* Dubrovnik, Yugoslavia, Aug. 28-Sept. 1 (1978) 305-314.

[85] Servais, S. B., J. G. Webster, H. J. Montoyo. Estimating human energy expenditure using an accelerometer device. *IEEE Frontiers of Engineering in Health Care* (1978) 309.

[86] Wong, T. C. *et al.* Portable accelerometer device for measuring human energy expenditure. *IEEE Trans. on Biomed. Eng.* 28 (1981) 467-471.

[87] Humen, D. P., K. Anderson, D. Brumwell, S. Huntley, G. J. Klein. A pacemaker which automatically increases its rate with physical activity. Proceedings of the VIIth World Symposium on Cardiac Pacing, edited by K. Steinbach, in *Cardiac Pacing.* Darmstadt: Steinkopff Verlag (1983) 259-264.

[88] Humen, D. P., W. J. Kostuk, G. J. Klein. Activity sensing rate responsive pacing. Improvement in myocardial performance with exercise. *PACE* 8 (1985) 52-59.

[89] Schaldach, M., M. Hubmann, A. Weikl, R. Hardt. Bewegungsenergie als Steuergröße für die Anpassung der Stimulationsfrequenz. *Biomedizinische Technik* 34 (1989) 191-196.

[90] Res, J. C. J., T. J. M. de Boer. Evaluation of a new body motion sensing rate response sensor (Ergos 2), using a multiphase exercise protocol. *PACE* 13 (1990) 1207.

[91] Weisswange, A., G. Csapo, W. Perach, E. Kannegieber. Frequenzsteuerung von Schrittmachern durch Bluttemperatur. *Verh. Ges. Kreislaufforsch.* 44 (1978) 152.

[92] Aschoff, J., B. Gunther, K. Kramer. *Energiehaushalt und temperaturregulation*. München, Berlin, Wien: Urban und Schwarzenberg (1971).

[93] Fearnot, N. E., M. L. Evans. Heart rate correlation, response time and effect of previous exercise using an advanced pacing rate algorithm for temperature-based rate modulation. *PACE* 11 (1988) 1846-1852.

[94] Griffin, J. C., K. R. Jutzy, J. P. Claude, J. W. Knutti. Central body temperature as a guide to optimal heart rate. *PACE* 6 (1983) 498-501.

[95] Schaldach, M. Hybrids, an essential for high reliability human implant electronics. *ISHM Proceedings International Symposium on Microelectronics* (1991) 120-124.

[96] Isbruch, F. M., Th. Koch, H. Greve, H. Dittrich, H. Heuer. Die zentralvenöse Bluttemperatur als Sensor für ein frequenzadaptives Schrittmachersystem. *Biomedizinische Technik* 33 (1988) 295-299.

[97] Koch, Th., F. M. Isbruch, B. Frenking, H. Greve, H. Gülker, H. Heuer. Long-term results with the temperature-controlled pacemaker Thermos 01. *PACE* 11 (Cardiostim 88) 807.

[98] Laczkovics, A., G. Simbrunner, U. Losert *et al.* Temperaturmessungen zur Steuerung der Herzfrequenz in der Schrittmacherchirurgie. *Kongressbericht Osterr. Ges. Chir.* 23 (1982) 119.

[99] Laczkovics, A., G. Laufer, Th. Oehner, W. Schlich. First clinical results with a temperature-guided rate-responsive pacemaker. *PACE* 10 (1987) 1216.

[100] Naumann d'Alnoncourt, C., F. Schnabel, B. Baumann, H. P. Helwing. Temperaturgesteuerter Herzschrittmacher THERMOS 01 nach His-Bündel Ablation. *Herzschrittmacher* 7 (1987) 168-174.

[101] Schaldach, M., H. Hutten. A physiological approach to different concepts of rate-adaptive cardiac pacing. *Medical Progress Through Technology* 16 (1990) 235-246.

[102] Wassermann *et al.* Interactions of physiological mechanisms during exercise. *J. Appl. Phys.* 22 (1966) 71-85.

[103] Rickards, A. F., J. Norman. Relation between OT interval and heart rate. New design of physiologically-adaptive cardiac pacemaker. *Br. Heart J.* 45 (1981) 56.

[104] Kubicek, W. G., R. P. Patterson, D. A. Witsoe. Impedance cardiography as a noninvasive method of monitoring cardiac function and other parameters of the cardiovascular system. *Annals of the New York Academy of Sciences* 170 (1970) 724-732.

[105] Webster, J. G. *Electrical Impedance Tomography.* Bristol and New York: Adam Hilger (1990).

[106] Valentinuzzi, M. E. Intracardiac conductance volumetry. *Automedica* 11, Gordon and Breach Science Publishers (1989) 4.

[107] Salo, R. W., B. D. Pederson, A. L. Olive, W. C. Lincoln, T. G. Wallner. Continuous ventricular volume assessment for diagnosis and pacemaker control. *PACE* 7 (1984) 1267-1272.

[108] McGoon, M. D., J. E. Shapland, R. Salo, B. Pederson, A. Olive. The feasibility of utilizing the systolic pre-ejection interval as a

determinant of pacing rate. *JACC* Vol. 14, No. 7 (December 1989) 1753-8.

[109] Morris, III, J. J., G. L. Pellom, D. P. Hamm, C. T. Everson, A. S. Wechsler. Dynamic right ventricular dimension: Relation to chamber volume during the cardiac cycle. *J. Thorac. Cardiovasc. Surg.* 91 (1986) 879-887.

[110] Schwiep, F., S. S. Cassidy, M. Ramanathan, R. L. Johnson, Jr. Rapid in vivo determinations of instantaneous right ventricular pressure and volume in dogs. *Am. J. Physiol.* 254 (1988) H622-H630.

[111] Linker, D. R., W. E. Moritz, A. S. Pearlman. A new three-dimension echocardiographic method of right ventricular volume measurement: in vitro validation. *JACC* 8 (1986) 101-106.

[112] Salo, R. W., T. G. Wallner, B. D. Pederson. Measurement of ventricular volume by intracardiac impedance: theoretical and empirical approaches. *IEEE Trans. Biomed. Eng.*, BME-33 (1986) 189-195.

[113] Czegledy, F., N. Aebischer, A. Tamari, A. Tortolani. A new mathematical model for right ventricular geometry. *Proc. 12 Annual Intern. Conf. IEEE Eng. in Med. & Biol.* (1990) 1813-195.

[114] Aebischer, N. M., F. P. Czegledy. Determination of right ventricular volume by two-dimensional echocardiography using a crescentric model. *J. Amer. Soc. of Echocard* 2 (1989) 110-118.

[115] Schaldach, M., M. Schaldach. Volume conductor model of the right ventricular impedance plethysmography. *Medical & Biological Engineering & Computing* 29 Supplement (1991) 343.

[116] Schaldach, M. Automatic adjustment of pacing parameters based on intracardiac impedance measurements. *PACE* 13 (1990) 1702-1710.

[117] Schaldach, M. Evoked response recognition in cardiac pacing. *Proc. 2nd Int. Conf. on Pacemaker Electrodes*, Ferrara (1991).

[118] Urbaszek, A., H. Hutten, M. Schaldach. Modell des menshlichen Blutkreislaufs und der Herzfunktion mit Schwerpunkt auf kurzzeitigen Regulationsvorgängen. *Biomedizinische Technik* 36 (Ergänzungsbank 1) (1991) 260-261.

[119] Furman, S., G. Robinson. The use of an intracardiac pacemaker in the correction of total heart block. *Surg. Forum* 9 (1958).

[120] Mindt, W., M. Schaldach. Electrochemical aspect of pacing electrodes, in *Pacemaker Technology,* edited by M. Schaldach, S. Furman. New York: Springer-Verlag (1975) 297.

[121] Schaldach, M., M. Hubmann, A. Weikl, R. Hardt. Sputter-deposited TiN electrode coatings for superior sensing and pacing performance. *PACE* 13 (1990) 1891-1895.

[122] Irnich, W. The fundamental law of electrostimulation. *Proc. of the 11th Ann. Int. Conf. of the IEEE/EMBS* (1989) 84-85.

[123] Breme, J. Titanium and titanium alloys, biomaterials of preference. *Proc. Sixth World Conf. on Titanium,* France (1988).

[124] Schaldach, M., J. Breme. Porous heart pacemaker electrodes of TiAL Fe 2.5 alloy. *Proc. of Int. Conf. on Titanium Products and Applications,* Orlando (1990) 604-622.

[125] MacGregor, D. C., G. J. Wilson, W. Lixfeld, R. M. Pilliar, J. D. Bobyn, M. D. Silver, S. Smardon, S. L. Miller. The porous-surface electrode. A new concept in pacemaker lead design. *J. Thorac. Cardiovasc. Surg.* 78 (1979) 281.

[126] Avnir, D. *The Fractal Approach to Heterogenous Chemistry.* Chichester: Wiley & Sons (1989).

[127] Thornton, J.A. The microstructure of sputter-deposited coatings. *J. Vac. Sci. Technol.* 6 (1986) 3059-3065.

[128] Schaldach, M. New pacemaker electrodes. *Trans. Amer. Soc. Artificial Int. Organs* 17 (1984) 134.

[129] Lerner, J., *et al.* Miniature implantable tantalum/tantalumoxide stimulating electrodes. *IEEE Trans.* BME 29, 290.

[130] Schaldach, M. Bioelectric phenomena in cardiac pacing. *Proc. IEEE Ninth Annual Conference of the Engineering in Medicine and Biology Society* (1987) 139.

[131] Schaldach, M., R. Thull. Das elektrochemische Verhalten gesinterter Metallelektroden. *Biomed. Technik* 31 (Ergänzungsband) (1988) 333.

[132] Thull, R. Eigenschaften implantierbarer Elektroden für die physiologische Stimulation des Herzens, in *Physiologische Stimulation des Herzen,* edited by A. Weikl. Erlangen: Perimed-Fachbuch (1984) 89-106.

[133] Thull, R., M. Schaldach. Electrochemistry of after-pacing potentials on electrodes. *PACE* 9 (1986) 1191.

[134] Mindt, W., P. Racine. Stimulating electrode with low energy consumption. *Med. Biol. Eng.* 11 (1973) 659.

[135] Hirshorn, M. S., L. K. Holley, D. K. Money, M. Spector, F. A. Young, J. R. S. Hales. Histological evaluation of porous titanium cardiac pacemaker electrode tips. *J. of Biomed. Mater. Res.* 18 (1984) 47.

[136] Mitamura, Y., T. Yuta, T. Mikamit. Ceramic heart valve: high tech ceramics, edited by P. Vincensini. Amsterdam: Elsevier (1987) 127.

[137] Mitamura, Y., *et al.* Ceramic heart valve of cardiac prostheses progress in artificial organs, edited by Y. Nosé, C. Kjellstrand, P. Ivanovich. Cleveland: ISAO Press (1986) 508.

[138] Albrektsson, T., P. J. Branemark, H. A. Hansson, B. Kasemo, K. Larsson, I. Lundstrom, D. H. McQueen, R. Skalak. The interface zone of inorganic implants in vivo: titanium implants in bone. *Ann. of Biomed. Eng.* 11 (1983) 1.

[139] Elmquist, H., H. Schüller, G. Richter. The carbon tip electrode. *PACE* 6 (1983) 436.

[140] Mugica, J., B. Duconge, L. Henry, B. Atchia, B. Lazarus. Clinical experience with new leads. *PACE* 11 (1988) 1745.

[141] Mund, K. Prüfen und Bewerten von Beschichtungen für Implantentate durch electrochemische Messungen. Prüfen und Bewerten von Oberflächenschutzschichten, Tagung Hagen, 8. u. 9 Nov. 1988 *VDI-Ges. Wertestofftechnik*. Düsseldorf: VDI (1988) 387-406.

[142] Mund, K., *et al*. Electrochemical properties of platinum, glassy carbon, and pyrographite as stimulating electrodes. *PACE* 9 (1986) 1225.

[143] Shigemitsu, T., G. Matsumoto, S. Tsukahara. Electrical properties of glassy-carbon electrodes. *Med. Biol. Eng. Comp.* 17 (1979) 465-470.

[144] Albrektsson, T., P. J. Brånemark, H. A. Hansson. The interface zone of inorganic implants in vivo: Titanium implants in bone. *Ann. of Biomed. Eng.* 11:1 (1983).

[145] Pesch, H. J., K. Glückert, H. P. Tümmler, R. Thull. Tissue reactions of titanium and its alloys. Titanium, science and technology, edited by G. Lütjering, U. Zwicker, W. Bunk. *Proc. of the 5th Int. Conf. on Titanium,* Munich 1985. Deutsche Gesellschaft für Metallkunde e. V., Oberursel 1985.

[146] Bolz, A., H. Matlok, M. Still, M. Schaldach, M. Hubmann, R. Hardt. Langzeitstabilität des Detekins-und Reizschwellenverhaltens von Titannitrid-Herzschrittmacher-Elektroden. *Biomedizinische Technik* 35 (Ergäzungsband) (1990) 131-133.

[147] Schaldach, J., A. Bolz, J. Breme. Porous heart pacemaker electrodes of TiA15Fe2.5 and TiN. *Proc. 1st European Conf. on Biomed. Eng.* Nice (1990) 312-313.

[148] Bolz, A., M. Schaldach, M. Hubmann. Interface aspects of stimulating electrodes. *Proceedings of the Annual Int. Conf. of the IEEE Eng. in Med. and Biology Society* 13 (1991) 714-716.

[149] Schaldach, M., M. Hubmann, R. Hardt, A. Weikl. Titannitrid-Herzschrittmacher-Elektrode. *Biomed. Technik* 34 (1989) 185-190.

[150] Schaldach M. M., M. A. Bolz. Long-term stability of TiN. To be published in the *Proc. of the Conf. "Bioceramics and the Human Body"* (IRTEC-CNR)(1991) Faenza.

[151] Rubin, L. R. *Biomaterials in Reconstructive Surgery.* St. Louis, Toronto, London: The C. V. Mosby Company (1983) S 1017.

[152] Frank, E., M. Zitter. *Metallische Implantate in der Knochenchirurgie.* Berlin, Heidelberg: Springer-Verlag (1971).

[153] Winter, G. D., *et al. Evaluation of Biomaterials.* Chichester, New York, Brisbane, Toronto: John Wiley & Sons Bk. 1 (1980) S 553.

[154] Matloff, J. M. *Cardiac Valve Replacement-Current Status.* Boston, Dordrecht, Lancaster: Martinus Nijhoff Publishing (1985) S 310.

[155] Callow, A. D. Historical overview of experimental and clinical development of vascular grafts, in *Biological and Synthetic Vascular Prostheses,* edited by I. C. Stanley, *etal.* New York, San Francisco, London: Grune & Stratton, Inc. (1982) 11-26.

[156] Hufnagel, C. A. History of vascular grafting, in *Vascular Grafting: Clinical Applications and Technique,* edited by C. B. Wright, *et al.* Boston: John Wright, PSG, Inc. (1983) 1-12.

[157] Temple, L. J., J. T. M. Wright. Implants in the cardiovascular and respiratory systems, in *Implants in Surgery,* edited by D. F. Williams and R. Roaf. London: W. B. Saunders Company Ltd. (1973) 481-536.

[158] Schaldach, M. M., M. A. Bolz. Biocompatibility optimization of materials by hybrid structuring. *Medical & Biological Engineering & Computing.* Vol. 29 Supplement (1991) 134.

[159] Williams, D. F. *Definitions in Biomaterials*. Amsterdam, Oxford, New York, Tokyo: Elsevier Science Publishers B.V. (1987) S 21.

[160] Han, C. D. *Rheology in Polymer Processing*. New York: Academic Press (1976).

[161] Middleman, S. *Fundamentals of Polymer Processing*. New York: McGraw-Hill (1977).

[162] Miller, E., ed. *Plastics Design Handbook: Part B, Process and Design for Processes*. New York: Dekker (1983).

[163] Billmeyer, F. W., Jr. *Textbook of Polymer Science*. 3rd ed. New York: John Wiley & Sons (1984).

[164] Rembaum, A., M. Shen. *Biomedical Polymers*. New York: Dekker (1971).

[165] Shalaby, S. W. Polymeric materials, in *Encyclopedia of Medical Devices and Instrumentation*, edited by J. G.Webster. New York: John Wiley & Sons (1988) 2324-2335.

[166] Szycher, M. *Biocompatible Polymers, Metals, and Composites*. Lancaster: Technomic Publishing Co. Inc. (1983) S 1071.

[167] Mohtashemi, M., G. L. Hines. Tissue response to permanently implanted pacemaker generàtors and electrodes, in *Biomaterials in Reconstructive Surgery*, edited by L. R. Rubin. St. Louis, Toronto, London: The C.V. Mosby Company (1983).

[168] Ratner, B. D. Surface Characterization of Biomaterials. *Progress in Biomedical Engineering* 6. Amsterdam, Oxford, New York: Elsevier Science Publishers (1988).

[169] Cazenave, J. P., *et al.* Biological Principles Underlying Haemocompatibility with Artificial Materials. In: *Blood-Surface Interactions*. Amsterdam, Oxford, New York: Elsevier Science Publishing Co., Inc. (1986).

[170] Semlitsch, M., H. G. Willert. Metallic materials for artificial hip joints, in *Encyclopedia of Medical Devices and Instrumentation*, edited by J. G. Webster. New York, Toronto: John Wiley & Sons (1988) 137-149.

[171] Fraker, A. C., C. D. Griffin. Corrosion and degradation of implant materials. *Second Int. Symposium on Corrosion and Degradation of Implant Materials 1983*. Philadelphia: ASTM, Bd. 859 (1985) S 457.

[172] Breme, J. Titanium and titanium alloys, biomaterials of preference. *Proc. Sixth World Conference on Titanium,* France, 1988.

[173] Zitter, H., H. Plenk, Jr. The electromechanical behaviour of metallic implant materials as an indicator of their biocompatibility. *J. Biomed. Mater. Res.* 21 (1987) 881.

[174] Fraker, A. C., A. W. Ruff, P. Sung, A. C. van Orden, K. M. Speck. Surface Preparation and Corrosion Behaviour of Titanium Alloys for Surgical Implants. *Ti '80 Science and Technology*. Plenum Press (1980) 2447.

[175] Mears, D. C. Dissimilar metals in orthopaedic surgery. *J. Biomed. Mater. Res.* 6 (1975) 133.

[176] Brown, S. A. Biomaterials, corrosion and wear of corrosion. In *Encyclopedia of Medical Devices and Instrumentation*, edited by J. G. Webster. New York, Toronto: John Wiley & Sons (1988) 351-361.

[177] Zitter, H. The suitability of metals for surgical implants, in *Advances in Artificial Hip and Knee Joint Technology, Engineering in Medicine Bd.* 2, edited by M. Schaldach, D.Hohmann. Berlin, Heidelberg, New York: Springer-Verlag (1976) 227-241.

[178] Higham, P. A. Ion implantation as a tool for improving the properties of orthopaedic alloys. *Proc. Conf. Biomed. Mat.*, Boston, Dec. 1985 (1986) 253.

[179] Kubaschewski, O., E. C. Evans, C. B. Alcock. *Metallurgical Thermochemistry*. London: Pergamon Press (1967).

[180] Steinemann, S. G., S. M. Perren. Titanium as metallic biomaterials. *Ti '84 Science and Technology*, edited by G. Lütjering, U. Zwicker, W. Bunk. (1984) 1327.

[181] Breme, J., G. Heimke. Corrosion fatigue test of TiAl5Fe2.5 hip implant under high stresses. *Ti '84 Science and Technology*, edited by G. Lütjering, U. Zwicker, W. Bunk (1984) 1351.

[182] Semlitsch, M., F. Staub, H. Weber. Development of a vital, high strength wrought Ti-6Al-7Nb alloy for surgical implants. *5th Europ. Conf. on Biomaterials*, Paris (1985).

[183] Steinemann, S. G., P. A. Mäusli. Titanium alloys for surgical implants-biocompatibility from physicochemical principles. *Sixth World Conference on Titanium*, France (1988).

[184] Albrektsson, T., P. I. Brånemark, H. A. Hansson, B. Kasemo, K. Larsson, I. Lundström, D. H. McQueen, R. Skalak. The interface zone of inorganic implants in vivo: titanium implants in bone. *Annals of Biomedical Engineering* 11 (1983) 1-27.

[185] Lemm, W., *et al*. Biodegradation of some biomaterials in vitro. *Proc. Europ. Soc. Art. Organs* 7 (1980) 86.

[186] Imai, Y., *et al*. Biodegradation of polymeric materials. *Trans. Soc. Biomat.* 3 (1979) 84.

[187] Nossel, H. C. Assessment of activation of coagulation and platelets in vivo. In *Interaction of the Blood with Natural and Artificial Surfaces*, edited by E. W. Salzman. New York: Marcel Dekker (1981) 171-184.

[188] Andrade, J. D. Principles of protein adsorption. In *Surface and Interfacial Aspects of Biomedical Polymers*, edited by J. D. Andrade. New York, London: Plenum Press, Bk. 2 (1985) 1-80.

[189] Park, K., S. J. Gerndt, H. Park. Patchwise adsorption of fibrinogen on glass surfaces and its implication in platelet adhesion. *Journal of Colloid and Interface Science* 125 (1988) 702-711.

[190] Mosher, D. F. Influence of proteins on platelet-surface interactions. In *Interaction of the Blood with Natural and Artificial Surfaces,* edited by E. W. Salzman. New York: Marcel Dekker (1981) 85-102.

[191] Brash, J. L. Protein interactions with artificial surfaces. In *Interaction of the Blood with Natural and Artificial Surfaces,* edited by E. W. Salzman. New York: Marcel Dekker (1981) 37-60.

[192] Baurschmidt, P., M. Schaldach. Alloplastische materialien für den herzklappenersatz. *Biomedizinische Technik* 25 (1980) 89-95.

[193] Morrissey, B. W. The adsorption and conformation of plasma proteins: a physical approach. *Annals of the New York Academy of Science* 283 (1977) 50.

[194] Vanholder R, S. Ringoir. Bioincompatibility: An overview. *The International Journal of Artificial Organs* 12 (1989) 356-365.

[195] Anderson, G. H., J. D. Hellums, J. L. Moake, C. P. Alfred. Platelet lysis and aggregation in shear fields. *Blood Cells* 4 (1978) 499-507.

[196] Benninghofen A., W. H. Werner, F. G. Rudenauer. *Secondary Ion Mass Spectrometry.* New York: John Wiley (1987).

[197] Andrade, J. D. X-ray Photoelectron Spectroscopy. In *Surface and Interfacial Aspects of Biomedical Polymers,* Bk. 1, edited by J. D. Andrade. New York, London: Plenum Press. (1985) 105-196.

[198] Whalen, R. L., D. L. Jeffrey, J. C. Norman. A new method of in vivo screening of thromboresistant biomaterials utilizing flow measurement. *Trans. Amer. Soc. Int. Organs* 19 (1973) 19.

[199] Gott, V. L., A. Furuse. Antithrombogenic surfaces, classification and in vivo evaluation. *Fed. Proc.* 30 (1971) 1679.

[200] *Guidelines for Blood-Material Interactions.* Chap. 8: Species effects in testing materials and cardiovascular devices in experimental animals. Report of the National Heart, Lung, and Blood Institute

Working Group. NIH Publication No. 85-2185, Revised Edition (1985).

[201] Baurschmidt, P., M. Schaldach. The electrochemical aspects of the thrombogenicity of a material. *Journal of Bioengineering* 1 (1977) 261.

[202] Bolz, A., M. Schaldach. Entwicklung einer amorphen, halbleitenden Beschichtung zur Verbesserung der Blutverträglichkeit kardiovaskulärer Implantate. *Biomedizinische Technik* 33 Ergbd. (1988) 175-176.

[203] Sung Wan Kim, C. D. Ebert, J. Y. Lin, J. C. McRea. Nonthrombogenic polymers: pharmaceutical approaches. *J. Am. Soc. Artif. Intern. Organs* 6 (1983) 76-87.

[204] Yasuda, H. Plasma for modification of polymers. *Journal of Macromolecular Sci.-Chem.* A10 (1976) 383-420.

[205] Schaldach, M. Kardiovasculäre implanttechnologie. *Biomedizinische Technik.* (Ergäzungsband 1) (1991) 1-6.

[206] Bolz, A., M. Schaldach. General relationship for the growth of passivation layers on biomaterials. *Proceedings of the Annual Int. Conf. of the IEEE Eng. in Med. and Biology Society* 13 (1991) 1474-1476.

[207] Ruben, S. R. U.S. Patent No. 2,422,045 (1947).

[208] Brennen K., B. Owens. *Medical Batteries: Lithium Battery Technology,* edited by H. V. Venkatasetty. New York: Wiley (1984) 139-158.

[209] Salkind A., S. Ruben. *Batteries for Implantable Biomedical Devices,* edited by B. B. Owens. New York: Plenum Press (1986) 261 ff.

[210] Ruetschi, P. *Jour. Electrochem. Soc.* 127 (1980) 1667.

[211] Schneider, A. A., J. Moser. U.S. Patent No. 3,674,562 (1972).

[212] Greatbatch, W., J. Lee, W. Mathias, M. Eldridge, J. Moser, A. Schneider. *IEEE Trans. BME* 18-5 (1971) 317.

[213] Lehmann, G., J. P. Gabano. U.S. Patent 3,853,627 (1974).

[214] Harney, M. J., S. Brown. *Proceedings of the Symposium on Power Sources for Biomedical Implantable Applications and Ambient Temperature Lithium Batteries,* edited by B. B. Owens, N. Margalit. Pennington, New Jersey: The Electrochemical Society (1980) 102 ff.

[215] Untereker, D. Lithium primary cells for power sources. *Batteries for Implantable Biomedical Devices,* edited by B. B. Owens. New York: Plenum Press (1986) 51.

[216] Ibid., p. 64.

[217] Holmes C. F. Lithium/halogen batteries. *Batteries for Implantable Biomedical Devices,* edited by B. B. Owens. New York: Plenum Press (1986) 133.

[218] Ibid., p.147.

[219] Schneider, A., G. Bowser, L. Foxwell. U.S. Patent No. 4,148,975.

[220] Mead, R., W. Greatbatch, F. Rudolph, N. W. Frenza. Lithium-Iodine Cell, U.S. Patent 4,210,708 (1980).

[221] Mead, R. T., C. F. Holmes, W. Greatbatch. *Proceedings of the Symposium on Cell Design and Optimization,* edited by S. Gross. Pennington, New Jersey: The Electrochemical Society (1979) 327-334.

[222] Wilson Greatbatch Ltd. Cumulative Survival of WGL Lithium Iodine Cells, *Reliability Evaluation Report* (1990).

[223] Verrier R. L. Neurogenic aspects of cardiac arrhythmias. *Cardiac Pacing and Electrophysiology, 3rd Edition,* edited by N. El-Sherif, P. Samet. (1991) 77-92.

[224] Kulbertus, H. E., G. Franck, eds. *Neurocardiology.* Mt. Kisco: Futura Publishing Co. (1988).

[225] Natelson, B. H. Neurocardiology. An interdisciplinary area for the 80s. *Arch. Neurol.* 42 (1985) 178.

[226] Schwartz, P. J., H. L. Stone. Effects of unilateral stellectomy upon cardiac performance during exercise in dogs. *Circ. Res.* 44 (1979) 637-645.

[227] Corr, P. B., K. A. Yamada, F. X. Witkowski. Methanisms controlling cardiac autonomic function and their relation to arrhythmogenesis. *The Heart and Cardiovascular System,* edited by H. A. Fozzard, *et al.* New York: Raven Press (1986) 1343-1403.

[228] Schwartz, P. J. The rational basis and the clinical value of selective cardiac sympathetic denervation in the prevention of malignant arrhythmias. *Eur. Heart J.* Suppl. A 7 (1986) 107-118.

[229] Schwartz, P. J., M. Stramba-Badiale. Parasympathetic nervous system and malignant arrhythmias. *Neurocardiology,* edited by H. E. Kulbertus, G. Franck. Mt. Kisco: Futura Publishing Co. (1988) 179-200.

[230] Schwartz, P. J. Manipulation of the autonomic nervous system in the prevention of sudden cardiac death. *Cardiac Arrhythmias: Where To Go From Here?* Mt. Kisco: Futura Publishing Co. (1987) 741-765.

[231] Hutten, H., M. Schaldach. Rate-adaptive cardiac pacing considering closed-loop control. *Proceedings of the Annual Int. Conf. of the IEEE Eng. in Med. and Biology Society* 13 (1991) 2111-2113.

[232] Schaldach, M. Microelectronics in implantable cardiovascular devices. *Proceedings of the Annual Int. Conf. of the IEEE Eng. in Med. and Biology Society* 13 (1991) 2120-2122.

INDEX

A

B

D

E

Q

R

S

T

U

V

W

Z

MIX
Papier aus verantwortungsvollen Quellen
Paper from responsible sources
FSC® C105338

FSC
www.fsc.org

If you have any concerns about our products,
you can contact us on
ProductSafety@springernature.com

In case Publisher is established outside the EU,
the EU authorized representative is:
Springer Nature Customer Service Center GmbH
Europaplatz 3, 69115 Heidelberg, Germany

Printed by Libri Plureos GmbH
in Hamburg, Germany

SpringerBriefs in Economics

More information about this series at http://www.springer.com/series/8876

Fikret Čaušević

The Global Crisis of 2008 and Keynes's General Theory

Springer

Fikret Čaušević
School of Economics and Business
University of Sarajevo
Sarajevo
Bosnia-Herzegovina

ISSN 2191-5504 ISSN 2191-5512 (electronic)
ISBN 978-3-319-11450-7 ISBN 978-3-319-11451-4 (eBook)
DOI 10.1007/978-3-319-11451-4

Library of Congress Control Number: 2014950674

Springer Cham Heidelberg New York Dordrecht London

Printed on acid-free paper

Springer is part of Springer Science+Business Media (www.springer.com)

The outstanding faults of the economic society in which we live are its failure to provide for full employment and its arbitrary and inequitable distribution of wealth and incomes.

John Maynard Keynes, *The General Theory*, Chapter 24

I conceive, therefore, that a somewhat comprehensive socialisation of investment will prove the only means of securing an approximation to full employment; though this need not exclude all manner of compromises and of devices by which public authority will co-operate with private initiative. But beyond this no obvious case is made out for a system of State Socialism which would embrace most of the economic life of the community. It is not the ownership of the instruments of production which it is important for the State to assume. If the State is able to determine the aggregate amount of resources devoted to augmenting the instruments and the basic rate of reward to those who own them, it will have accomplished all that is necessary. Moreover, the necessary measures of socialisation can be introduced gradually and without a break in the general traditions of society.

John Maynard Keynes, *The General Theory*, Chapter 24

In telling people how to read The General Theory, I find it helpful to describe it as a meal that begins with a delectable appetizer and ends with a delightful dessert, but whose main course consists of rather tough meat. It's tempting for readers to dine only on the easily digestible parts of the book, and skip the argument that lies between. But the main course is where the true value of the book lies.

Paul Krugman, July 2006

Preface

When Richard Nixon, the US president of the day, took the US dollar off the gold standard on 15 August 1971, it produced major disturbances on national and global financial markets, and also marked the beginning of the end for what had up until then been the dominant intellectual influence on official economic policy-making in the largest world economy, Keynesian economic thought, or so it seemed. Definite confirmation that the system of fixed exchange rates had been abandoned in favour of a freely floating US dollar came in March 1973. As the most important global currency began to suffer major volatility, it meant not just the end of the international financial system based on fixed exchange rates, but also the start of a series of major disruptions on global and national markets for goods, services and financial assets.[1]

The first markets for financial derivatives were established in the same year as the United States formalised its transition to a freely floating dollar. That year also saw the first oil crisis, when the price of oil practically quadrupled in just 2 months, a reaction on the part of the oil-producing countries that was both prompted by the fall in the dollar and represented a coordinated approach to limit the supply of this key fuel. The following year, 1974, the Basel Committee for Banking Supervision was created. At the time, 9 of the 10 largest banks in the world were American and the most important oil producers kept their deposits with them. In 1974, the developing countries mooted a proposal to establish new global economic relations, to be called the New Economic Order. Their intention was to respond to the urgent problems caused by rising oil prices, problems financing postcolonial recovery and attempts to re-establish the rules for international trade in goods and services on a new basis.

Chapter 1 of this book presents the international context and some of the reasons that led to this weakening influence of Keynesian economic thought at the

[1] This book has been translated by a native speaker, Desmond Maurer, MA.

beginning and, more especially, during the second half of the 1970s, and the subsequent strengthening of the intellectual influence of the New Classical macroeconomics. It also presents certain Keynesian economic responses offered by circles of economists who belonged (and still belong) to the neo-Keynesian and new Keynesian schools of economic thought.

Because of the intellectual influence previously enjoyed by Keynes' *General Theory of Employment, Interest, and Money*, an influence in large part recovered during the current global financial and economic crisis (to such a degree, indeed, that between 2008 and 2014, it has dominated economic policy-making in the most developed and largest economies of the world, particularly the United States and Japan), Chap. 2 of this book is dedicated to a commentary on the Master's great work. This decision to offer a concise interpretation of the *General Theory* stems from the fact that, although without doubt one of the most significant works of economic science, it nonetheless leaves unresolved a whole series of questions to which Keynes, whether because of his own lack of time or because of his primary focus on dealing with internal imbalances under given technological conditions (in the short-term), either provided no answer or provided answers which served in the 1930s his goal of securing an exit from the immediate trough of the business cycle, but fail to provide clarity now, in an environment of globalisation and very high international mobility of capital, as to the impact of the economic policy measures applied during the global crisis, even though they were almost entirely based on his immediate recommendations for a combination of expansionary monetary and expansionary fiscal policy in the *General Theory*.

Chapter 3 deals with the impact of financial liberalisation on the efficiency of economic policy of major economies in the world, from one side, and its impact on the cost structure in production of globally integrated manufacturing companies. The international capital mobility arising from the financial liberalisation measures implemented in developing countries, particularly the most populous ones like China and India, brought about a sharp reduction in the costs of production, compared to the same costs on the national markets of developed countries. Consequently, one of the fundamental assumptions of both the new classical model and the new Keynesian model of production in developed market environments, that is, the assumption of growing marginal costs and the consequent preoccupation with inflationary pressures, ceased to be a key problem in the period from 1990 to 2010 in the globally connected major economies.

On the other side, the measures of financial liberalisation adopted during the 1990s and in the first 5 years of this century created a situation in which the money supply was predominantly endogenously determined, that is, determined on the basis of the business policies and profit motives of banking groups which unified the operations of commercial and investment banking, as well as those of trading in financial derivatives on rapidly growing and, between 2000 and 2009, almost entirely deregulated over-the-counter markets. Given a US monetary policy that was, during the periods in which financial bubbles were being created, powerless (or uninterested) to step in, through determined measures to increase the interest rate, the enormous growth in lending activity from 2002 to 2008, particularly on the interbank market, and given the multiple systems for ensuring through the issue and sale of financial derivatives that risk transferred *de facto* onto the public budget, a situation was created which is best described in theoretical terms in the works of the post-Keynesian economists who developed the monetary circuit theory.

Sarajevo, Summer 2014 Fikret Čaušević

Contents

Chapter 1
Economic Theory and Economic Policy Since the Seventies: Keynesians Versus New Classical Economists

Abstract This chapter begins with the analysis of causes that led to the weakening of the intellectual influence of Keynesian economic thought at the beginning of the seventies of the last century and to the strengthening of the impact of the new classical economists led by Lucas, Sargent and Wallas. The theoretical response of the new Keynesians to the criticism, was based on the introduction of sticky prices in macroeconomic models in the works of Phellps, Fischer, Taylor and Dornubsch in the late seventies. The author also presents the role of modern financial theory based on the efficient market theory, portfolio theory and the capital market theory, and the criticism of these theories presented in the works of Mandelbrot, Schiller and Kahneman. In explaining the causes of the global crisis of 2008, the author pays special attention to the post-Keynesian monetary circuit theory and the Minsky's financial instability hypothesis and its relevance for the analysis of major factors that led to the global financial crisis. This chapter ends with the author's comparison of the effects of macroeconomic policies in the United States under the administrations led by the last three US presidents: Clinton, Bush and Obama. By presenting the data on the trends in unemployment, interest rates, inflation and changes in the market capitalization on the major capital markets, the author shows that the economic policy measures implemented during the global economic crisis in the U.S., Europe and Japan are based on the recommendations suggested by John Maynard Keynes in the General Theory regarding the simultaneous use of expansionary monetary and fiscal policy.

Keywords Keynes · Keynesianism · New classical economics · Modern financial theory · Monetary-circuit theory · Financial instability hypothesis · Economic policy · The global crisis

1.1 Keynes's Economic Thought on the Defensive

Oil shortages, a falling dollar and sharply rising oil prices during the second half of the 70s were accompanied by a marked growth in inflation in the United States, as in all the other developed and developing countries. This increase in the general

F. Čaušević, *The Global Crisis of 2008 and Keynes's General Theory*,
SpringerBriefs in Economics, DOI 10.1007/978-3-319-11451-4_1

level of prices in the leading world economy was at least partly due to the negative supply-side shock of sharply rising oil prices, but also to the conducting of expansionary monetary and fiscal policies, as recommended by economic policy-makers in the United States, whose economic programs were based on Keynesian modelling. John Maynard Keynes had himself recommended an expansionary monetary policy and, if required, cutting the interest rate to zero (or close to zero) as a remedy for maintaining actual employment close to full employment, as we shall see in the second part of this essay.

Keynesian economic thought, as initiated with the publication of its best-known work, *The General Theory of Employment, Interest and Money*,[1] dominated the first two decades after the Second World War—both in academic circles and in economic policy-making. Economic policy measures themselves and forecasting of their possible impact were, however, based on the Hicks IS–LM model[2] or the Mundell–Fleming IS–LM–BP[3] model of an open economy, in spite of the fact that the author of the IS–LM model later confessed that it was primarily useful for teaching purposes and did not provide an adequate basis for economic policy-making. Keynesian economic thought branched out, in the post-war period, in three directions: post-Keynesian, neo-Keynesian and new Keynesian. It is therefore worth noting that it was the neo-Keynesians and the neoclassical economic synthesis championed by Paul Samuelson that exerted the greatest influence on economic policy-making in the 50s and 60s.

Even though Richard Nixon had declared, on the day he took the dollar off the gold standard, "I am now a Keynesian",[4] the implementation of Keynesian recipes during the 70s, and particularly its second half, did not yield good economic results. Inflation was not under control, nominal interest rates were lower than inflation, and real interest rates were, as a result, negative.

This combination of expansionary monetary and expansionary fiscal policy between 1974 and 1976 (after Nixon's resignation over Watergate, when Gerald Ford took his place as US president) was marked by increasing unemployment and sharply falling real interest rates (particularly 1974/75).

These trends on money and labour markets strikingly contradicted the results of studies by William Phillips,[5] from which he had derived his recommendations for economic policy-making, represented on the Phillips curve. Investigating the relationship between the cost of labour and the unemployment rate in Great Britain, he had found that economic policy recommendations should rely on an expansionary monetary policy, which would facilitate, amongst other things, an increase in the price of labour and so in the inflation rate, leading directly to lower unemployment. The data from Tables 1.1 and 1.2 indicate that, in spite of negative real

---

[1] Reference [1]. http://www.marxists.org/reference/subject/economics/keynes/general-theory/.

[2] Reference [2].

[3] References [3, 4].

[4] See: http://www.ontheissues.org/celeb/Richard_Nixon_Budget_+_Economy.htm.

[5] Reference [5].

Table 1.1 Real interest rates in the United States 1974–1978

Year	Federal funds rate	Rate of Inflation	Real interest rate on inter-bank market
1974	10.51	11.03	−0.52
1975	5.82	9.20	−3.38
1976	5.05	5.75	−0.70
1977	5.54	6.50	−0.96
1978	7.94	7.62	+0.32

Source Federal Reserve System—http://www.federalreserve.gov/releases/h15/data.htm

Table 1.2 Real interest rates and the unemployment rate in the US: 1974/1978

Year	Real interest rate on interbank market in %	Unemployment rate in %
1974	−0.52	7.2
1975	−3.38	8.2
1976	−0.70	7.8
1977	−0.96	6.4
1978	+0.32	6.0

Source http://www.davemanuel.com/historical-unemployment-rates-in-the-united-states.php. http://www.federalreserve.gov/releases/h15/data.htm

interest rates and falling inflation, the unemployment rate rose between 1974 and 1975 and did not fall significantly during 1976, while inflation growth in 1977–1978 was accompanied by a growth in the federal funds rate, which switched from a real negative to a real positive rate, however small, accompanied by a reduction in unemployment from 7.8 % (1976) to 6 % (1978).

In a paper published in 1976, Robert Lucas[6] explained why the Keynesian models could not provide answers to the evident problems of inflation and built-in inflationary expectations, which had given rise to a phenomenon directly opposite to neo-Keynesian predictions based on the Phillips curve. The Phillips curve describes a relatively simple trade-off between inflation and unemployment. As we have already seen from the data in Tables 1.1 and 1.2, however, after the transition of the US dollar to a free float and the first oil shock which followed, certain economic players (companies and trade unions primarily) closed ranks and built their inflationary expectations into the prices of their products and labour. The result was stagflation.

Lucas demonstrated that the Keynesians, relying on either the Hicks IS-LM or Mundell-Fleming IS-LM-BP model and so exclusively on macroeconomic relations, had left entirely out of their models how key economic players, i.e. firms and households, actually react. In other words, one of his fundamental criticisms was

[6] Reference [6].

that the Keynesian models do not contain clearly specified goal functions describing how firms and households react when the government is changing its economic, and particularly its monetary and fiscal policy. Insofar as they are well informed and their expectations rational, economic policy measures will not, according to Lucas, have any impact on real variables. Econometric models based on past information do not provide a reliable basis for economic policy-making.

The rational expectations school thus appeared during a period of growing inflation, developing its conditions of "fitness of purpose" in economic policy-making by testing models for forecasting future prices. Based on a model for forecasting the inflation rate, adherents of the rational expectations school took the view that there was no need to use formula, except as an analytical condition for eliminating systematic forecasting error.[7] The analytical condition sufficient to eliminate systematic errors in forecasting may be presented in rudimentary form by the following equation:

$$P^{e}_{t+j} = E\left(P_{t+j}/\Omega_t\right)$$

Accordingly, the analytical condition of the rational expectations school boils down to the claim that the expected rate of inflation is based on estimation of the future level of prices, P(t + j), and a set of information available to all actors at time (t) during the decision-making process, or rather the time when expectations are being formed. For expectations to be rational, they must meet the condition that the deviation of real prices at a future time P(t + j) from expected price EP (t + j) is equal to zero, or that any eventual deviation is the result of the action of unforeseeable events (the random variable).[8]

Expectations are rational insofar as subjective expectations are consonant with objective expectations, which depends on the available set of information (Ω). Objective expectations represents the average value of the distribution of conditional probabilities of the variable P(t + j) for the given available information at time (t). It is a condition of rational subjective behaviour that all mistakes from the past (systematic mistakes or errors in forecasting) be avoided, so that there is no discrepancy between real and expected values:

$$E(P(t + 1) - Pe(t + 1)) = 0$$

Application of the theory of rational expectations assumes a democratic organisation of the society and so a transparent economic programme for the conduct of economic policy. Consequently, so long as the government publishes an economic programme with all the important information on which implementation will be

---

[7] Reference [7].

[8] The condition of rational behaviour is for subjective expectations formed by market actors to be the same as the average value of the distribution of probabilities of the variable being predicted, for a given range of available information.

based, rational market actors can forecast all future actions taken in the name of economic policy, reducing significantly any room for economic policy to actively influence GDP growth over the short term and eliminating it entirely over the middle or longer term.

Given this conclusion, changes in monetary and fiscal policy, insofar as they are foreseeable and foreseen, serve no purpose in essence. It follows from this that Keynesian models do not provide any valid intellectual basis for tackling economic problems. A year before Lucas published his critique of Keynesianism and of the effectiveness of any economic policy based on it, Thomas Sargent and Neil Wallace had published a paper outlining the intellectual basis for strengthening the impact of the new classical economic teachings (new classical economics).[9]

The academic response to the challenge issued to the Keynesians by Lucas, Sargent, Robert Barro and Milton Friedman, was contained in a number of papers published in the mid-1970s by Rudiger Dornbusch, Stanley Fischer, Edmund Phelps and Robert Taylor. In a work from 1976, Dornbusch offered a theoretical model to explain major volatility in exchange rates over the short term.[10] His "overshooting model" was a significant theoretical contribution to explaining the causes of major changes in exchange rates over the short term, at a time when the world of international finance had experienced a major transformation, with the transition to the system of floating exchange rates. The Dornbusch model combines elements from the monetary model and the Mundell Fleming model with his own original contribution, associated with analysis of how the exchange rate behaves over the short term. He started from the assumption that the prices of goods and services are rigid over the short term (the assumption of "sticky prices"), but they gradually adjust over the medium run, and are fully flexible in the long run as a result of changes in the supply of money and demand for it. On the other hand, financial assets prices (foreign exchange included) are flexible in the short run, and changes in the money markets are caused, primarily, by unanticipated changes in the money supply.

Unanticipated growth in the money supply over the short run causes a change in the nominal exchange rate. This change in the exchange rate is, however, greater in percentage terms than the change in the money supply. It is this more intensive exchange rate growth that forms the core of Dornbusch's model, whence its name, the "overshooting model." The exchange rate rise faster than the quantity of money in circulation because of (over)heated expectations of changes required in it over the coming period to establish a new exchange rate equilibrium. Since Dornbusch assumed that the prices for goods, labour and services are rigid over the short run (do not change), changes in the nominal exchange rate presuppose change in the real exchange rate of the same percentage (the prices of the goods of trading partners are also assumed not to change in the short run). Growth in the real exchange rate in the short run (real depreciation of the domestic currency) stimulate

---

9 Reference [8].

10 Reference [9].

Table 1.3 The federal funds rate, the inflation rate and the unemployment rate in the United States

Year	Inflation rate	Unemployment rate
1980	13.5	7.2
1981	10.3	8.5
1982	6.2	10.8
1983	3.2	8.3
1984	4.3	7.3

Sources http://www.usinflationcalculator.com/inflation/historical-inflation-rates/. http://www.davemanuel.com/historical-unemployment-rates-in-the-united-states.php

exports, which is to say they have an impact on real variables—output and employment. In the long run, however, the prices of goods are flexible, so that changes in the exchange rate are based on the assumptions of the monetary model of exchange rates.

Phelps and Taylor's paper from 1977[11] and Fischer's paper published the same year[12] share certain common characteristics with the Dornbusch model. In all three papers, the prices for goods, labour and services are rigid in the short run, i.e. they don't change (prices are sticky—sticky prices models), from which it follows that, given rational expectations, changes to monetary and fiscal policy (shifting of the IS and LM curves) will have an impact on real variables—output and employment, so long as they are unannounced and, accordingly, unanticipated by market actors.

Regardless of this intellectual response on the part of the Keynesians, their influence over the conduct of practical economic policy, at a time of major changes at the head of the Fed (the appointment of Paul Volcker as chair of the Fed in 1979) and Ronald Reagan's victory in the 1980 elections, fell to its lowest level in the post-war period. The newly elected president's personal adviser was Milton Friedman himself, while real influence on the US Council of Economic Advisers was concentrated in the hands of representatives of the new classical economics, which was based on the theory of rational expectations. The monetarist recipe of focusing on control of the monetary base and, indeed, sharp contraction of it, as the key instrument in the struggle against inflation and to eliminate built-in inflationary expectations in the prices of labour and goods, as applied by Paul Volcker and his colleagues from the FOMC between 1980 and 1982, did provide results in the short run (Table 1.3).

The monetary tightening aimed at eliminating inflationary expectations actually resulted in pushing both the nominal and real interest rates up to their highest levels since the Second World War. A steep recession followed, provoked by this exceptionally restrictive monetary policy. In spite of Ronald Reagan's promises to balance the US budget by the end of his first mandate, the cost of this restrictive monetary policy was so great that to pursue it at the same time as a restrictive fiscal

---

[11] Reference [10].

[12] Reference [11].

policy would have meant steeper recession and maybe even depression. Instead of balancing the budget, Reagan and his administration had, by the end of their first mandate, recorded the highest budget deficit in the US for three decades.[13] While unemployment rose sharply over the first 2 years of Reagan's first mandate, however, inflation had been dealt with as the key problem. Inflationary expectations had been eliminated, but the trade-off between inflation and unemployment described by the Phillips curve had been confirmed. Thus, while between 1980 and 1982 inflation was cut from 13.5 to 6.2 %, unemployment had risen from 7.2 to 10.8 %. On the other hand, the fall in inflation in 1983, compared to 1982, was accompanied by a simultaneous fall in unemployment from 10.8 to 8.3 %.

Nonetheless, the strengthening intellectual influence of the authors of the new classical macroeconomics and rational expectations theory in the late 1970s, their predominance over the circles of economic advisers during the 1980s, and the views of the supply-side economists that cutting corporate and income tax rates would lead to a re-equilibration of the economy and lay the groundwork for a new American prosperity and return to domination, entailed a sharp drop in the popularity of Keynesian economic ideas.

Supply-side economics drew its fundamental idea of stimulating production through low tax rates, supposedly leading to lower production costs, productivity growth, increased supply and increased consumer utility, from the ideas of Alexander Hamilton, the first US Secretary of the Treasury and one of the founders of the American institutional school of economics, as well as from the ideas of Adam Smith, Robert Mundell, and Arthur Laffer. Herbert Stein, Chair of the Council of Economic Advisers under two presidents, Nixon and Ford,[14] was the first to introduce the term "supply-side economics," the heart of which was the view that cutting income taxes would boost consumption, thanks to greater disposable income, while cutting profit taxes would boost profit potential.

Cutting the rate of income tax reduces the cost of labour and so inflationary pressures, while boosting potential profits and leading to a growth in net profit, so long as it is accompanied by a cut in the rate of profit tax. As net profit increases, the value of equity rises and there is a stabilisation of the economy, based on increased supply at lower costs, assuming the elimination of inflationary expectations. The Keynesian methods of managing the economy based on managing aggregate demand could not yield good results, according to the advocates of the new classical economics, monetarism and supply-side economics, as they were based on incorrect assumptions, which had contributed directly to the formation of inflationary expectations as the key problem for reproduction of the system founded on private ownership.

[13] Data on US budget balances for 1940–2013 are available on http://www.davemanuel.com/history-of-deficits-and-surpluses-in-the-united-states.php.

[14] Herbert Stein was Chair of the Council of Economic Advisors from 1972 to 1974.

The key economic decisions, according to the advocates of rational expectations theory, relate to maximisation of profit, on the supply-side (firms), and maximisation of utility, on the demand side (households). Consequently, in an environment of perfect information, or at least information sufficiently close to the ideal, companies and households maximise their goal functions, reducing severely the room for manoeuvre in economic policy, assuming prior publication of economic programmes. Consequently, economic policy plays no active role, since rational market actors are quite capable of making decisions about changes in quantity and price in the short, medium and long run, which will "clear the market" at the level of full employment.

1.2 The Keynesians' Theoretical Response and the Rise of the New Keynesianism

In the preceding section of the text, I have already noted that economists of Keynesian orientation responded to the new classical economists' and the Chicago school's objections regarding the ineffectiveness of either monetary or fiscal policy on output and employment. By introducing assumptions about sticky prices in the short run the new Keynesians demonstrated that monetary and fiscal policy could have a significant role to play in affecting the direction of the business cycle in the short run, even conceding the assumption of perfect information or at least access to all the information required (for rational expectations). Already in 1970, Edmund Phelps had edited a book in which the fundamental microeconomic theories of inflation and unemployment had been set out.[15] The influence of expectations on economic decision-making were also explained in the book, but, in contrast to the later dominant the rational expectations school, Phelps' approach was based on the autonomous expectations of actors in the economic system, which, in his own view, could not be identified with rational expectations, and which therefore produced significantly different outcomes. Together with his colleague Roman Frydman, Phelps edited a new collection of papers last year (2013)[16] which included what he and his co-authors considered a key distinction for understanding the poor record of all macroeconomic models based on rational expectations in forecasting the economic outcomes of market actors' decisions.

The critique contained in these texts does not relate only to the adherents of the new classical economics, but also to economists of the New Keynesian school of thought, who ground their analysis on rational expectations models and assumptions, albeit under conditions of sticky prices. Thus, one of the major currents amongst the New Keynesians bases its analysis of macroeconomic decision-making on the theory worked out by Kenneth Rogoff and Maurice Obstfeld in a series of

15 Reference [12].

16 Reference [13].

works published in 1994 and 1995. This pair of authors published the *Redux model* of exchange rate formation, basing their macroeconomic analysis on the assumptions of sticky prices for goods, labour and services and of integrated and clearly specified microeconomic goal functions on the part of the key actors in the economic system—households and firms.[17] This new open economy macroeconomics, with integrated goal functions for households and firms based on rational expectations, was, in effect, a definitive restatement of the New Keynesians' theoretical response to one of the major criticisms levied at Keynesians by Lucas and his fellow adherents of the new classical macroeconomics nearly two decades earlier.

The Redux model and the new open economy macroeconomics are based on imperfect competition. Market actors act under conditions of monopolistic competition, in which the number of households is equal to the number of producers. That is to say, households are at the same time consuming and productive units. The prices for goods, labour and services are sticky. Market actors have access to all the relevant information and as a result their subjective expectations are rational. In order to deal with the problem of specifying the utility function for households, Obstfeld and Rogoff based their original analysis on the representative agent (household) model. Since use of this model to approximate the behaviour of all households eliminates any possibility of differential behaviour by groups of consumers, the authors based the mathematical expression of maximisation of the household utility function on Gorman's polar (linear) form of the utility function, which provided them with an acceptable mathematical basis for maximisation of the household utility function. Household utility over the life cycle is determined by the following equation:

$$\mathrm{Ut} = \sum \beta^{s-t}\left[(\sigma/(\sigma-1))\mathrm{Cs}^{(\sigma-1)}\Big/\sigma + (\chi/(1-\varepsilon))(\mathrm{Ms}/\mathrm{Ps})^{(1-\varepsilon)} - (k/\mu)\mathrm{y},\ \mathrm{z}^{\mu}\right],$$

where β represents[18] the discount factor, the utility function is positively correlated with consumption (the first expression in the square brackets—Cs) and the real money supply (the second expression in the square brackets), but negatively correlated to the labour supply (the third expression), since working longer means less recreation and less enjoyment, so that more time spent at work reduces household utility but increases total output. The utility function is intertemporal. The utility of the representative household is maximised by a standard form of dealing with the optimisation of consumer intertemporal choice, i.e. maximising consumer utility in both major economies.

As this suggests, Obstfeld and Rogoff developed their model for two big economies,[19] both with floating exchange rates, so that changes in the money

---

[17] References [14, 15].

[18] Reference [14].

[19] As an example of two major economies, we may take Europe or the Eurozone and the United States, although at the time the authors were writing the redux model paper, the Euro had still to be introduced.

supply in one affect the potential for utility maximisation in the other. Consumer preferences are similar in both economies (because of high levels of income and similar economic structures), while the products produced by households are relatively homogenous on the national markets but differ, that is represent imperfect substitutes, in trade between the two economies. One of the key assumptions is the constant elasticity of substitution of goods from one country for goods from the other (the CES function), taken over from earlier work on monopolistic competition published in 1977 by Avinash Dixit and Joseph Stiglitz.[20] Any change to the interest rates and the exchange rates as a result of unanticipated changes in the money supply leads to the stabilisation of the initial equilibrium.

Gregory Mankiw and Ricardo Reis supplemented these New Keynesian models, based on sticky prices, and the New Keynesian Phillips curve derived from them in a paper of theirs[21] written at the beginning of the current century (2001/2002) and dealing with sticky information. They identified the key factor for economic policy's impact on real categories (output and unemployment) in the short run not as sticky prices for goods, labour and services in themselves, but sticky information and changes in the economic parameters derived from them, as a result of which the key actors in the economic system (firms and households) adapt with a certain time lag. Mankiw and Reis proposed replacing the New Keynesian Phillips curve based on sticky prices with a new New Phillips curve based on sticky information. They stressed that their paper included three significant characteristics differentiating their model from the New Keynesian one of the sticky prices Phillips curve, namely:

- Anti-inflationary programs (disinflation) always entail contraction in economic activity (a recession), which will, however, be milder if the anti-inflationary programme is known in advance.
- Monetary policy shocks have their maximum impact on inflation with a substantial delay.
- Changes in inflation are positively correlated with the level of economic activity.

Mankiw and Reis's paper was based on the previously mentioned paper by Stanley Fischer from 1977. Fischer was the first author to introduce the concept of "sticky information," in addition to "sticky prices," referring to the contractual relations of trade unions and employers, whereby a certain segment of the information from contracts agreed at some previous date remains unchanged and provides grounds for retaining prices in the current period. In this way, information from the preceding period has a direct impact on business decisions and economic outcomes during the current period, opening up room for monetary and fiscal policy to have an impact on output and employment.

What the Mankiw-Reis model and the Keynesian models based on sticky prices, and indeed most of the new classical models, have in common is the assumption of

---

[20] Reference [16].

[21] Reference [17].

the operation of the law of diminishing returns and so a pro-inflationary impact of expansionary monetary policy, particularly under conditions when actual output exceeds potential output (over-heated economy). The key factor on which this causality rests is the assumption of rapidly rising marginal costs, largely due to growing costs of labour whose marginal physical productivity is falling sharply as a result of the law of diminishing returns.

The 1990s and 2000s (the last two decades) differ from the period when these models based on sticky prices and sticky information were being developed, or indeed when the alternative models based on the new classical macroeconomic analysis and of the monetary circuit theory, however, in that they have seen steep growth in foreign direct investment in the countries of Southeast Asia and, more particularly, in China, i.e. in countries where labour costs are many times lower than in developed countries. Growing investment in the mass production of consumer goods, the very significant fall in labour costs thanks to the flow of capital to the Far East and the import of the resulting goods into the United States, Europe and other developed countries have resulted in a marked fall in the marginal cost of their production, with a major impact on eliminating inflationary pressures in developed countries, which had been such a problem for anti-cyclical economic policy during the 1970s and, to a lesser extent, the 1980s.

Globalising capital flows, financial deregulation and liberalisation, the liberalisation of international trade in goods and services, the opening up of a large number of countries to international trade and capital flows (China, India, and the former socialist countries) significantly increased the degree of international cooperation and, as a result, interdependence in the global economy.

1.3 The Dominant Financial Theory and Its Criticism

Modern financial theory or financial economics, as some call it, is based on the assumptions of free market activity and the efficient market theory. One of the three winners of the Nobel Prize for Economics in 2013 was Eugene Fama. In contrast to his colleague, Robert Shiller, who also awarded the Prize in the same year, Fama is one of the founders of the efficient market hypothesis. Shiller's position is diametrically opposed in theoretical terms, at least with regard to explaining how financial markets function. He was awarded the Nobel Prize for his behavioural theory of finance, which allows no room for Fama's interpretation of how financial markets operate on the basis of rational expectations and their consequent efficiency.

Historically speaking, the dominant financial theory did develop out of the efficient market theory, which has its roots in the work of Louis Bachelier,[22] who used fluid mechanics to provide the theoretical groundwork for understanding

[22] Reference [18].

changes in the price of financial assets. Bachelier analysed changes in the price of government bonds on the Paris stock exchange, concluding that changes in the price of financial assets tend to accord with a model of random deviation. His work was, consequently, fundamental to the development of writings about the efficiency of capital markets during the 1930s and 40s in works published by Working and Cowles. However, the founding father of the efficient market theory and its definite establishment was made by Eugene Fama in 1970.[23] The fundamental conclusion of this theory is that financial markets provide an effective mechanism for determining the price of financial assets, assuming a stable and developed institutional and legal framework in which all (or all the relevant) information for pricing securities is available and accessible. In such an environment, according to this theory, there is no systematic relationship between the return on securities and their price at some time in the past and their current market price. Their current price is exclusively a function of the information available now, which excludes any possibility of manipulating the market or prices and so of extracting extra profits on the basis of market asymmetries.[24]

Harry Markowitz developed portfolio theory in the 1950s,[25] which Lintner,[26] Sharpe[27] and Treynor[28] took as a normative framework for their model for pricing capital assets and the capital market theory. Their theoretical contribution was to consider decision-making about investment in financial asset portfolios and individual financial assets *under conditions of risk and uncertainty*. The Markowitz model of optimal portfolio choice relies on a two-parameter criterion for decision-making (the E-V criterion, where E is expected return on the portfolio and V the variance of the return on portfolio as a measure of investment risk). It rests upon the following assumptions:

- The investors' subjective views regarding the likelihoods of particular rates of return on the portfolio are mutually consistent. In other words, the distribution of the likelihoods of expected return on individual securities and portfolios are the same (objectively given) for all investors.
- The distribution of likelihoods of expected returns have a normal distribution.
- All investors share a single time horizon of choice, that is their choices relate to the same time period.
- The interest rates on invested and borrowed funds are equal to each other and they equal to the risk-free rate. Moreover, there is unlimited potential for investment and for taking on debt at the risk-free rate. By unlimited investment at the risk-free rate, it is meant unlimited potential or opportunity to buy risk-free securities

[23] Reference [19].

[24] Eugene F. Fama, Op. cit., see the sections under C (the Random Walk Model) and D (Market Conditions Consistent with Efficiency).

[25] Reference [20].

[26] Reference [21].

[27] Reference [22].

[28] Reference [23].

(government bonds), while by unlimited potential for taking on debt at the risk-free rate it is meant a possibility for selling borrowed securities (short sales).
- There are no transaction costs.
- The capital market is in equilibrium.

In a book published in 2004, Benoit Mandelbrot explained that modern financial theory is based on mistaken assumptions, or rather on assumptions that are directly contrary to how those markets actually function.[29] In the fourth chapter, entitled "The Case against the Modern Theory of Finance," Mandelbrot claims that the assumptions on which orthodox financial theory, as he calls it, rests are absurd, insofar as they do not reflect the behaviour of key market actors. He focuses particularly on four "shaky" or, in his words, absurd assumptions:

- The first assumption—that human beings are rational and that their only goal is to become rich. Rationality of choice is based on utility theory. Mandelbrot points out that people simply do not think in the terms of any theory of utility that can be measured in dollars and cents, nor are they always rational, anymore than they always operate in their own best interest.
- The second assumption—all investors are identical, i.e. have the same goals when investing, the same time horizon of investment, and make the same decisions on the basis of the given information. According to Mandelbrot, in reality, people are not equal and they do not behave identically, even when they have the same level of wealth and equal access to all information. Even under such circumstances, people do not make the same decisions, because their systems of preferences are not identical.
- The third assumption—changes in prices are in practice continuous. According to financial theory, the prices of stocks and bonds do not change steeply or precipitately over short time periods, but are subject to small and continuous changes. Continuity of this sort is characteristic of physical systems subject to inertia. In his book, Mandelbrot demonstrates that financial asset prices quite clearly do change precipitately over very short time periods, followed by periods of lesser change and discontinuity.
- The fourth assumption—that changes in financial asset prices follow Brownian motion. This term was borrowed from physics and refers to the motion of molecules in a medium at uniform heat. Bachelier suggested that this process could be used successfully to describe changes in securities prices. Brownian motion is based on three assumptions: first, that the magnitude of change during the present period is independent of change in the preceding period; second, on the statistical stationarity of prices; and third, on a normal distribution. Mandelbrot stresses that life is quite simply more complex than that and that the third assumption (the normal distribution of the expected returns on securities) is in the most marked contradiction to reality.[30]

---

[29] Reference [24].

[30] Benoit B. Mandelbrot and Richard L. Hudson (2004), Op. cit., pp. 79–87.

Since not just portfolio theory, but the capital market theory and the later Black-Scholes model[31] for determining the prices of options and derivative financial instruments all rely upon the assumption of the normal distribution of expected returns, in a section entitled "Pictorial Essay: Images of the Abnormal" Mandelbrot presented the results of his own investigations into changes in exchange rates and the prices for securities included in the Dow Jones Industrial Average on the basis of the theory of fractals—a theory founded by Mandelbrot himself. He points out that under the Brownian model most changes (68 %) should be small in scale and place within one standard deviation of the expected (average) value. In a normal distribution, 98 % of changes are found within three standard deviations either side of the average value, so that changes outside this range are very rare. Mandelbrot showed that changes to the Dow Jones Industrial Average over the period from 1929 to 2003 sometimes ranged, on a given day, as far as 10 standard deviations from the average, while on 1 day in October 1987 the change was 22 standard deviations.[32] This can be expressed in the terms of probability theory as a ratio of $1:10^{50}$! In other words, such deviations are not even incorporated into the standard Gauss table, because they are basically theoretically impossible. Nonetheless, they have, as Mandelbrot himself points out, obviously happened in practice.

Mandelbrot's analyses in his book finish with time series of data up to 2002, since the book was published in 2004. Major changes in securities prices took place during the global financial crisis of 2007–2009. According to the data on the Seeking Alpha website, there was a record high on 9 October, 2007, in the total value of world market capitalisation, amounting to US$61.264 trillion.[33] In March 2008, this total had fallen to approximately US$51.5 trillion, while by September of the same year it was down to approximately US$40 trillion.[34] World market capitalisation bottomed out on 9 March, 2009, at US$25.597 trillion.[35] Thus, over these 15 months world market capitalisation fell some 58.2 % (Fig. 1.1, Table 1.4).

One of the best critiques of the theory of the perfect rationality of market actors, other than Mandelbrot's, was that offered by Daniel Kahneman, winner of the Nobel Prize for Economics in 2002. In his book, *Thinking—Fast and Slow*,[36] Kahneman explains how he and Amos Twersky developed their theory and arrived at the following conclusions after many experiments and much research:

---

[31] Reference [25]. http://www.cs.princeton.edu/courses/archive/fall09/cos323/papers/black_scholes73.pdf.

[32] Benoit B. Mandelbrot, and Richard L. Hudson (2004), Op. cit, p. 93.

[33] Data from: http://seekingalpha.com/article/194972-world-market-cap-at-46_8-trillion.

[34] See: http://en.wikipedia.org/wiki/Market_capitalization.

[35] See: http://seekingalpha.com/article/194972-world-market-cap-at-46_8-trillion.

[36] References [26, 27].

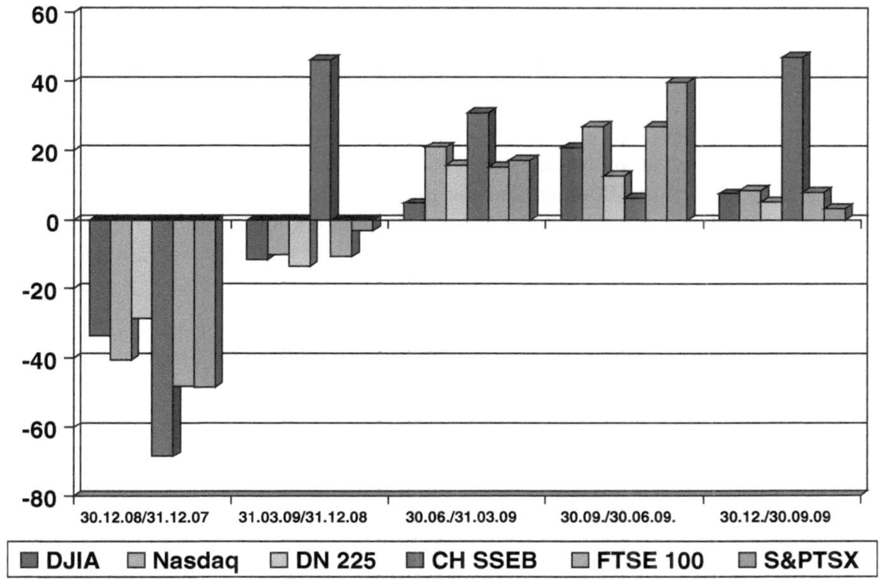

Fig. 1.1 Percentage change in the capital market indices on selected representative global capital markets. *Source* The Economist, various issues

Table 1.4 Percentage change in selected major indices for securities on global capital markets: 31st of December 2007 to 30th of December 2009

State/Index	30.12.2008/ 31.12.2007	31.03.2009/ 31.12.2008	30.06./ 31.03.2009	30.09./ 30.06.2009	30.12./ 30.09.2009
United States DJIA	−33.5	−11.6	+4.8	+20.9	+7.6
United States Nasdaq	−40.5	−10.2	+21.0	+26.9	+8.6
Japan Nikkei 225	−28.5	−13.5	+15.9	+12.8	+5.2
China SSEB	−68.4	+46.3	+30.9	+6.3	+47.1
Britain FTSE 100	−48.1	−10.8	+15.3	+27.0	+8.1
Canada S&P TSX	−48.3	−3.1	+17.3	+ 39.8	+3.3

Source The Economist, various issues

> We retained utility theory as a logic of rational choice but abandoned the idea that people are perfectly rational choosers. We took on the task of developing a psychological theory that would describe the choices people make, regardless of whether they are rational. In prospect theory, decision weights would not be identical to probabilities.[37]

In the 27th chapter of his book, Kahneman presents results of his work into the impact of reference values on decision-making about buying and selling, as well as into the very idea of indifference and the theory of utility derived from the map of indifference curves, on which the theory of rational choice is based. He draws the following conclusion regarding indifference curves and the theory of rational choice based on maximising utility through a map of indifference curves:

> The representation of indifference curves implicitly assumes that your utility at any given moment is determined entirely by your present situation, that the past is irrelevant, and that your evaluation of a possible job does not depend on the terms of your current job. These assumptions are completely unrealistic in this case and in many others.[38]

1.4 The Post-Keynesian Approach to Financial Theory: The Monetary Circuit Theory and Minsky's Financial Instability Hypothesis

In contrast to the classical and neoclassical theories of interest and money, on which monetarism and the new classical macroeconomics draw and which start from the assumptions that the money supply is exogenously determined and consequently given and that regular economic actors in the private economy cannot affect the money supply, the post-Keynesian economic school has built on Marx's work on the production circuit cycle and Keynes's theory of money as developed in the *General Theory* to develop a theory of the monetary circuit. Under this theory, money is created primarily endogenously as a consequence of the central role of private banking institutions in creating credit. This theory of the endogenous supply of money based on the lending activities of banks was established by the post-Keynesian, Roy Harrod, Nicholas Caldor, followed by Paul Davidson, Basil Moore, Victoria Chick and Sheila Dow.[39]

Further development of the monetary circuit theory resulted in the appearance of three currents, differing with regard to their authors' views of the role of the central bank in determining the money supply, that is to say what, if any, purpose it serves insofar as the banking sector plays the key role in creating the supply of liquidity. Keynesian fundamentalists believe that the money supply is entirely determined by the banking sector lending and that the only point of the central bank is to adjust for

---

37 Daniel Kahneman, Op. cit., p. 345.

38 Daniel Kahneman, Op. cit., p. 318.

39 For a concise overview of the development of this theory, see Ref. [28].

the need for additional liquidity in periods when the key actors in the system, that is firms, banks and households, have to "close" the circuit and need liquid resources to do so. Authors from this stream therefore tend to advocate the thesis that the money supply should be derived exclusively from the needs of economic actors (and so is entirely endogenously determined).[40]

In contrast to the "fundamentalists," the structuralists are closer to Keynes's original theory of interest and money based on liquidity preference, on the one hand, and the role central banks play in the process of creating money, on the other. This stream takes the position that the money supply is endogenously determined by the lending activity of banks, but that the overall money supply comprises both the supply of money based on lending (the endogenous supply which makes up the largest part of the total money supply) and the supply of funds in the reserves set by the central bank, which allow banks to close the monetary circuit (the exogenous component of the money supply).

Cavalieri explains the basic assumptions on which the monetary circuit theory and its primary endogeneity are based, which are themselves a product of the link between real and financial sectors:

- Capitalist production entails the existence of three groups of economic actors: firms (production units), banks and waged employees.
- The levels of production and of employment are set by business decisions made by banks and firms. Interest rate levels affect business decisions. Firms avail of loans to finance investments and pay their employee wages.
- The cause-and-effect chain in the economic system is based on demand for credit which is shaped by businesses in line with their investment plans. Banks approve loans and enable firms to undertake investment activities. Firms in return have unlimited access to additional liquid resources, on the basis of the loan approved, so long as they remain creditworthy.
- The cause-and-effect chain with regard to the money supply starts for the banking system with the loans that create new deposits, as against the interpretation of the classical economists for which the supply of credit was exclusively a function of deposits or savings. Thus, in classical economic theory, the supply of credit is a function of saving. In the monetary circuit theory, bank lending does not depend on savings. Rather it is lending that determines the level of deposits and of savings and, consequently, is of key significance in shaping the business cycle.
- The monetary circuit takes place in time, i.e. it assumes the passage of time. When firms receive loans, the impact of the money invested becomes apparent only after a certain lapse of time. Firms use loans to invest and pay salaries. Employees use their income (salaries) to buy goods produced by and securities issued by firms. In this way, households become the source of financial resources that "close" the circuit of money, i.e. the money the firms need to pay off their bank loans.[41]

---

[40] For an excellent overview of Monetary Circuit Theory see Ref. [29].

[41] Duccio Cavalieri (2004), Op. cit. http://mpra.ub.uni-muenchen.de/43769/7/MPRA_paper_43769.pdf, p. 7.

Hyman Minsky made a major contribution to the development of the monetary circuit theory over the period from the 1960s to the mid-1990s. He did so by developing his financial instability hypothesis. In a work from 1992, Minsky explained that he had based his theory of money and market function on Keynes's *General Theory* and Schumpeter's theory of the role of money and credit in the entrepreneurial economy.[42] Minsky further explained that his theory was not based on equilibrium conditions of the sort introduced into economics by Adam Smith and Leon Walras. The history of frequent episodes of financial crisis had shown that the economic system was not an inherently stable one, and that the money supply was not exogenously determined.

Minsky emphasised that the definition of economics as the allocation of available resources for alternative uses is not what lies at the heart of his theory. Rather it is Keynes's statement that the essence of the problem of economics is contained in the focus on the development and accumulation of capital. The formation of capital in a capitalist economy is based on the exchange of existing for future capital, i.e. on committing capital now in the hope of increasing it at some future period. In other words, the development of capital is based on the process of investing in the present to earn a profit at some point in the future, facilitating the enlargement of available resources (capital).

Minsky looked at the overall structure of the economic units in an economy and developed a threefold classification, depending on how economic activities were financed.[43] The first group of economic units includes those with access to stable sources of financing and satisfactory share capital, which represent a buffer or equity reserve during periods of crisis that can ensure these units' solvency, as well their liquidity (hedged economic units). During periods of crisis, such institutions can withstand external financial shocks, because they are sufficiently capitalised. A further characteristic of such economic units is that they are capable of regularly obtaining funds through bond and share issues. This group of institutions can make regular repayment of both the prinicipal and the interest on the loan out of their regular cash flows. This type of debt allows financial institutions to conduct their business in a regular manner and make a profit on the basis of a stable credit portfolio.

The second group is made up of speculative units, which secure their sources of financing by issuing short and long-term bonds (money market and capital market instruments). These institutions typically finance their purchases of assets by borrowing funds whose term is shorter than that of the assets. When these debts come due, the institutions must raise new sources of financing through new debts with a similar term (either by renewing their obligations or by re-issuing debt). As a result, such speculative economic units don't base their operations on returning the principal of the debt, focusing instead on regularly servicing the interest on their sources of funding.

---

[42] Reference [30]. http://www.levyinstitute.org/pubs/wp74.pdf.

[43] Hyman P. Minsky (1992), Op. cit., p. 7.

The third group of institutions includes financial institutions which base their operations on Ponzi schemes. In other words these institutions operate on the basis of very high levels of leverage (borrowings), and instead of paying back the funds financing their operations (the principal), they focus on paying off the interest on their borrowings by issuing new instruments and transferring the risk to third parties. Such institutions are, like the speculators, a source of systemic risk, and their operations become particularly risky when central banks pursue an anti-inflationary policy, reducing the money supply. Minsky stressed that liquidity shock is a particular problem for such institutions, as the majority of financial institutions in the second group (speculative financial institutions), suddenly unable to renew their leverage sources of financing, are left not only without access to funds to pay off the principal, but, unable to find new sources by renegotiating due obligations, can't even pay off regularly the interest on their borrowings and transform into Ponzi institutions. Minsky considered this phase particularly dangerous and a key source of financial crashes:

> In particular, over a protracted period of good times, capitalist economies tend to move from a financial structure dominated by hedge finance units to a structure in which there is large weight to units engaged in speculative and Ponzi finance. Furthermore, if an economy with a sizeable body of speculative financial units is in an inflationary state, and the authorities attempt to exorcise inflation by monetary constraint, then speculative units will become Ponzi units and the net worth of previously Ponzi units will quickly evaporate. Consequently, units with cash flow shortfalls will be forced to try to make position by selling out position. This is likely to lead to a collapse of asset values.[44]

Under such circumstances, Minsky believes that central banks have to act to prevent the whole system from collapsing. What is required, however, is for the central bank to conduct an expansionary monetary policy in order to prevent collapse of the system on the basis of pyramid structures via an accelerated spread of illiquidity from the speculative institutions to institutions which had sufficient capital up until that point.

In the same article, Minsky stresses that his theory of financial instability is based on two theorems. The first theorem is that for every economic system there exist financial systems under which they can function stably and financial systems under which they will display instability. The second theorem is that economies which have been enjoying prosperity over a longer period of time have a tendency to transition from stable systems of financing towards systems of financing in which economic units which belong to the second and third groups tend to dominate—that is speculative economic units and economic units which function on the basis of Ponzi schemes.[45] A predominance of such economic units necessarily leads to financial crisis, threatening profits, which are the basis for reproduction of the capitalist system.

---

[44] Hyman P. Minsky (1992) Op. cit., p. 8.

[45] Hyman P. Minsky (1992) Op. cit., p. 8.

The relevance of Minsky's analysis of financial cycles and the problems that arise from the inherent instability of speculative and Ponzi scheme financial institutions received very pointed confirmation from the financial shocks in the United States during the 2007–2009 crisis. Viral Acharya, Nirupama Kulkarni and Matthew Richardson provided an explanation, in a paper published in the book *Regulating Wall Street*,[46] for how the largest American investment banks became illiquid and insolvent, threatening the entire system. The data from their article provides very clear confirmation of Minsky's analysis and his hypothesis on financial instability. Namely, in the period from 1992 to 2007, the United States of America experienced a stable rate of growth, with the second Clinton mandate the most successful in three decades. During these 15 years, but particularly after the repeal of the Glass-Steagall Act in 1999 and denial of effective control over OTC markets for financial derivatives by the SEC in 2000, investment banks, institutional investors and hedge funds came to dominate the American financial system. These institutions belong to Minsky's second and third groups of financial institutions, defined as speculative and Ponzi respectively. These institutions were sufficiently powerful, in combination with financial groups from a number of the other financially most important countries in the world, to succeed, through lobbying the Basel Committee for Banking Supervision, in securing the adoption of Basel II,[47] which allowed them to apply internal risk management models and specific formula based on them in calculating their capital adequacy ratios.

Acharya, Kulkarni, and Richardson explain how the largest investment banks in the United States then used regulatory arbitrage based on internal models (Basel II), in combination with the rating agencies. A typical way for the major US investment banks to finance themselves during the first seven and a half years of this century was to issue asset-backed commercial papers—commercial papers with a typical maturity of 7 days.[48] These instruments were bought by money market funds. However, to buy them, the funds required asset backed commercial papers to be highly liquid and to have the highest rating (AAA). The investment banks' commercial papers were "secured" (asset-backed) by the banks' assets. A considerable part of the banks' assets thus consisted of MBS—mortgage-backed securities or bonds issued on the basis of mortgage loans. The investment banks tipically purchased "Triple-A" rated tranches of such securities.

As this form of "security" was not sufficient to give these commercial papers, with which the investment banks were financing themselves, the highest rating, their issuers also issued guarantees for them to money market funds. Thus, commercial papers were issued with a typical maturity of 7 days and, accordingly, every 7 days, as they matured, new commercial papers with the same maturity were issued (a renewal of obligations or paying commercial papers by issuing new commercial papers, which is a basic characteristic of speculative institutions under

---

46 Reference [31].

47 The Basel Committee for Banking Supervision adopted Basel II in June 2004.

48 Reference [32].

the Minsky model), guaranteed by guarantees issued by the investment banks with typical maturity of 364 days. These guarantees therefore had to be renewed every year. The investment banks treated these guarantees as off-balance sheet items and, accordingly, did not put aside capital reserves to cover them. Because of treating them in this way, the total issue of securities "backed" by guarantees rose from US $600 billion in 2004 to US$1.2 trillion at the end of June in 2007.

According to the official balance sheets of the five largest investment banks in the United States on 15 September, 2008, they were all sufficiently capitalised, with an average capital adequacy ratio of 6.2 %. Within just 2 days, as the money market funds holding investment bank securities, fast becoming illiquid, in their portfolios called in the guarantees for payment, the investment banks were forced to transfer their guarantee obligations to their balance sheets. Because of the large number of activated guarantees, it had become clear in just 2 days (17 September, 2008) that the largest investment banks in the United States had in fact average capital adequacy ratios of under 2 %. Table 1.5 shows the total value of the resources written-off and credit portfolio losses, to a large extent based on the purchase of securities "backed" by loans to purchase property (mortgage-backed securities):

George W. Bush and his administration had made a nominal commitment to the principles on which free market economy is based. Greg Mankiw was Chair of the Council of Economic Advisers during this administration. Keynes' authentic

Table 1.5 The amounts of the largest write-offs and losses by American financial institutions (June 2007–March 2010)

Company name	Written off resources and losses on loans (billions of USD)	Return on share capital (June 2007–December 2008)
Fannie Mae	151.4	−98.14
Citigroup	130.4	−82.46
Freddie Mac	118.1	−97.98
Wachovia	101.9	−88.34
Bank of America	97.6	−67.79
AIG	97.0	−97.57
JPMorgan	69.0	−31.51
Merrill Lynch	55.9	−85.16
Wells Fargo	47.4	−10.77
Washington Mutual	45.3	−99.95
National City	25.2	−94.29
Morgan Stanley	23.4	−75.99

Source Reference [32]

teachings were discarded or ignored, along with the positions taken by the post-Keynesians in economic theory and in particular those of the founders of the monetary circuit theory. It is an irony of the laws of economics and, more particularly, of the relevance of the ideas and teachings of the post-Keynesians, and more particularly of Hyman Minsky, that in the final year of his second mandate, George W. Bush ended his presidential mandate by intervening in a way entirely in line with the recommendations of John Maynard Keynes. This urgent intervention to save a collapsing financial system was a direct consequence their having tolerated a situation that provided evident confirmation in fact and reality of everything Minsky had written and described in his theory of financial instability — that at a high level of development in financially developed environments speculative and Ponzi financial institutions tend to dominate.

In order to save the financial system from de facto bankruptcy, the US president was forced to intervene by emergency procedure with a fiscal package in the amount of US$700 billion.[49] Moreover, in the second half of 2008, the Fed increased the money supply by about 124 %, which had not happened since the great crisis of 1929. These measures of monetary and fiscal policy provide a marked example of return to Keynes's original teachings, as presented in the *General Theory*.

1.5 The Global Financial and Economic Crisis and the Return in the Major Economies of Economic Policy Based on Keynes' Recommendations from the *General Theory*

During the first 7 years of this century, the American economy achieved good results, measured by the rates of economic growth and unemployment. In mid-2007 the unemployment rate in the United States was 4.6 %, which may be considered approximately close to the natural rate of unemployment, in terms of the theoretical concept developed by Milton Friedman and Edmund Phelps. The inflation rate in the United States was also low, so that measured by the popular, if superficial indicator of the misery index, George W. Bush's administration at that time seemed at first glance to be achieving good economic results.

The data from Table 1.6, particularly when supplemented by the data from Table 1.7, show that this "economic success" was essentially underwritten by an expansionary monetary policy, that is a monetary policy that allowed the taking on debt on the interbank market at what was in reality a negative interest rate from 2003 to 2005, and at an effective zero real interest rate during 2002.

The average level of the federal funds rate in the last year of George W. Bush's second mandate was significantly lower than the average inflation rate, resulting in

[49] The Emergency Economic Stabilisation Act—which entered into force on October 4, 2008.

Table 1.6 The federal funds rate, the rate of inflation and the rate of unemployment in the United States

Year	Inflation rate	Unemployment rate
2000	3.38	4.0
2001	2.83	4.7
2002	1.59	5.8
2003	2.27	6.0
2004	2.68	5.5
2005	3.39	5.1
2006	3.24	4.6
2007	2.85	4.6
2008	3.85	5.8

Source http://www.federalreserve.gov/releases/h15/data.htm. http://www.statista.com/statistics/193290/unemployment-rate-in-the-usa-since-1990/

Table 1.7 The federal funds rate, the rate of inflation and the rate of unemployment in the United States

Year	Federal funds rate	Inflation rate	Real interbank market interest rate
2000	6.24	3.38	+2.86
2001	3.88	2.83	+1.03
2002	1.67	1.59	+0.08
2003	1.13	2.27	−1.14
2004	1.35	2.68	−1.33
2005	3.22	3.39	−0.17
2006	4.97	3.24	+1.73
2007	5.02	2.85	+2.17
2008	1.92	3.85	−1.93

Source http://www.federalreserve.gov/releases/h15/data.htm

the lowest real interest rate on the US interbank market in 33 years (since 1975). Compared to the last year of Bill Clinton's administration (2000), when unemployment in the US economy was at its lowest level for three decades (4 %) and the real positive interest rate on the interbank market was 2.86 %, the economic results of the Bush administration take on a whole other aspect.

In analysing the economic results achieved by these two American presidents (Bill Clinton and George W. Bush), one must first stress the fact that even though the second Clinton mandate saw the US economy achieve its best results in three decades, and better results than the Bush administration as well, the creation of the dot-com bubble in 1997–2000 and the consequent implosion of the bubble did nonetheless result in a slowing down of economic activity and in financial shocks which spilled over from capital markets to the real sector of the economy in the second half of 2000 and during 2001. The US economy was faced, precisely because of this implosion of the financial bubble that had been created, with serious problems regarding deflationary trends in financial asset prices and consequent falling investment. The terrorist attacks on the United States (September 11, 2001) further exacerbated these economic problems, but the Fed, acting decisively, cut the federal funds rate in thirteen consecutive sessions, from 6.24 % at the end of 2002 to 1.13 % in late 2003.

In the last 2 years of the Clinton administration, however, two laws were passed which would contribute to the creation of major financial problems, in combination with the above-mentioned, much lobbied-for adoption of new international banking standards (Basel II), which allowed large banks to use internal models for risk management and regulatory arbitrage. Comparison of data on changes in the DJIA and NASDAQ indexes between 1992 and 2000, and percentage changes in US nominal GDP reveals that during those 8 years the Dow Jones rose 226.8 %, the NASDAQ 265 %, but nominal GDP just 57.4 %.[50] In fact, the changes in the NASDAQ index had been far stronger, as is clear from Fig. 1.2, rising nearly 500 % between 31 December, 1992 and 10 March, 2000, when it reached a record 5048 index points. In just a year (10 March 1999—10 March 2000), the index rose 109.8 % (from 2406 to 5048 index points).

Alan Greenspan's speech and annual lecture at the American Enterprise Institute for Public Policy Research in December 1996 show that Greenspan and his colleagues at the Fed were, naturally, quite conscious of the danger of "financial bubbles":

> But how do we know when irrational exuberance has unduly escalated asset values, which then become subject to unexpected and prolonged contractions as they have in Japan over the past decade? And how do we factor that assessment into monetary policy? We as central bankers need not be concerned if a collapsing financial asset bubble does not threaten to impair the real economy, its production, jobs, and price stability. Indeed, the sharp stock market break of 1987 had few negative consequences for the economy. But we should not underestimate or become complacent about the complexity of the interactions of asset markets and the economy. Thus, evaluating shifts in balance sheets generally, and in asset prices particularly, must be an integral part of the development of monetary policy.[51]

Even though the Fed did increase the federal funds rate between 1998 and 2000, from 4.75 to 6.5 %,[52] this increase was neither timely nor large enough to show the money markets and particularly the capital markets that the Fed was ready to take quick action to "toughen" the conditions for accessing funds for primary liquidity on the interbank market and so indirectly affect the interest rates and the yield on

---

[50] Data on the values of the DJIA, Nasdaq and US GDP are available on the following websites: http://www.fedprimerate.com/dow-jones-industrial-average-history-djia.htm. http://www.google.com/finance/historical?cid=13756934&startdate=Mar+10%2C+1999+&enddate=Mar+20%2C+2000&num=30&ei=N2QsU7j2CsmpwAPnSg. http://data.worldbank.org/indicator/NY.GDP.MKTP.CD?page=4.

[51] See: Ref. [33]. Available on the following web site: http://www.federalreserve.gov/boarddocs/speeches/1996/19961205.htm.

[52] See: http://www.fedprimerate.com/fedfundsrate/federal_funds_rate_history.htm.

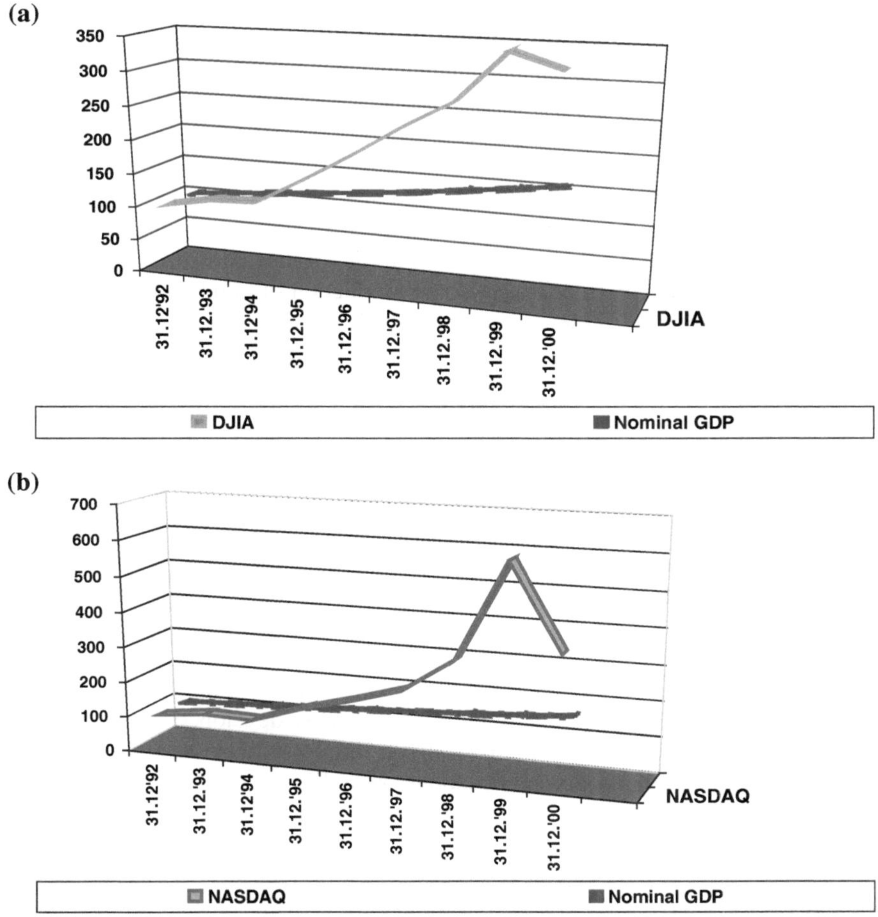

Fig. 1.2 Percentage changes in US GDP, the Dow Jones Industrial Average, and the NASDAQ index during the Clinton administration: 1992–2000. **a** Changes in nominal US GDP and the Dow Jones Industrial Average: Clinton administration 1992–2000 (1992 = 100). **b** Changes in nominal US GDP and the NASDAQ index: Clinton administration 1992–2000 (1992 = 100). *Sources* http://www.fedprimerate.com/dow-jones-industrial-average-history-djia.htm. http://www.google.com/finance/historical?cid=13756934&startdate=Mar+10%2C+1999+&enddate=Mar+20%2C+2000&num=30&ei=N2QsU7j2CsmpwAPnSg

corporate shares and bonds, particularly those of technology, media and telecommunications (TMT) companies. The view of so-called new economy gurus, who alleged that the impressive productivity growth in US manufacturing was due to increasing use of information technology, which was driving productivity growth in the United States more generally, was only in part correct. In a paper published in 1999, Robert Gordon from Northwestern University showed that the increase in productivity between 1995 and 1999 had indeed been impressive, but only in the production of computers (41.7 % annually). In private sector non-agricultural manufacturing productivity growth was at a lower level during this period (2.15 % annually) than it had been in 1950 to 1972, though higher than in 1972 to 1995. Similarly, productivity over those 4 years in the perishable non-agricultural goods sector was considerably lower than it had been between 1950 and 1972.[53]

In other words, the claims by advocates of the so-called new economy that the law of diminishing returns had become obsolete in an environment of new information technologies and that these technologies allowed ever increasing returns, rapid productivity growth, and reduced production costs, in turn allowing production to increase without inflationary pressure on the production of goods, were simply not confirmed. According to these economists, it followed from these trends that the entire increase in share prices was "justified" by productivity growth, in other words that it was based on positive supply shocks. There was no need for central banks to react to positive supply shocks, because the rapid growth in financial assets prices "was closing" the gap between rapidly-growing supply and the temporary shortage in aggregate demand. Frank Smets was one of the first to present a model of the role of asset prices in monetary policy-making, including both the price of the financial assets and the target function of the monetary authorities.[54]

In their 2010 book,[55] Howard Davis and David Green presented an analysis of the causes of the problem of the central banks' late reaction to changes in the prices of assets and its major impact on the real economy, as made manifest in the global economic crisis. This pair of authors stress that the central banks were slow to react to the accumulated problems of the first 8 years of the century and that, if they really want to maintain financial stability in the most developed countries, they must expand their focus from a narrow one on money markets to a wider one on assets markets, particularly property markets, including them in the price indexes which they use as the basis for monitoring inflation. Even though the most important central banks have, in the meantime, since the book was published, significantly changed how they deal with these issues, focusing on unconventional measures of monetary policy, they still have not included changes in property prices as a reference value in determining the target rate of inflation.

Davis and Green, in fact, did explain some of the key problems regarding the appearance and development of this bubble. Thus, in contrast to the period of the Clinton administration in the US, when a financial asset bubble was created, during that of the Bush administration, it was the fiscal policy of cutting taxes and incentives for residential building for families without regular income, combined with the Fed's monetary policy, again lagging behind events during the period of 2004–2006, that led to the creation of a bubble on the property market and speculation in securities issued on the basis of loans for residential building (mortgage-backed securities).

[53] Reference [34]. Available on: http://research.stlouisfed.org/conferences/workshop/gordon.pdf.

[54] Reference [35]. http://www.bis.org/publ/work47.pdf.

[55] Reference [36].

Figures 1.3 and 1.4 show that in contrast to the financial assets bubble created during the Clinton period, financial asset prices, as captured by the Dow Jones and NASDAQ indexes, grew somewhat more slowly than nominal GDP under the Bush administration. A different type of bubble was created on the property market, however, which had a direct impact on the steep growth of the balance-sheets of American and European banks, as well as institutional investors actively participating on American capital markets. These transactions during the 2002–2007 period were largely based on speculative activities, as illustrated in the above text through the example from the book on *Regulating Wall Street.*

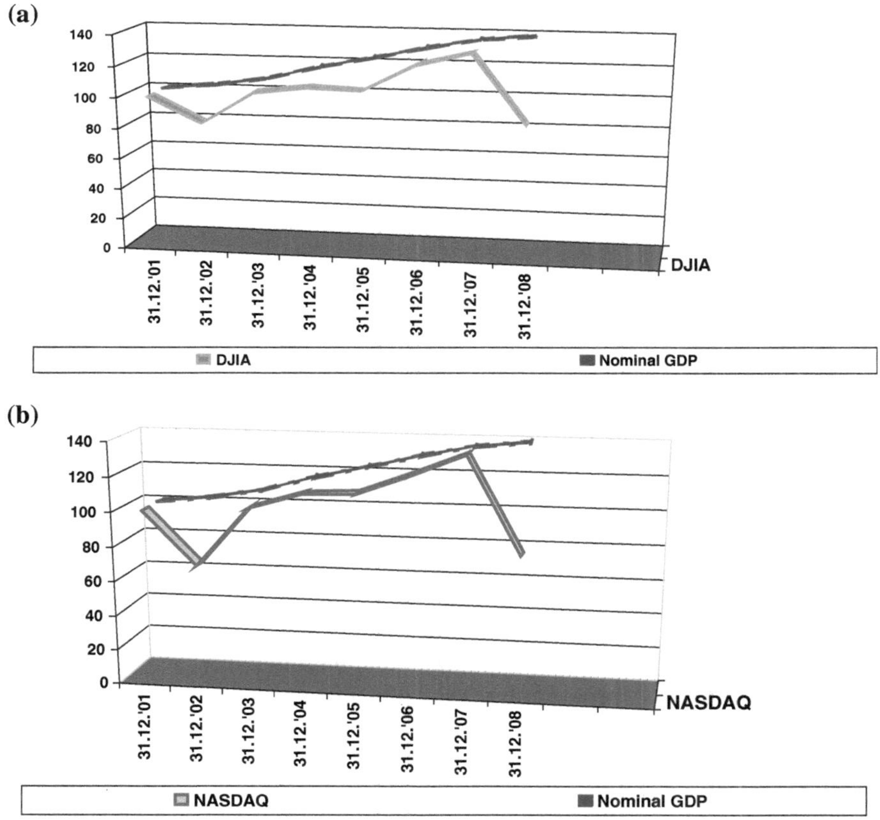

Fig. 1.3 Percentage changes in US GDP, the Dow Jones Industrial Average and the NASDAQ. Under the Bush administrations 2001–2008. **a** Changes in US GDP and the Dow Jones Industrial Average: the George W Bush administration, 2001–2008 (2001 = 100). **b** Changes in US nominal GDP and the NASDAQ index: the George W Bush administration, 2001–2008 (2001 = 100). *Sources* http://www.fedprimerate.com/dow-jones-industrial-average-history-djia.htm. http://www.google.com/finance/historical?cid=13756934&startdate=Mar+10%2C+1999+&enddate=Mar+20%2C+2000&num=30&ei=N2QsU7j2CsmpwAPnSg

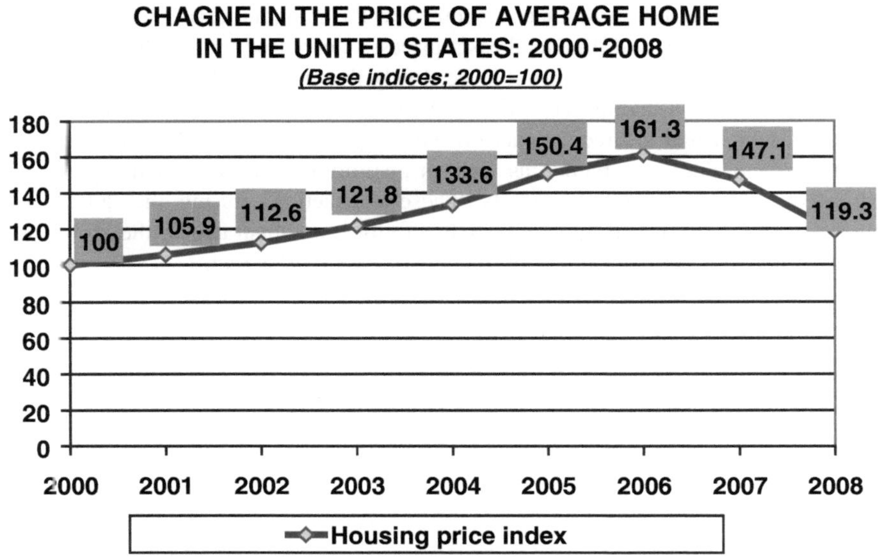

Fig. 1.4 Changes in house prices under the George W Bush administration. *Source* Robert Shiller, Yale University—published in [42]

As a consequence of *The Commodity Futures Modernisation Act,*[56] the SEC was stripped of any effective capacity to monitor and control the contracting of derivative financial contracts, opening up room for major investment banks and dealers to increase their trade in such instruments sharply, which in turn significantly reduced the Fed's capacity to control the interest rate effectively, given that interest swaps were one of the most intensively traded instruments on these markets. The growth in trade of derivatives between 2000 and 2008 was spectacular. The total notional value of contracts for which financial derivatives were agreed grew from $98 trillion in 2000 to $668 trillion in mid-2008 (Fig. 1.5).[57]

The almost absolute financial deregulation of the market in derivatives, along with the system enabled by Basel II of using internal models to determine the rating of internationally active banks (and to determine their minimum capital adequacy ratios) caused a series of problems and led to the financial collapse of 2007 to 2008. Paul Krugman, in a book from 1999, reprinted in an extended edition in 2008 called the creation of this alternative system of financing through investment banking, hedge funds and institutional investors, the "shadow banking system."[58] In fact,

[56] Signed by William Clinton in December 2000.

[57] Data available on the web page of the Bank for International Settlements: http://www.bis.org/statistics/dt1920a.pdf.

[58] Reference [37]. Chapter 8: Banking in the Shadows (pp. 153–164).

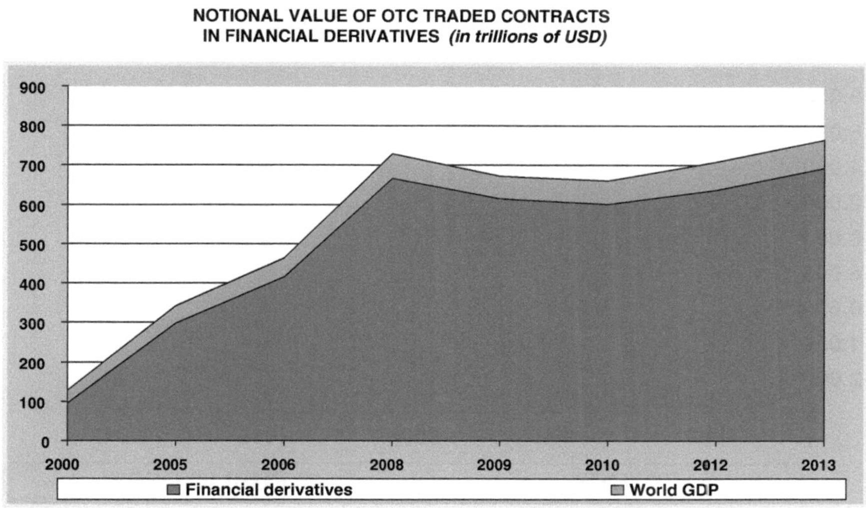

Fig. 1.5 The notional value of contracts on the OTC market: 2000–2013. *Source* The Bank for International Settlements

Krugman's analysis only confirmed how right Hyman Minsky had been in developing his financial instability hypothesis—the endogenous and integrated motive of the financial sector which, innovating in the area financial services, produces a transformation of economic systems from stable sources of financing into ones in which speculative activity and Ponzi schemes of financing dominate and have a major negative impact on systemic risk.

1.6 The Return of Keynes to Economic Policy

In his book, *The Return of the Master*, the best-known biographer of Keynes, Lord Skidelsky[59] describes the importance of Keynes's economic and social ideas, and particularly the relevance of counter-cyclical actions during periods of crisis, that is the need for timely action to reduce or absorb external shocks to economic stability. In Skidelsky's view, the post-Keynesians are the closest of the three currents of Keynesianism to Keynes's original and authentic teaching and his stress on the importance and role of uncertainty in economics. In a book published in 2009, Paul Davidson, one of the most prominent post-Keynesians and the founder of *The Journal of Post-Keynesian Economics*, explains Keynes's original teachings and the confusions of most neo- and New Keynesians, as well as of the advocates of the

[59] Reference [38].

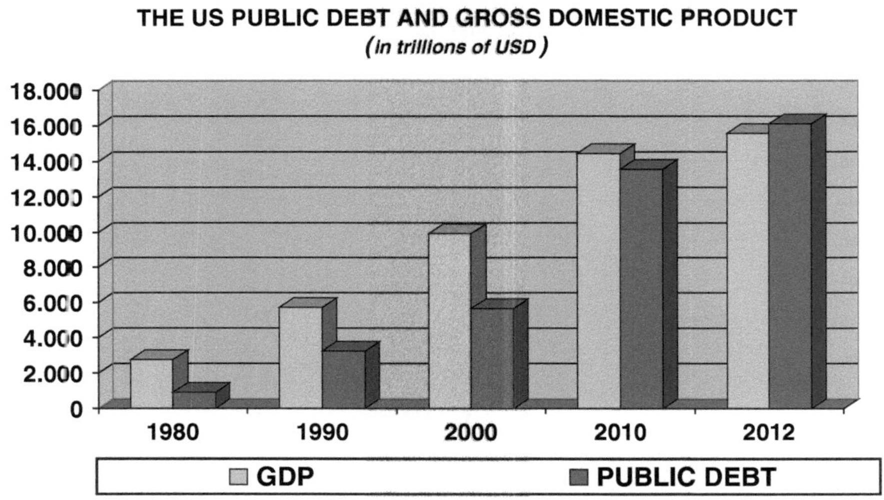

Fig. 1.6 US public debt and GDP: 1980–2012. *Source* The Federal Reserve of the United States

new classical economics and monetarists, over the interpretation of Keynes's recommendations on how to exit a depression, and particularly their confusions regarding the interpretation of sticky prices over the short term.[60]

During the period from 1989 to 2007, the total public debt of the United States increased *by* US$3.03 trillion. Over the next 5 years (2008–2013) the country's public debt grew US$6.47 trillion. Thus, over those 5 years, American public debt grew 2.13 times more than it had in the previous eighteen. Although in absolute terms the US public debt increased considerably more over those 5 years than it had in the previous eighteen, the percentage increase in the public debt between 2002 and 2008 was, however, 75.5 %, compared to 82 % between 2008 and 2012. Between 1990 and 2010, the nominal GDP of the United States grew 170.6 %, while during the first decade of this century it grew only 45.7 %. Comparing growth in nominal GDP in the US in the first decade of this century with the growth of the total public debt, we thus arrive at the following ratio: 139 % as against 45.7 %, in favour of public debt (Fig. 1.6).[61]

Between 17 September and 31 December, 2008, the Fed increased the money supply from US$ 995 billion to US$ 2.239 trillion, that is by 123 %.[62] Such an increase in the money supply over just three and a half months had not been recorded in four decades. Along with this massive increase in the money supply,

[60] Reference [39].

[61] Reference [40].

[62] *Source* Federal Reserve of the USA, available on the following webpages: http://www.federalreserve.gov/monetarypolicy/bst_recenttrends.htm; http://www.federalreserve.gov/monetarypolicy/bst_recenttrends_accessible.htm.

there came a decision to cut the federal funds rate to a target range of between 0 and 0.25 %. The FOMC made this decision at a session in December 2008. This increase in the monetary supply by 123 % and cut in the interest rate to a de facto zero level did not result in an increase in economic activity, inflation, or employment in 2009. In fact the opposite occurred. As Tables 1.8 and 1.9 show, the expansionary monetary and fiscal policy failed to prevent recession during 2009. The unemployment rate went up from 5.8 to 9.3 %, while inflation, which had been a serious problem in the first half of 2008, gradually began to fall in the second half of the same year, becoming deflation during 2009.

The major psychological shocks associated with a sharp rise in investor pessimism, on the one hand, and a sharp fall in demand for final consumer products and homes, because of the previous enormous expansion in lending in the US, Europe and the Far East, resulted in a sharp rise in unemployment. Thus, the markedly expansionary monetary and fiscal policies were not accompanied by an increase, but a sharp fall in employment in 2009, contradicting the new classical model, the neo-Keynesian model and its Phillips curve, and even the New Keynesian model based on the New Keynesian Phillips curve. Not even the model offered at the very beginning of this century by Mankiw and Reis, with its new New Phillips curve

Table 1.8 The real interest rate in the US: 2007–2013

Year	Federal funds rate	Inflation rate	Real interest rate
2008	1.92	3.8	−1.88
2009	0.16	-0.4	+0.56
2010	0.18	1.6	−1.42
2011	0.10	3.2	−3.10
2012	0.14	2.1	−1.96
2013	0.11	1.5	−1.39

Source Federal Reserve System—http://www.federalreserve.gov/releases/h15/data.htm. http://www.usinflationcalculator.com/inflation/historical-inflation-rates/

Table 1.9 The inflation rate, the real interest rate and the rate of unemployment in the United States (2007–2013)

Year	Inflation rate	Real interest rate	Unemployment rate
2008	3.8	−1.88	5.8
2009	−0.4	+0.56	9.3
2010	1.6	−1.42	9.6
2011	3.2	−3.10	8.9
2012	2.1	−1.96	8.1
2013	1.5	−1.39	7.4

Source Federal Reserve System—http://www.federalreserve.gov/releases/h15/data.htm. http://www.usinflationcalculator.com/inflation/historical-inflation-rates/

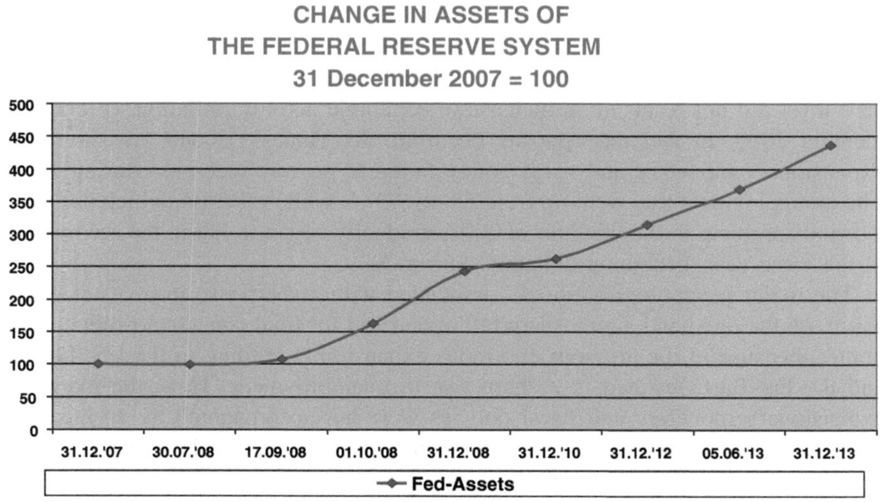

Fig. 1.7 Monetary expansion in the United States of America 2007–2013. *Source* The Federal Reserve of the United States of America

based on sticky information, was any more successful in explaining the major disturbances that had taken place on the largest world markets. As it is, the Mankiw-Reis model claims that anti-inflationary policy, that is a combination of monetary and fiscal restriction, necessarily brings about a fall in employment, because of the sticky information contained in contracts agreed earlier between trade unions and employers, which reduce the profit potential under conditions of restrictive macroeconomics policy. Between 2008 and 2009, the sharply *rising* money supply and fiscal expansion came to be combined with steep growth in unemployment, because of major falls in the prices of securities, market capitalisation (between October 2007 and March 2009), the consequent steep rise in the cost of financing investment and the associated risks (Fig. 1.7).

The winner of the United States presidential elections in November 2008 was the Democratic Party candidate, Barack Obama (a success he repeated in the November 2012 elections). Obama's administration started its first mandate with a proposal for urgent passage of an anti-recession law, *The American Recovery and Reinvestment Act* (ARRA).[63] The law was passed in Congress and signed by President Obama on 17 February, 2009. The proposal put together by Obama's cabinet had been to approve a fiscal stimulus worth US$820 billion. As actually passed, the law authorised fiscal intervention to the tune of US$787 billion. With this law, the Obama administration made clear that its economic policy exit from the crisis was firmly grounded in Keynes's teachings and Keynes's recommendations for getting out of

[63] US Department of the Treasury, recovery act, http://www.treasury.gov/initiatives/recovery/Pages/recovery-act.aspx.

Table 1.10 The value and breakdown of US GDP: 2008–2013

Year	GDP Billions of US$	Household consumption (% of GDP)	Gross private domestic investment (% of GDP)	Value of the net exports of goods and services (% of GDP)	Government consumption of goods and services (% of GDP)
2008	14,720	68.0	16.5	−4.9	20.4
2009	14,418	68.3	13.0	−2.7	21.4
2010	14,958	68.2	14.1	−3.5	21.2
2011	15,534	69.0	14.4	−3.7	20.3
2012	16,245	68.6	15.2	−3.4	19.5
I-VI 2013	16,668	68.6	15.8	−3.0	18.7

Source Board of Governors of the Federal Reserve System, *Financial Accounts of the United States—Flow of Funds, Balance Sheets, and Integrated Macroeconomic Accounts*, Second Quarter 2013. Note: Data for the first half of 2013 have been seasonally adjusted to the annual level

crises. Although Obama's administration faced constant pressure from Republican Party representatives to limit the fiscal expansion, particularly evident in August 2011 and August 2013 in the context of setting the debt limit, the administration's economic policy was very largely led by Keynes's economic logic. The role of fiscal policy in avoiding a double-dip recession, or indeed preventing the 2009 recession becoming a depression, was unquestionably great. Table 1.10 shows a breakdown of US GDP for the past 5 years, indicating that the key factor in preventing a considerably greater fall in US GDP in 2009 was the increase in the role of public spending in answer to the marked fall in gross private domestic investment.

Monetary and fiscal expansion as counter-cyclical economic policy measures were not limited to the United States. Although the European Central Bank (ECB) was, because of legal obstacles in the European Union and the Eurozone member countries, rather limited at the beginning of the crisis in terms of how quickly could react to the emerging conditions, it was, after a series of legal adjustments in the EU and Eurozone member countries, able to increase the money supply, cut interest rates and intervene on money markets and the markets for government securities in the countries of the southern Eurozone and Ireland (Italy, Spain, Portugal, Greece and Ireland). The result was an increase in the assets of the Euro system from €1.1 trillion at the end of December 2006 to €1.8 trillion at the end of 2009 and €2.9 trillion at the end of 2012. Thus, the ECB's assets grew by as much as 163.6 % between December 2006 and December 2012 and by 61.1 % during the period of crisis from 2009 to 2012.[64] On the other hand, data regarding the ECB-Euro system

[64] Data on changes to the ECB-Euro system on the ECB webpage: https://www.ecb.europa.eu/press/pr/wfs/2008/html/fs080326.en.html.

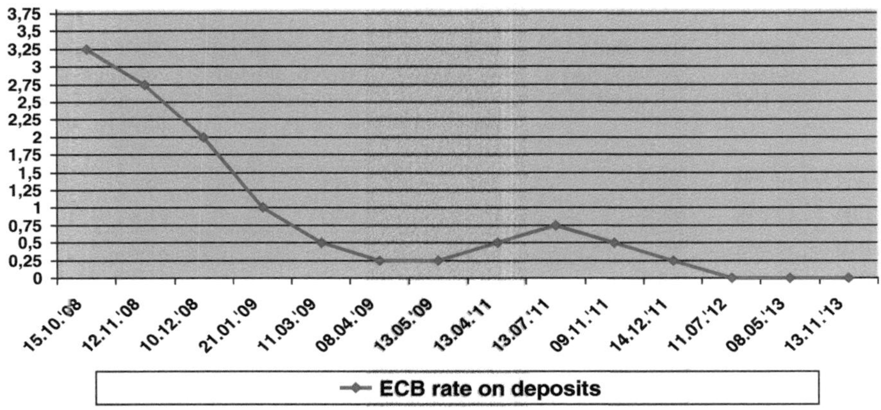

Fig. 1.8 European Central Bank: deposit interest rate: October 2008–December 2013. *Source* European Central Bank

monetary aggregates show that M1 increased by 34.5 % between 2008 and 2013, while M2 and M3 grew rather more slowly over the same period—15.5 and 5.4 %, respectively.[65] The expansionary nature of the Eurozone countries' monetary policy is well illustrated by data on this institution's interest-rate policy, as shown in Fig. 1.8.

The interest rate paid on deposits by the ECB and the central banks of Euro system countries was cut from 0.25 % to zero on 11 July, 2012, and it remained at that level up to the point when this text was written (mid-May 2014). At a session of the Board of Governors of the ECB on 8 May, 2014, the possibility was mooted that the deposit interest rate might go negative in the second half of 2014 (−0.10 %). At the same time, between 13 July, 2011, and 13 November, 2013, the key ECB short-term lending rate to commercial banks (the main refinancing operation rate) was cut from 1.5 to 0.25 %. At the above-mentioned session of the Board of Governors, it was announced that the basic reference interest rate might be cut during the second half of 2014 to 0.10 %, if the economic recovery of Eurozone countries did not reach expected levels, as a strengthening euro against the dollar had reduced the competitiveness of exporters from Eurozone countries.

The application of Keynes's recommendations for using fiscal policy and particularly public spending to combat recession and reduce unemployment are illustrated for Eurozone countries by Fig. 1.9. According to Eurostat data, between 2006 in the first half of 2013, the public debt of all the Eurozone countries except Luxembourg exceeded the criteria set under the Maastricht Treaty and the 1994 the Stability and Growth Pact. The criterion for the ratio of total public debt to GDP had been set at a level of 60 %. During the period of crisis, starting in 2008 and

[65] Data on changes to the ECB-Euro system monetary aggregates on the ECB webpage: http://www.ecb.europa.eu/stats/money/aggregates/aggr/html/hist.en.html.

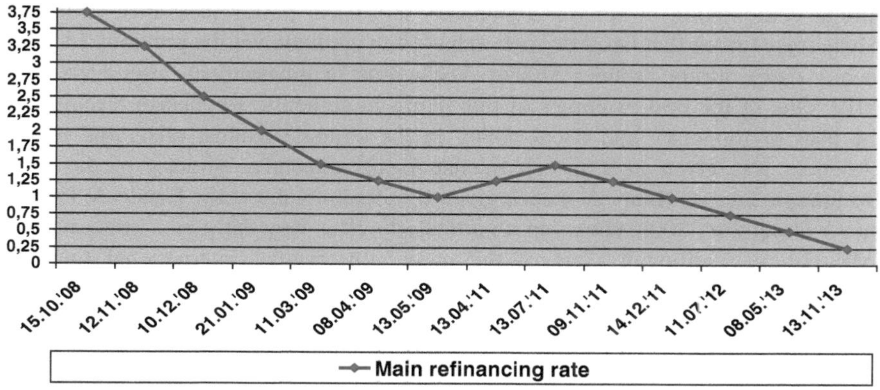

Fig. 1.9 European Central Bank: main refinancing rate. October 2008–December 2013. *Source* Euroepan Central Bank

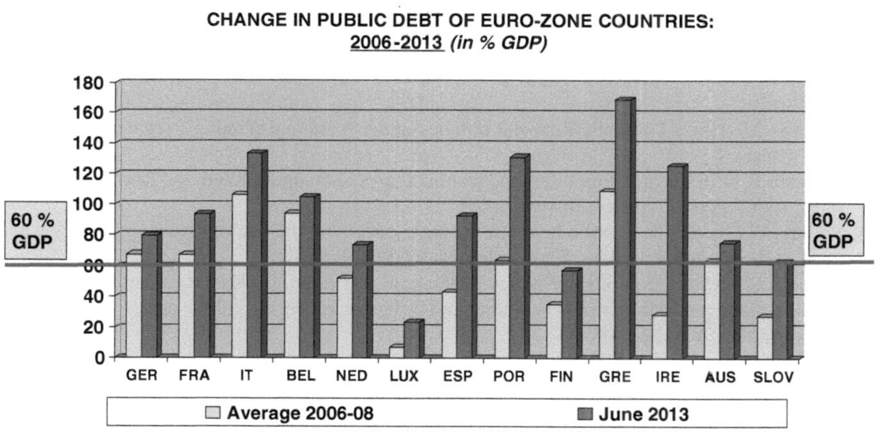

Fig. 1.10 Change in public debt in Eurozone countries. *Source* Eurostat—http://epp.eurostat.ec.europa.eu/cache/ITY_PUBLIC/2-23102013-AP/EN/2-23102013-AP-EN.PDF

continuing up until mid-2013, the public debt of Ireland, Greece and Spain grew enormously. The public debt of the other Eurozone countries, with the exception of Germany and Austria, also rose significantly over this period of just 5 years (Fig. 1.10).

Recourse to a combination of a highly expansionary monetary and equally expansionary fiscal policy had also been practicised in Great Britain during the final 2 years of Gordon Brown's tenure as British Prime Minister and Mervyn King's tenure as Chairman of the Bank of England. The government of Japan applied a similar combination of agressive monetary and fiscal policy measures, known as Abenomics (given that fiscal expansion had been applied over the previous 15 years in any case) during Premier Shinzo Abe's time in government, after December 2012. One of the best examples of this return to authentic Keynesian teaching on the uses of monetary and fiscal policy in the struggle against crisis and unemployment was given in a speech by the former executive director of the IMF, Dominique Strauss Khan. A speech given by Strauss Kahn at the Banco de España on 15 December, 2008, shows in a particularly clear way the major turnaround in IMF policy, traditionally one of the more conservative international financial institutions when it comes to use of expansionary fiscal policy. Strauss Kahn recommended marked monetary expansion and, in particular, marked fiscal expansion for all the developed and major developing economies, particularly those whose public debts had up until that point been at relatively low levels.[66]

References

1. Keynes JM (1936) The general theory of employment, interest and money. Macmillan & Co., London
2. Hicks JR (1937) Mr. Keynes and the classics—a suggested interpretation. Econometrica 5:147–159
3. Fleming M (1962) Domestic financial policies under fixed and floating exchange rates. IMF Staff Papers 9
4. Mundell R (1963) Capital mobility and stabilization policy under fixed and flexible exchange rates. Can J Econ Polit Sci 29(4):475–485
5. Phillips W (1958) The relation between unemployment and the rate of change of money wage rates in the United Kingdom. Economica 25(100):1861–1957
6. Lucas R (1976) An econometric policy evaluation: a critique. In: Karl B, Allan M (eds), The Phillips curve and labor markets. Carnegie-Rochester series on public policy 1. American Elsevier, New York, pp 19–46
7. McCallum B (1989) Monetary economics—theory and policy. Macmillan Publishing Company, London
8. Sargent T, Wallace N (1975) Rational expectations, the optimal monetary instrument, and the optimal money supply rule. J Polit Econ 83(2):241–254
9. Dornbusch R (1976) Expectations and exchange rate dynamics. J Polit Econ 84(6):1161–1176
10. Phelps E, Taylor JB (1977) Stabilizing powers of monetary policy under rational expectations. J Polit Econ 85(1):173–190
11. Fischer S (1977) Long-term contracts, rational expectations, and the optimal money supply rule. J Polit Econ 85(1):191–205
12. Phelps E (ed) (1970) Microeconomic foundations of employment and inflation theory. Macmillan, London
13. Phelps E, Frydman R (eds) (2013) Rethinking expectations: the way forward for macroeconomics. Princeton University Press, Princeton
14. Obstfeld M, Kenneth R (1994) Exchange rate dynamics redux. NBER Working Paper Series. Working Paper No 4693, Cambridge, MA
15. Obstfeld M, Rogoff K (1995) Exchange rate dynamics redux. J Polit Econ 103:624–660

[66] Reference [41]. http://www.imf.org/external/np/speeches/2008/121508.htm.

16. Dixit AK, Stiglitz JE (1977) Monopolistic competition and optimum product diversity. Am Econ Rev 67(3):297–308
17. Mankiw G, Ricardo R (2002) Sticky information versus sticky prices: a proposal to replace the new Keynesian Phillips curve. Qurarterly J Econo 117:1295–1328
18. Bachelier L (1964) The random character of stock market prices. MIT Press, Cambridge
19. Fama EF (1970) Efficient capital markets: a review of theory and empirical work. J Finan 25 (2):383–417
20. Markowitz H (1952) Portfolio selection. J Finan 7(1):77–91
21. Lintner J (1965) The valuation of risk assets and the selection of risky investments in stock portfolios and capital budgets. Rev Econ Stat 47(1):13–37
22. Sharpe WF (1970) The portfolio theory and capital markets. McGraw-Hill, New York
23. Treynor J (1962) Toward a theory of market value of risky assets, The MIT finance seminar
24. Mandelbrot BB, Hudson RL (2004) The (mis)behaviour of markets—a fractal view of risk, ruin, and reward. Profile Books, London
25. Fischer B, Myron S (1973) The pricing of options and corporate liabilities. J Polit Econ 81 (3):637–654
26. Kahneman D (2011) Thinking—fast and slow. Farrar, Strauss and Giroux, New York
27. Daniel K (2013) Misliti, brzo i sporo, Mozaik knjiga, Zagreb
28. Toporowski J (2012) The monetary theory of Kalecki and Minsky. SOAS department of economics working paper series, no. 172. The school of oriental and African studies, London
29. Cavalieri D (2004) On some equilibrium and disequilibrium theories of endogenous money: a structuralist view. Hist Econ Ideas 12(3):49–81
30. Minsky HP (1992) The financial instability hypothesis. Working paper no. 74. The Jerome Levy Economics Institute of Bard College (May)
31. Acharya VV, Cooley TF, Richardson MP, Walter I (eds) (2011) Regulating wall street—the Dodd-Frank act and the new architecture of global finance. Wiley, New Jersey
32. Acharya VV, Kulkarni N, Richardson M (2011) Capital, contingent capital, and liquidity requirements. In: Acharya VV, Cooley TF, Richardson MP, Walter I (eds) Regulating wall street—the Dodd-Frank act and the new architecture of global finance. Wiley, New Jersey, pp 148–149 (Chapter 6)
33. The American Enterprise Institute for Public Policy Research, Remarks by Chairman Alan Greenspan at the Annual Dinner and Francis Boyer Lecture of The American Enterprise Institute for Public Policy Research, Washington, 5 Dec 1996
34. Gordon RJ (1999) Has the 'New Economy' rendered the productivity slowdown obsolete? Northwestern University and NBER, 14 June
35. Smets F (1997) Financial asset prices and monetary policy: theory and evidence. BIS working paper, no. 47, Basel, Sept
36. Davis H, Green D (2010) Banking on the future—the fall and rise of central banking. Princeton University Press, Princeton
37. Krugman P (2008) The return of depression economics and the crisis of 2008. Penguin Books, London
38. Skidelsky R (2010) The return of the master. Penguin Books, London
39. Davidson P (2009) The Keynes solution—the path to global economic prosperity. Palgrave Macmillan, New York
40. Board of Governors of the Federal Reserve System: Financial Accounts of the United States—Flow of Funds, Balance Sheets, and Integrated Macroeconomic Accounts, 25 Sept 2013
41. Strauss-Kahn D (2008) The IMF and its future. Speech at the Banco de Espana, 15 Dec
42. Cauchon D (2008) Why home values may take decades to recover. USA Today, December 12

Chapter 2
The General Theory of Employment, Interest and Money: An Overview with Commentary

Abstract Due to the intellectual influence previously enjoyed by Keynes' *General Theory*, this chapter is dedicated to a commentary on Keynes's great work, whose influence was largely recovered during the current global financial and economic crisis. The author points out the great importance and relevance of the book in solving the problems that occur in the periods of high involuntary unemployment. In this analysis, special attention is paid to Keynes's theory of cost structure based on the distinction between the primary cost (the user cost as its part), and the supplementary cost. In addition, attention is paid to the importance of changes in the user cost for the analysis of inflationary pressures, in the long run. Keynes's theory of capital and investment process, based on the relationship between the marginal efficiency of capital and the interest rate, and his position that the investment process is sufficiently attractive for entrepreneurs as long as the capital is scarce, is presented as one of the most contradictory parts of the General Theory in this book, taking into account that Keynes explicitly recommended expansionary monetary policy in order to keep the interest rate at a very low level when the actual output is close or equal to the potential output. Since Keynes did not write a single chapter about changes in the cost structure in the long run, caused by the changes in technology, the author ends this chapter by integrating Keynes's theory of capital with Horvat's model of economic growth based on the contributions of the sectors producing capital goods and final consumer goods to the average growth rate, and changes in the cost structure in the two sectors in the long run. Thereby the author presents that Keynes's theory of capital scarcity combined with an expansionary monetary policy does not necessarily cause inflationary pressures as long as the marginal productivity in the second sector (production of final consumer goods) increases up to the level necessary to maintain price stability.

Keywords Keynes · General theory · User cost · Capital scarcity · The employment function · Theory of wages · Demand for money · Theory of prices

F. Čaušević, *The Global Crisis of 2008 and Keynes's General Theory*,
SpringerBriefs in Economics, DOI 10.1007/978-3-319-11451-4_2

2.1 The Starting Point for Analysis

Up until halfway through the 1930s, the dominant economic doctrine was that of the neoclassical economists. According to them, supply always creates its own demand—whatever producers produce on the market is realised. A glut in supply is simply followed by a reduction in prices and trade continues unhindered. When supply is less than demand, prices rise and again an equilibrium will be established. In this way, supply and demand are always balanced. Just as there are no problems in realising production, neither is there a problem of unemployment. There is a certain (small) percentage of unemployed persons who do not wish to work at the wage offered, as they do not consider it adequate compensation for the labour to be expended. The neoclassical economists consider such unemployment to be voluntary. There is no such thing as involuntary unemployment.

Actual conditions in the economies of Western Europe and the United States at the end of the third and beginning of the fourth decades of the 20th century differed quite significantly from their interpretation of reality. Hundreds of thousands of workers were seeking job, even at wages lower than being offered on the labour market. The warehouses of the captains of industry were full of unsold finished products. Prices fell, but there was no effective demand. The industrialists let the workers go. There was a need in practical economic life to take radical measures for recovery and in theoretical economics to write books and articles to explain what was happening and what needed to be done to ensure that the economies of the most developed countries could recover.

John Maynard Keynes was a student and then a follower of the ideas of Alfred Marshall for nearly two and a half decades. In analysing the causes of these major disturbances to real economic life, he understood that they could not be explained on the basis of a theory that denied what the major players in economics—people—actually based their behaviour on, whether as producers or consumers. So, Keynes conducted a rigorous examination of the fundamental views of the classical writers, amongst whom he included the followers of David Ricardo, like Marshall, Mill, Edgeworth and Pigou. Keynes concluded that the assumption that only voluntary unemployment existed, or rather that involuntary unemployment did not, did not correspond to reality. Nor was it a realistic assumption that supply automatically creates its own demand on the market. For Keynes, classical economic theory was valid only in a special case—when full employment actually exists. That condition is, however, the exception and not the rule.

Reality is different and rests on the following series of causal relations. By selling products, entrepreneurs earn income. As income rises, so does consumption, but by a smaller percentage than income. The difference is made up by savings. If the savings are not put to work through investment activities, an imbalance arises between supply and demand. The entrepreneurs' expectations regarding future revenues become pessimistic. This is why they do not create new jobs and may indeed get rid of existing ones. Unemployment begins gradually to increase, while demand for products and services falls.

Thus, there is a need to increase aggregate demand. The gap between income and consumption has to be bridged. Keynes found the solution in additional investment, which has to be made more attractive. Investment becomes more attractive or lucrative the lower the interest rate. The interest rate is a function of liquidity preference for money and of the money supply, and not of the supply of and demand for capital. The monetary authorities (central bank) must increase the money supply and, so, reduce interest rates. Then, securities prices can begin to grow and entrepreneurs find themselves in a more favourable situation. Investment becomes more attractive or profitable and expectations of future revenues more favourable. As a result of this growth in investment, unemployment reduces. The propensity to consume must be kept at a higher level and the propensity to save within reasonable boundaries. The government becomes an active player within the economy. Through effective fiscal policy, the government can provide relief for activities in sectors operating under less favourable conditions, while gathering resources from more efficient players. By issuing government securities, the Ministry of Finance collects financial resources, which are greater than those collected on the basis of taxation. One part of these resources so invested contributes to increasing aggregate demand. The government subsidies industrialists working in sectors with lower levels of efficiency, and so increases the marginal efficiency of capital and stimulates investment in those sectors too.

What follows is a very concise introduction to presenting the *General Theory of Employment, Interest and Money*,[1] following the order of the chapters as Keynes laid them out. Accordingly, in the *General Theory*, Keynes analyses the phenomena of economic life that determine total employment and total revenue, which are aggregate quantities for a given economy. The fundamental variables on which Keynes built his theory are: the psychological propensity to consume, the marginal efficiency of capital and the interest rate. None of these three variables can be directly derived from the other two—in this regard they are independent of each other.

The *General Theory* is without doubt one of the greatest works in the history of economic thought. Its great significance resides in the fact that the theoretical answers put forward in the work did actually find practical application. For two and a half decades after the Second World War, Keynes' prescriptions were followed by the governments of the United States and Great Britain. The results of applying the Keynesian approach were evident. By the beginning of the 1970s, however, application of the approach was yielding ever weaker results. As the United States switched to a system of floating exchange rates, against a background of steeply rising oil prices, the short-term focus of the approach became itself an obstacle to effective economic policy in the struggle against inflation. A significant source of

---

[1] The first edition of *The General Theory of Employment, Interest and Money* was published in 1936, in London, by Macmillan & Co. For this commentary, an electronic version available on the ISN ETH Zurich webpage (www.isn.ethz.ch) has been used. The author has also used the most recent version of the Collected Writings of John Maynard Keynes published by the Royal Economic Society [3].

problems in the approach lay in the fact that Keynes himself was contradictory in certain matters, and that he did not give a valid answer as to how to control inflation in the long-run. What are these problems and contradictions? The commentary on the *General Theory* that follows is intended for those who want to engage in more detail with the work, and with both its strong and weak points.

2.2 The Principle of Effective Demand

Effective demand is represented by the point where the aggregate supply curve and the aggregate demand curves meet. The price of aggregate supply is determined by the sum of factor cost, user cost and profit. The factor cost is the amounts which the entrepreneur pays out to the factors of production (labour services and other factors) exclusive of other entrepreneurs products and services. The user cost is the amounts the entrepreneur pays out to other entrepreneurs for what he has to purchase from them together with cost of depreciation. In other words, the user cost is the sum of intermediate goods and depreciation. If the price of aggregate supply is less than the expected demand, then the entrepreneur will have an interest in providing additional employment, up to the point where these two quantities are equal. This point of equilibrium of the supply and the demand, therefore, represents effective demand. The importance of the entrepreneur's expectations is clear from the following quote:

> It follows that in a given situation of technique, resources and factor cost per unit of employment, the amount of employment, both in each individual firm and industry and in the aggregate, depends on the amount of the proceeds which the entrepreneurs expect to receive from the corresponding output.[2]

The following short presentation of Keynes's economic logic may serve as an introduction into the mode of thought presented in the *General Theory*. Given a particular set of techniques, resources and costs, income depends upon the quantity of employment. The income so realised is partly spent, partly saved. What amount is spent depends on the propensity to consume. The amount of employment entrepreneurs decide to create depends on expected expenditure on this investment and on consumption, i.e. on effective demand. Effective demand determines the quantity of employment, while the quantity of employment determines the total income. The level of employment in equilibrium depends on the following three variables: the aggregate supply, the propensity to consume, and on investment. As employment grows, so does the income of the community, and therefore consumption. However, consumption grows at a lower rate than income itself. The growing gap between income and consumption should be covered by growing investment, in order to keep the economy in a state of equilibrium, so that real demand can always be equal to expected demand.

[2] See Ref. [2], p. 20.

2.3 The Definition of Income, Savings and Investment

In Keynes's theory, entrepreneurs face three forms of cost: the factor cost of production, user cost, and supplementary costs. The primary cost is the sum of factor cost and user cost. Factor cost is expenditures paid by the entrepreneur to hired employees, and other factors of production exclusive of what he pays to other entrepreneurs. The entrepreneur's expenses on hired labour represent income for the owners of that labour—the employees. In the same way the entrepreneur's expenses on services of other factors of production (capital, natural resources) represent income for the owners of these factors. The user costs are the sum of the entrepreneur's expenditures for securing resources required from other entrepreneurs, that is the resources to be used up in producing the entrepreneur's product (intermediate products), and the costs of investment in, maintenance of and improving the equipment (depreciation and ongoing maintenance of equipment). Keynes specifies the structure of costs as follows:

> We have defined the user cost as the reduction in the value of the equipment due to using it as compared with not using it, after allowing for the cost of the maintenance and improvements which it would be worth while to undertake and for purchases from other entrepreneurs. It must be arrived at, therefore, by calculating the discounted value of the additional prospective yield which would be obtained at some later date if it were not used now. Now this must be at least equal to the present value of the opportunity to postpone replacement which will result from laying up the equipment; and it may be more.[3]

The entrepreneur's income is the difference between total revenues and primary costs. Supplementary costs are expenditures the entrepreneur has out of its income —that is spending which is not related to regular production, but may arise as a result of excessive use of equipment, careless management, fire, or similar unforeseen circumstances.

Net income is the difference between income and supplementary costs. Aggregate consumption represents the difference between the sum of the value of all sales and the sum of the value of sales between entrepreneurs (the value of sales of intermediate products).

The fundamental psychological law according to Keynes is the relative lag in consumption as income grows:

> The fundamental psychological law, upon which we are entitled to depend with great confidence both a priori from our knowledge of human nature and from the detailed facts of experience, is that men are disposed, as a rule and on the average, to increase their consumption as their income increases, but not by as much as the increase in their income.[4]

For the economy to function properly the difference between the increase in the value of aggregate supply and the slower growth in aggregate consumption must be covered by net investment. Keynes points out the danger of excessive financial

---

[3] John Maynard Keynes, Op. cit., p. 41.

[4] John Maynard Keynes, Op. cit., p. 52.

caution on the part of entrepreneurs, when it comes to increased spending on depreciation and maintenance of equipment. Where that is the case, net investment does not increase efficiently and cannot cover the difference between the value of aggregate supply and the slower growth in consumption, with the result that there is a shortfall in effective demand, and as a result of that a fall in economic activity.

The following quotes are characteristic of Keynes's views on the relationship between investment and consumption:

> New capital-investment can only take place in excess of current capital-disinvestment if *future* expenditure on consumption is expected to increase. Each time we secure today's equilibrium by increased investment we are aggravating the difficulty of securing equilibrium tomorrow. A diminished propensity to consume today can only be accommodated to the public advantage if an increased propensity to consume is expected to exist some day.[5]

So, one cannot increase consumption without investment—in the absence of an increase in aggregate consumption, a fall in economic activity will follow, that is to say there will be a recessionary trend. Consumption increases, as real wages increase, and for real wages to increase, one must reject the assumption of the unchangeable (given) nature of technology and equipment. Investment should be directed towards new technology, to improve productivity (increase in the marginal product of labour). Still, Keynes did not build into his theory the effects of technical or technological progress, even though he did later assume them implicitly. As a result, the dynamic of the economic system is eliminated from his analysis, or, to be more precise, Keynes presented an analysis of the potential for arriving at full capacity utilization, that is of increasing production to the level of the potential output, but did not incorporate into his theory factors of economic growth that lead to a shifting of the long-run aggregate supply curve as a consequence of technical progress or improvements.

Analysing factors affecting consumption, Keynes listed the following incentives amongst the subjective factors of consumption: precaution, foresight, calculation, improvement, independence, enterprise, pride, and avarice. These factors rarely change much over the short term and are of no particular importance for short-term analysis. In addition to these incentives for the individual, there are also incentives to government institutions. Keynes lists amongst such motives the following four ones: the motive of enterprise, the motive of liquidity, the motive of improvement, and the motive of financial prudence. The motive of financial prudence is particularly interesting in terms of introducing (for the first time in his book) the term of technical change. Keynes' definition of the motive is following:

> The motive of financial prudence and the anxiety to be 'on the right side' by making a financial provision in excess of user and supplementary cost, so as to discharge debt and write off the cost of wastage and obsolescence, the strength of this motive mainly depending on the quantity and character of the capital equipment and the rate of technical change.[6]

---

[5] John Maynard Keynes, Op. cit., p. 56.

[6] John Maynard Keynes, Op. cit., p. 58.

2.4 The Marginal Propensity to Consume and the Multiplier

The marginal propensity to consume represents the relationship between that part of the increase in income which is consumed and the increase as a whole. Consequently, the value of the marginal propensity to consume may be between zero and unity. According to Keynes, it is a fundamental psychological law that as income increases the marginal propensity to consume falls, and vice versa, as income decreases, the marginal propensity rises. The marginal propensity to consume tells us, accordingly, by what ratio a given increase in income will be split between consumption and savings or investment. We may represent the increase in income using the following expression:

$$\mathbf{Y_w = k \times I_w},$$

where (k) is the investment multiplier, a variable that depends on the marginal propensity to consume (dC/dY = 1 − (1/k)). Yw and Iw are income and investment expressed in the wage unit (w). There is therefore a direct positive relationship between the marginal propensity to consume and the investment multiplier—as the marginal propensity to consume increases, so does the investment multiplier. The converse is also true, naturally. The investment multiplier represents the relationship between the change in income and the change in investment. In other words, the investment multiplier shows by how much the national income will change, if investment changes to the tune of a single monetary unit.

Keynes provisionally equated his investment multiplier with Richard Kahn's employment multiplier. Kahn's employment multiplier represents the relationship between the change in total employment and the change in employment in industries which produce capital equipment [capital goods]:

$$\mathbf{\Delta N = k \times \Delta N2}$$

Insofar as the multiplier is positively correlated with changes in the marginal propensity to consume, it follows that its value is determined by psychological changes, like consumption itself. In an economy at a higher degree of development, which presupposes employment at a level close to full employment, it therefore follows that a reduction in the marginal propensity to consume will bring about significantly higher levels of investment, but relatively little change in employment. This is understandable, insofar as investment in such economies is in highly productive technologies which require a higher degree of capital outfitting for labour and greater financial resources than investment in developing countries.

2.5 The Marginal Efficiency of Capital

The marginal efficiency of capital depends on expectations of future incomes and costs, as well as on present interest rates, which affect the calculation of the present value of the investment. Keynes describes the high degree of uncertainty in the process of investment as follows:

> ***When a man buys an investment or capital-asset, he purchases*** the right to the series of ***prospective returns, which he expects to obtain from selling its output***, after deducting the running expenses of obtaining that output, during the life of the asset. This series of annuities $Q1$, $Q2$, … Qn it is convenient to call the *prospective yield* of the investment.[7]

In the above quotation, the words "when… he purchases" and "prospective returns which he expects to obtain from selling its output" have been put in bold in order to stress the fundamental problem of investing implicit to Keynes' approach. Thus, when a given individual or firm buys specific equipment, resources have been quite definitely invested, and whether that investment of resources, which is now a done deal, will pay off depends on the prospective, but nonetheless uncertain revenues for a certain number of future periods.

The marginal efficiency of capital represents the relationship between the prospective yield brought by investment in any additional unit of capital equipment and the replacement cost of that equipment. Keynes stressed that his definition of marginal efficiency of capital was identical to 'the rate of return over costs'—a variable introduced into analysis by Professor Irving Fisher in his book *The Theory of Interest*. One can determine for each form of equipment a lower limit of efficiency, below which investing in it does not pay off. This limit determines demand for investment in that type of capital equipment. In the same way, the demand functions can be determined for other forms of capital equipment. One may use the sum of these individual functions to determine the general function of the marginal efficiency of capital, the function of demand for investment.

The lower limit of the efficiency of investment is determined by the interest rate. As long as the marginal efficiency of capital is greater than the interest rate, it pays off to invest. In other words, there is a positive differential between the yield from investment and the average interest rate, which represents the stimulus to entrepreneurs to take on additional risk and accept insecurity or uncertainty with regard to future income from investments. Once the marginal efficiency of investment is equal to the interest rate, additional exposure to risks and uncertainties loses any economic sense and there is no longer any incentive to take on additional risk over and above the incentive for relatively secure income in the form of interest on deposited savings.

The following two forms of risk have a major impact on the level of investment:

1. The first is investor or entrepreneurial risk with regard to the likelihood of actually realising the expected profit. The entrepreneur is most often a borrower,

[7] John Maynard Keynes, Op. cit., p. 69.

excluding those cases of ability to finance investment from own resources or share capital.
2. The second is creditor risk and it relates to the possibility that the debtor may not be able to settle the debt on deadline and so return the borrowed funds and associated interest, whether for subjective or objective reasons.[8]

Both these forms of risk are calculated on investing. Firstly, they are built into the level of the interest rate on any loan approvals, while secondly they are built into the price being offered by the investor.

In addition to these two forms of risk, Keynes singles out a third, which is related to the possibility of a change in the monetary regime which may affect the expectations of either the entrepreneur-borrower or the financial institution-creditor.

2.6 The State of Long-Term Expectations

The marginal efficiency of capital and the interest rate are what determine the scale of investment. The marginal efficiency of capital depends on the relationship between projected revenues or income and the supply price and the type of capital, i.e. replacement costs. The projected revenues depend on investor expectations, and expectations are in turn a function of existing facts, which are more or less certain, and of future events which can be forecast with a greater or lesser degree of certainty. Keynes included amongst existing facts the existing quantity of all capital goods, capital equipment taken as a whole and the strength of consumer demand for products whose production entails the possession of relatively expensive capital equipment.

Under future uncertain events, Keynes meant changes in the type and quantity of capital goods, changes in consumer taste, changes in effective demand during the lifetime of the investment and changes in the wage unit. This constellation of psychological expectations Keynes named jointly the state of long-term expectations.

The scale of investment can grow, even at unchanging interest rates, if the price of long-term securities (shares and bonds) itself grows on the capital markets for particular branches of industry. Keynes was entirely aware of the major role which well-organised capital markets had in the investment cycle in developed economies, even at the time of the writing of the *General Theory*. Keynes, however, stressed that important problems existed on capital markets, which could render investment in significant social projects ineffectual. Keynes included the following under this form of problem[9]:

- The significant percentage of total invested capital belong to shareholders who do not dispose of sufficient information and are not well informed regarding the prospects of their future expectations.

[8] John Maynard Keynes, Op. cit., p. 73.
[9] John Maynard Keynes, Op. cit., p. 78.

- Seasonal fluctuations in the profits of companies whose secondary activities may result in a significant part of capital being invested in securities.
- The formation of a herd behaviour under the influence of large numbers of uninformed individual financial investors, leading to major fluctuations in opinion related to factors which have no significant impact in reality on the investment.
- The very significant relationship between speculators and entrepreneurs. Speculators specialise in studying herd behaviour, with a view to achieving rapid and large-scale profits. A predominance of speculation on capital markets may lead to major changes in securities prices and to a crowding out of investment in significant social projects. Entrepreneurs are oriented towards viewing utility over the long-term—so long as entrepreneurial motives dominate the markets, investment markets can be a significant factor for stabilisation of the economy and of investment activities.
- The degree of confidence of financial institutions approving funds in those whose funding they are approving.

Keynes was personally involved in financial transactions and understood the functioning of capital markets very well at two levels: he was himself an investor of his own funds in financial assets (structuring his own portfolio), on the one hand, while also being the Chariman of the National Mutual Life Assurance Society (1921–1938), in which capacity he was involved in decision-making regarding the structuring of the portfolio of financial assets of that institution. Consequently Keynes's understanding of the impact of psychological factors on the behaviour of financial investors and financial markets played a major role in how he analysed the impact of these markets on investment activities in the real sector, as well is having a major impact on his theory of demand for money, in which he gave great importance to speculative demand for money.

2.7 Keynes' General Theory of the Interest Rate

Depending on the propensity to consume, an individual spends part of their income on purchasing goods and services. The remainder may be invested in the bank (a savings account) or in securities (financial assets), or held in the form of cash (cash money as asset). What part of the remaining income will be kept in the form of cash depends on the degree of insecurity over the level of future interest rates. Thus, insecurity over the future interest rate is a necessary condition for a propensity to hold cash to arise. Keynes called this propensity the liquidity preference. The liquidity preference and the quantity of money remaining on satisfaction of the propensity to consume determine the level of the interest rate. The interest rate is not the price which equalises supply and demand for capital, but is rather a price derived from the supply of and demand for money. Keynes did not accept the definition of interest as the reward for refraining from consumption. According to

Keynes, the interest rate is the reward for refraining from liquidity, as the simple act of putting off consumption does not necessarily mean depositing money in the bank. An individual may save by retaining cash in their own possession, i.e. by keeping cash as a form of asset and therefore forgoing the interest rate (the interest rate is an opportunity cost for holding money as an asset).

The phenomenon of liquidity preference is psychological in nature. Keynes considered the liquidity preference an initial independent variable. Still, insofar as the liquidity preference comes about as a result of uncertainty over future interest rates, which are at least in part derived from current interest rates, it is only in cases where current interest rates coincide with past expectations that the liquidity preference can be considered an entirely independent variable of the system. If present interest rates differ significantly from previous expectations, the liquidity preference ceases to be an independent variable, as a certain number of other factors now become implicated in the causal chain, including, for example, measures taken by the monetary authorities, which may be pursuing a different policy than previous authorities.

> Liquidity preference is a potentiality or functional tendency, which fixes the quantity of money which the public will hold when the rate of interest is given; so that if r is the rate of interest, M the quantity of money and L the function of liquidity-preference, we have $M = L(r)$. This is where, and how, the quantity of money enters into the economic scheme.[10]

Keynes defines three types of need as having a crucial impact on the liquidity preference. These are the following three motives:

- The motive for exchange,
- The motive for security, and
- The motive for speculation.[11]

The quantities of money required for exchange (buying and selling goods and services) and security do not depend to any great degree on the interest rate. The liquidity preference, however, prompted by a speculative motive, can be a significant source of change in the interest rates. If the interest rate equalises the demand and the supply of money at the quantity reached once the motives for exchange and security have been met, then an equilibrium state is established. If an excessive motive for holding cash for speculative reasons arises, then there will be instability of the interest rates. This is why an increase in the quantity of money will affect growth in bond prices above all expectation, to the point at which further growth is not expected, and a tendency for their prices to fall will then appear. This fall in bond prices will bring supply and demand of money, which is to say the interest rate, back into balance.

---

[10] John Maynard Keynes, Op. cit., p. 84.

[11] John Maynard Keynes, Op. cit., p. 85.

2.8 The Classical Theory of the Interest Rate

Keynes agreed with the classical economists regarding the need for savings to equal investment. On the other hand, he did not accept the classical theory of the interest rate, that is the definition of the interest rate as the price which brings the supply of and demand for capital, or savings and investment, into equilibrium. The classical economists started from a given quantity of income, assuming which, they then analysed various relations of saving and investment. According to them, the interest rate grows, when demand for capital, that is investment grows. If investment is growing and investments are equal to savings, this means that savings must also be growing. Still, changes in savings and in investment entail a change in income—one cannot have changes in investment and saving at a set level of income. For Keynes, the classical economists overlooked this change in income because they were analysing relations in an economy enjoying full employment. Since a change in income is possible only on the basis of a change in employment, which the classical economists had excluded from their analysis, classical analysis was really reduced to just one special case of the functioning of a given economy.

Demand for investment is not determined by the interest rate, but by the relationship of the supply and demand for capital goods: their price. On the basis of the price for capital goods and the expected income from investment, one may arrive at the marginal efficiency of capital, and from the marginal efficiency and the interest rate at the scale of investment. The root of the mistake in the classical theory of the interest rate lies in supposing that interest is a reward for waiting (saving or deferring consumption) rather than for abstaining from liquidity.

2.9 Psychological and Business Incentives to Liquidity

Demand for money appears as a result of four forms of motive: the income motive, the business motive, the precautionary motive and the speculative motive.

- The income motive appears as a result of the need to “bridge the interval between the receipt of income and its disbursement.”
- The business motive appears as a result of the interval between paying for intermediary products ordered and work done, on the one hand, and charging for one’s own products and services sold, on the other (the time difference between income and outgoings of cash).
- The precautionary motive (holding cash for security) arises as a consequence of the possibility of unforeseen expenditures.
- The speculative motive is very important in Keynes’s system, because of the transmission of the effects of changes in the quantity of money in circulation,

which is the same as to say the effectiveness of the monetary policy transmission mechanism through financial markets onto the real sector.[12]

How homogenous securities markets are in predicting future interest rate movements is very important. If overall expectations are relatively homogenous, changes in the interest rate will have no very significant impact on changes in the quantity of money. If, however, expectations of the interest rate differ, i.e. are not homogenous, then bond and share prices can be subject to very intensive change. Accordingly, there will be a greater volume of financial transaction, which can cause the quantity of money and the interest rate to change.

In essence, Keynes distinguishes two categories of demand for money: transaction and speculative. The transaction demand for money depends on the level of income which is intended to meet our needs for exchange and security. The speculative demand for money depends on the interest rate and expectations. When additional money is issued, the amount of money in circulation increases and that increase will accrue to somebody as income, so that it follows that there will be an increase in income. Increased income will in part be allocated to the transaction demand and in part to satisfying the speculative motive. Part of the money allocated for meeting the speculative motive will push bond prices up, which will in turn lead to an increase in the volume of trade in financial assets. This increased volume of trade on capital markets will cause the interest rate to fall. As a result of this fall in interest rate, investment will rise and so will income.[13]

According to Keynes, it is of great importance for any economy what part of income is designated to meet speculative desires (the speculative motive), precisely because that quantity of money will play a decisive role in forming the interest rate. Changes in securities prices determine the volume of trade on capital markets and the level of the interest rate. Mass psychology is of great importance on such markets, as it determines expectations about future interest rates. If expectations conflict with monetary policy measures, the result will be major and undesirable changes in the interest rate and the quantity of money, which will influence the investment activity, the level of employment and income levels. If the policy of monetary authorities is in line with expectations formed on the securities markets, any interest-rate level will clear the market (represent an equilibrium). In the opposite case, there will be very undesirable and destabilising movements.

2.10 Sundry Observations on the Nature of Capital

In this chapter, Keynes tries to complete his theory of capital, or rather the effectiveness of capital over the long-term. *The basic assumption is that capital has to be sufficiently scarce to be effective over the longer term.* This is clear from the following quote:

---

[12] John Maynard Keynes, Op. cit., pp. 97–98.

[13] John Maynard Keynes, Op. cit., pp. 99–100.

> For the only reason why an asset offers a prospect of yielding during its life services having an aggregate value greater than its initial supply price is because it is *scarce*; and it is scarce because of the competition of the rate of interest on money. If capital becomes less scarce, the excess yield will diminish, without it having become less productive—at least in the physical sense.[14]

From this quotation, it may be concluded that capital is sufficiently scarce if its marginal efficiency is greater than the interest rate, since this means that the overall investment has still to reach its maximisation level, defined as equality of the marginal efficiency of capital and the interest rate. So long as the marginal efficiency of capital is greater than the interest rate, in Keynes's framework, i.e. without any analysis of the impact of international capital flows, investment will remain less than savings. Investments are equal to savings when the marginal efficiency of capital is equal to the interest rate.

The basic problem for investing lies in uncertainty over future effective demand, which gives rise to uncertainty over the rate of return on resources invested. Were saving to entail consumption on a specific day in the future, investment would become so much the more efficient, insofar as it would be possible to calculate effective demand and therefore the marginal efficiency of capital fairly exactly:

> In optimum conditions, that is to say, production should be so organised as to produce in the most efficient manner compatible with delivery at the dates at which consumers' demand is expected to become effective. It is no use to produce for delivery at a different date from this….[15]

Meanwhile, given that the simple act of saving does not mean consumption on a given day in the future, or even that those resources will ever be spent or that the owner will ever convert them into cash, investment depends to a very large degree on the condition of expectations.

In his theory of the multiplier, Keynes assumes that an initial increase in employment will come about in the sector manufacturing investment goods and that only later will a more rapid increase in employment in the manufacturing sector producing for consumption follow. For capital to be efficient, it must also be relatively scarce. In other words, it must represent a reward to entrepreneurs for taking additional risks and, in particular, for the uncertainty which is so important a characteristic of investment. The longer the production process, the more favourably it will augment demand for its products to a sufficient level for their price to grow and render investment efficient:

> … And conditions of equilibrium require that articles produced in less agreeable attendant circumstances (characterised by smelliness, risk or the lapse of time) must be kept sufficiently scarce to command a higher price. But if the lapse of time becomes an agreeable

14 John Maynard Keynes, Op. cit., p. 106.

15 John Maynard Keynes, Op. cit., p. 107.

> attendant circumstance, which is a quite possible case and already holds for many individuals, then, as I have said above, it is the short processes which must be kept sufficiently scarce.[16]

Accordingly, Keynes connects the idea of sufficient scarcity with higher prices or reduced supply of specific types of product, in comparison to demand for them. Additional employment is created on the basis of knocking down wages in the sector producing investment or capital goods. Production processes in the sector are sufficiently lengthy to meet the conditions Keynes considers key to stimulating the investment cycle. Because of their lengthiness and sufficient scarcity, the prices for these products should provide sufficient incentive (be sufficiently high) to increase both profits and employment, at least on the basis of lower real wages, as Keynes advocates.

If the growth in employment is to continue, after this initial increase prompted by investment in the production of capital goods, the investment cycle must transfer to the production of goods for mass consumption. Since these production processes are relatively short, it follows that they too must be made sufficiently scarce for the price of the products to be high enough to act as an incentive for investors ready to invest in producing them. This part of Keynes's theory about the nature of capital and how it should stimulate additional employment is, at least to some extent, contradictory and at the root of the unresolved problem in the *General Theory* of how to rein in accelerating inflation while keeping capital relatively scarce and translating employment growth induced by falling real wages into employment growth with higher real wages.

The theory of the scarcity of capital is in conflict with the theory of employment, and in particular with Keynes's assumption that higher employment with falling real wages can over time be replaced with higher employment with higher real wages. This leaving to one side of the issue of the structure of costs and how employment at lower real wages is to be magically transformed into employment with growing real wages results in an unavoidable tendency for increased inflationary pressure, as the economy approaches full capacity utilization.

2.11 The Essential Properties of Interest and Money

The interest rate is one of the key economic variables and plays a very major role in the economic system of any country, whether it has a large or small economy. By comparing the interest rate with the marginal efficiency of capital, we may determine the scale or volume of investment and so the level of employment and income. Keynes emphasises that it is possible to express the interest rate for any good in terms of itself. The interest rate is calculated by taking a money amount equivalent to the price of the good, putting it on deposit for a set period, and comparing it to

---

16 John Maynard Keynes, Op. cit., p. 107.

the future price of the good at the end of the same period. By multiplying this ratio by 100, we get a number that tells us what quantity of that good we would be able to buy after the set period with the resources we have at our disposal. If the number is less than 100, the interest rate on the good expressed in terms of itself is negative. On the other hand, if the number is larger than 100, the interest rate in terms of the good itself is positive. A good has a nil interest rate in terms of itself, according to this way of going about things, when the ratio of the two quantities is equal to the number 100.

Different assets entail utility and expense in different ratios, which depend on the following variables, as defined by Keynes:

1. Some assets provide a yield to their owner in the form of an additional product or some form of service—we may designate this return (q).
2. Almost all assets, other than money, entail carrying costs. Keynes designates these costs (c).
3. Different assets have different liquidity praemia. The liquidity premium (l) is the possibility that, at any given moment, one can use this asset to purchase some other asset. By definition, money is the asset with the highest liquidity premium, insofar as it is the universal medium of exchange and standard of value.[17]

The total utility of a given asset is determined by the sum of the yield and the liquidity premium minus the carrying costs ($q + l - c$). The liquidity premium is negligible, for most goods, compared to the carrying costs. Thus, it is the relationship between the return or yield and the carrying costs that determines the efficiency of many, if not most assets. So long as the returns are sufficiently high to cover the carrying costs, with an additional sum left over that is at least equal to the rate of interest on the cash deposit, then investment in the asset pays off. The interest rate on cash deposits is determined by the liquidity premium it secures. The characteristic features that give money its role as the universal medium of exchange and standard of value are as follows:

1. From the perspective of private entrepreneurs, money has zero elasticity of production. Elasticity of production is the term used for how much additional labour has to be invested in producing an additional unit, given an increase in the amount of labour commanded by it. This condition means that entrepreneurs cannot produce money and change its quantity.
2. The second feature that money has is that the elasticity of substitution of money by some other asset is or near to zero. This feature means as the exchange value of money rises there is no tendency to substitute another good for it, and it stems directly from money's role as a means of exchange and standard of value.
3. The determining characteristic of money, which is what gives it the role of universal means of exchange, is its high liquidity premium, the highest of all other goods, and it's very low, almost negligible, carrying costs.[18]

[17] John Maynard Keynes, Op. cit., pp. 112–113.

[18] John Maynard Keynes, Op. cit., pp. 115–116.

Determining the value of assets in any other asset than money would result in higher carrying costs, which in combination with the liquidity premium, would eat into the asset's value. Consequently the value of assets expressed in that asset would fall. The zero elasticities of production and substitution, like the high liquidity premium, are what secure money the privilege of being the measure or standard in which debts and wages are agreed, in other words what secure its stable value. It is because of these characteristics of money that the interest rate on money represents the fundamental measure of the efficiency of other assets.

It is, moreover, an important characteristic of money that its interest rate declines considerably more slowly than for other goods. In producing other forms of asset, entrepreneurs render those assets more abundant and therefore less valuable, so that the interest rate of those goods expressed in terms of themselves declines relatively fast, which is not the case of money, thanks to the zero elasticity of production of money for the entrepreneur. It is precisely the entrepreneurs' inability to produce money that makes money a sufficiently scarce good, allowing it to maintain a relatively stable value over the longer term.

2.12 The Underlying Logical Framework of the General Theory

In the 18th chapter of the General Theory, Keynes reviews once more the basic logic of his system of analysis and views on how an economy functions, or should do. Keynes explains the basic assumptions of the analysis as follows:

> We take as given the existing skill and quantity of available labour, the existing quality and quantity of available equipment, the existing techniques, the degree of competition, the tastes and habits of the consumer, the disutility of different intensities of labour and the activities of supervision and organisation, as well as the social structure, including the forces, other than variables set forth below, which determine the distribution of the national income.[19]

Keynes takes these factors as given, that is as unchanging, but only in the short run. In fact, the short-run is basically defined as the period in which these factors do not change. The following quantities in Keynes's system are exogenously determined:

1. Three variables which are psychological in character: the propensity to consume, the liquidity preference, and the state of expectations regarding yield on assets.
2. The level of the wage unit, as determined by negotiations between trade unions and employers.

[19] John Maynard Keynes, Op. cit., p. 122.

3. The quantity of money, as determined by the action of the central bank.[20]

Keynes offers the following as the independent variables of the model:

- The propensity to consume,
- The marginal efficiency of capital, and
- The interest rate.

There is no interdependence between these variables, insofar as none of them can be directly derived from either of the others, alone or in combination. The marginal efficiency of capital and the interest rates depend on psychological factors, like the state of expectations about yield on assets (the marginal efficiency of capital) and liquidity preference (interest rates).

The dependent variables and the Keynes' model are:

- The level of employment, and
- The output (gross domestic product).

Keynes expresses both these dependent variables in the wage-units, given that he built his theory of value on the foundations of the labour theory of value.

2.13 Changes in Money-Wages

In the 19th chapter of the General Theory, Keynes provides a critique of the analysis of the impact of changes in money wages on unemployment as applied by the classical economists. Their analysis had rested on the assumption that reducing money wages would increase demand by cutting prices, so that growth in the real wage could then bring about increased employment, due to a higher level of aggregate demand. The classical economists' analysis involved three phases. In the first phase, they looked to the quantities which can be placed on the market at a given price. Then they connect various quantities with various prices, and then on the basis of these demand curves, they deduce a labour demand curve, on the assumption that none of the other costs change (except those related to changing the output). In this way, the classical economists determined or deduced the demand function for labour at the level of an industry (branch of industry), which they then mechanically transferred to the economy as a whole.[21]

It was precisely this equation of a partial equilibrium, or rather transferring this logic for an individual branch of industry onto the economy as a whole, without any significant changes or analysis of the interaction between branches of industry and sectors that Keynes considered mistaken as an approach and an unacceptable way of analysing the general equilibrium. The second important shortcoming of the classical analysis of the impact of changes to wages was that the approach assumed

[20] John Maynard Keynes, Op. cit., p. 123.

[21] John Maynard Keynes, Op. cit., p. 127.

the overall supply of labour was equal to demand for it, from which it follows that there is no such thing as involuntary unemployment. Third, and for Keynes the most illogical position in classical economic analysis, was that every, even a very small, reduction in the real wage leads to a reduction in the supply of labour.

In his own analysis of changing wages and their impact on the relationship between the supply of labour and demand for it, Keynes started from the following two questions:

1. Will reducing the wage result in an increase of employment, so long as the marginal efficiency of capital, the propensity to consume and the interest rate remain unchanged?
2. Does reducing the wage have a probable impact on employment through a mediated influence on these three fundamental variables of Keynes's system?

Cutting wages has an impact on reducing employment, and so on changing expectations regarding the future level of aggregate demand, and therefore an impact on changing the marginal efficiency of capital, the propensity to consume, and the interest rate. Thus, if these variables are unchanged, no additional employment can come about, as future income would be less than the price of supply. The mediate impact of cutting wages on the propensity to consume, the marginal efficiency of capital and the interest rate is manifest in the following way[22]:

- A fall in wages will, to some degree, cut prices, which will mean the reallocation of actual income away from workers to other factors entering into marginal prime cost and away from entrepreneurs to rentiers. One result of this reallocation will be a reduction in the propensity to consume, which will have negative repercussions on the future level of employment.
- Insofar as the analysis is being conducted of an open economy, then cutting wages represents a reduction in comparison to abroad, which will be positive impulse for investment, because of the lower price of domestic goods abroad. This effect of reducing wages should be positive when it comes to increasing employment.
- While cutting wages may have a positive impact on the balance of payments, it can at the same time effect a worsening of the terms of trade. Worsened terms of trade will depress real wages, which will affect any increase in the propensity to consume.
- If the reduction in wages is temporary in character, that is if wages are expected to grow again soon and as a result demand to increase, there may be an increase in employment as a result of the expected growth in the marginal efficiency of capital, derived from the expected immediate increase in wages. If, however, even lower wages are expected in future, then such expectations will lead to a fall in the marginal efficiency of capital, a fall in investment and a consequential fall in employment.

---

[22] John Maynard Keynes, Op. cit., pp. 129–131.

- If the cut in wages is accompanied by a general reduction in the level of prices, this will have a positive effect on reducing the liquidity preference, which may result in lower interest rates. A fall in the interest rate would increase the marginal efficiency of capital and create the conditions for a more favourable investment climate, in other words increase investment. If this is a short-term phenomenon, the effect will be so much the less because of the expectation that interest rates will rise again soon, as a result of increasing prices and liquidity preference.
- A general reduction in wages may give rise to favourable expectations on the part of entrepreneurs about the future and in that way increase the marginal efficiency of capital. If, however, optimistic expectations of increasing wages in the near future are also prevalent amongst workers, then the optimistic views of the entrepreneurs may easily change into pessimistic ones and on that basis cause a reduction in the marginal efficiency of capital, investment and employment.
- A fall in wages will not provoke more favourable expectations in entrepreneurs who are heavily in debt, because of the expectation that it will lead to a fall in consumption and real wages, and consequently an inability to create their product.

Keynes then explained the shortcomings of the policy of flexible wages, which may be interpreted in brief as follows:

- There is no way to bring about universal change of wages for all employees and simply reduce wages for everyone by the same percent after a long period of time, which would have a series of negative effects on the marginal efficiency of capital.
- Because additional employment produces a fall in the marginal productivity of labour, the real wage has to be cut, not by cutting nominal wages, but by increasing the price of final goods at unchanged wages, which would maintain some sort of fairness in allocation between workers and factors whose income is determined on the basis of an unchanged amount.
- Increasing the quantity of money by reducing wages, increases the debt burden on entrepreneurs.
- If a fall in the interest rate is achieved by increasing the quantity of money and reducing wages (reducing the wage-unit), it results in negative trends regarding expectations of future effective demand and consequently a reduction in the marginal efficiency of capital.

On the basis of the foregoing interpretation of Keynes's views, it is clear that Keynes advocated a stable money wage (nominal wage), which would facilitate conducting a stable price policy. This view of Keynes is entirely logical for analysis of the short run, but it is contradictory in the long run, given that he never gave any answer as to how one might limit the workers' expectations or demands for higher wages, insofar as he advocated cutting real wages, as a consequence of preferring modest growth in the prices of final products in the hope of growth of the profit potential and investment. That is, given increased prices for final products and unchanged wages, the real wage would be reduced and in this way the real cost of labour and so marginal cost reduced.

Keynes thus claims in this chapter that price stability may be achieved by a policy of unchanged money wages, so that over the short term prices change only in a dependent relationship to employment, which is in contradiction with his view from Chapter 6 that the short-term price is formed out of marginal prime costs, from which one should not exclude the user cost. This is evident from the following quotation:

> Apart from "administrative" or monopoly prices, the price level will only change in the short period in response to the extent that changes in the volume of employment affect marginal prime costs; whilst in the long period they will only change in response to changes in the cost of production due to new techniques and new or increased equipment.[23]

2.14 The Employment Function

The employment function represents the relationship between the number in employment and effective demand measured by the wage-unit. By definition, this function represents an inverse form of the total supply function. The employment function for a given branch of industry is derived from two sets of curves:

- The supply function for the branch, and
- The supply function for the industry as a whole.

Let us suppose that all the other factors of the system are unchanged, but that effective demand changes as a result of increased investment. At any given level of effective demand, expressed in the wage-unit, there will be a corresponding level of employment for the industry as a whole, and, depending on how effective demand is distributed, there will be a certain level of employment for each industrial branch. Individual employment functions can be summed, because they are expressed in the same unit—the wage-unit. If we let **Nr** stand for the number employed in a branch of industry and **Dw** for effective demand in the industry as a whole, then **Nr** = **Fr** (**Dw**), while the total number of employed is equal:

$$\mathbf{N} = \sum \mathbf{Nr} = \sum \mathbf{Fr}(\mathbf{Dw})$$

Keynes defined three forms of elasticity with regard to effective demand, expressed in the wage-unit:

1. *The elasticity of employment* $\mathbf{e_{er}} = (\mathbf{dNr}/\mathbf{dWr}) \times (\mathbf{Dwr}/\mathbf{Nr})$ This elasticity shows the change in employment in a branch of industry resulting from the change in effective demand, expressed in wage-units, intended for purchasing a product from that branch of industry.

23 John Maynard Keynes, Op. cit., p. 134.

2. *The elasticity of output* $\mathbf{e_{or}} = (\mathbf{dOr/dDwr}) \times (\mathbf{Dwr/Or})$ The elasticity of output shows the percentage by which output will change if effective demand, expressed in wage-units, changes by 1 %.
3. *The elasticity of expected prices* $\mathbf{e_{pr}} = (\mathbf{dPwr/dDwr}) \times (\mathbf{Dwr/Pwr})$ The elasticity of expected prices shows by what percent prices of the product of a given industrial branch will change if effective demand in that branch changes by 1 %.[24]

Because the effective demand of a branch of industry is equal to a multiple of that branch's total output (**Or**) and the price, expressed in wage-units, (**Pwr**), it follows that:

$$\mathbf{OrPwr} = \mathbf{Dwr},$$

and from this relation, it follows:

$$(\mathbf{dOr/dDwr}) \times (\mathbf{Dwr/Or}) + (\mathbf{dPwr/dDwr}) \times (\mathbf{Dwr/Pwr}) = 1, \quad \text{that is}$$

$$\mathbf{e_{or}} + \mathbf{e_{pr}} = 1.$$

Thus, according to this law, an increase in effective demand is partially induced by an increase in the price and partially by an increase in output. The same holds true for the whole industry. Were we now to express the wage and the price of a unit of production in money terms, so that W is the money wage for a unit of labour, while (p) is the unit price of a product expressed in money, then we might designate with *ep* and *ew* the elasticities of money prices and the elasticities of money wages' independence from changes in effective demand, expressed, also, in money (D). From the foregoing relations, Keynes deduces the following relation[25]:

$$\mathbf{e_{pr}} = 1 - \mathbf{eo}(1 - \mathbf{ew}).$$

It is precisely from this relation that we may deduce Keynes's weakness in explaining how to transition from additional employment with a falling real wage to additional employment with a rising real wage. For the real wage to grow, *ew* has to be greater than *ep*, and, since by definition for Keynes $ep + eo = 1$, we may quite easily state that there is no value for which *ew* is greater than *ep* that can satisfy both equations.

If we reject the idea that there is only one way of allocating effective demand to industrial branches and accept the realistic assumption of a larger number of potential distributions of aggregate effective demand, then we may conclude that total employment does not depend only on the amount of total effective demand, but on how it is allocated. One consequence of this is situations in which different levels of employment are possible for the same level of effective demand. For

[24] John Maynard Keynes, Op. cit., pp. 140–142.
[25] John Maynard Keynes, Op. cit., p. 143.

example, where the lion's share of demand in the allocation is directed towards a branch with a lower elasticity of employment, then overall employment will be lower. Where a greater share of aggregate effective demand is directed towards a branch with a greater elasticity of employment, then total employment will be greater for the same quantity of effective demand.

If demand is directed towards products with a lengthier or more roundabout period of production, the number of additional people employed will be less, and so the elasticity of employment will be less. Keynes defined the period of production as the number of time units required for an announcement of a change in demand to produce the greatest elasticity of employment. Consumer goods have the longest period of production, determined in this way.

So, if effective demand rises as a result of higher investment, the elasticity of employment will be greater and more workers will be employed over the short term because of the shorter period of production. If this increase in demand is the result of higher consumption, however, it will in large part be absorbed by higher price, and to a lesser extent by an increase in the number of people employed. This means that a longer time period is required for the elasticity of employment to achieve a value close to one.

In other words, Keynes was suggesting in this way that the fastest way out of a depression in the short term, which was the primary motivation for writing the General Theory, will be investment in capital goods, that is investment in infrastructure and the production of goods for producing mass consumption goods. These investments have a larger multiplier of employment because of the shorter period of production and, particularly importantly, because they have a greater elasticity of employment than elasticity of price, so that the increase in aggregate demand (effective demand) is primarily absorbed by an increase in output and employment and to a lesser extent by an increase in prices. Thus, effective demand has a dominant impact on the real variables.

The foregoing presentation of Keynes's analysis and his recommendations for exiting depression relates to an economy that has a low or relatively low degree of capacity utilization (a spare capacity), so that it can react quickly in the short term by increasing production, without major changes in the cost structure. When, however, the economy reaches the level of full employment, the elasticity of employment will be zero and the entire increase in effective demand will be absorbed by price rises. Under Keynes's theory, it is only when prices rise under conditions of full employment that we have a situation of real inflation.

2.15 The Theory of Prices

At the beginning of the twenty first chapter, Keynes points out a shortcoming of classical economic theory in explaining how prices are formed, namely that, in dealing with the theory of value, the classical economists conclude that price is a result of the relationship of supply and demand, as well as of the movements of marginal costs and the elasticity of the short-run supply function. Once they have introduced money into the analysis, however, they treat the quantity of money as a factor independent of the relationship of supply and demand for goods and conclude that prices are determined by the quantity of money and the income velocity of money. Keynes claimed that money could not be considered an independent factor and, consequently, that the level of prices was the result of the interaction of supply and demand for goods and the quantity of money, while the quantity of money depends on the relationship of supply and demand for goods and on psychological expectations.

Keynes emphasised as a fundamental characteristic of money its role in connecting past and future:

> In a single industry its particular price level depends partly on the rate of remuneration of the factors of production which enter into its marginal cost, and partly on the scale of output. There is no reason to modify this conclusion when we passed to industry as a whole. The general price level depends partly on the rate of remuneration of the factors of production which enter into marginal cost and partly on the scale of output as a whole, i.e. (taking equipment and technique is given) on the volume of employment.[26]

Keynes stressed the importance of changes in demand and their impact on costs and the output, without neglecting how prices in the various branches of industry are related:

> It is on the side of demand that we have to introduce quite new ideas when we are dealing with demand as a whole and no longer with the demand for a single product taken in isolation, with demand as a whole assumed to be unchanged.[27]

Introducing money into the analysis of macroeconomic interdependencies, Keynes considers the effects of changes in the quantity of money on the level of prices as a consequence of the following influences:

1. Change in effective demand caused by changes in the quantity of money;
2. The impact of the law of diminishing returns;
3. The inability to substitute productive resources and inelasticity of supply in certain branches;
4. Changes to the wage-unit; and
5. Different rates of return on factors which enter into marginal cost.[28]

1. *Changes in effective demand* Keynes defined effective demand as the product of the quantity of money in circulation and the income velocity of money (MV = D). Changes in the quantity of money are reflected in the value of effective demand in the first phase through the impact on the interest rate. Changes in the interest rate, as a consequence of changes in the quantity of money, effect change in the liquidity preference, the marginal efficiency of

[26] John Maynard Keynes, Op. cit., p. 147.

[27] John Maynard Keynes, Op. cit., p. 147.

[28] John Maynard Keynes, Op. cit., p. 148.

capital and the investment multiplier. An increase in the money supply should cause a fall in the interest rate, an increase in the marginal efficiency of capital, an increase in investment and a consequent increase in employment. However, the distribution of effective demand will play the key role in determining the final effects of increasing the money supply and the fall in interest rates on the change in prices and production (nominal and real categories). If the increase in effective demand is directed for the most part to industrial branches with a high degree of capacity utilization, the increase in effective demand will result in a larger increase in prices than in production. If the lion's share of the increase in effective demand is oriented towards branches with a low degree of capacity utilization, the increase in prices will be less, while the largest percentage of the increase in effective demand will be absorbed through rising output and employment. Only in exceptional cases will an increase in the quantity of money be connected with a fall in effective demand.

2. *The impact of the law of diminishing returns* Because additional employment under conditions of existing techniques, equipment and technology leads to a fall in the marginal product of labour, as a result of the operation of the law of diminishing returns, in turn causing a fall in wages, given that the level of wages is derived from the marginal product of labour. Since Keynes advocated a fixed or moderately rising money wage, this means that an increase in output necessarily entails higher costs. For entrepreneurs to increase their profits, they must compensate for the increasing costs of production, as a consequence of the growth in the cost of labour, by putting up the price of their product. If money wages rise, then the rise in the price of products must be higher than the increase in money wages, the consequence of which is a fall in real wages.
3. *The impossibility of substituting for productive resources and the inelasticity of supply* As there is no way of substituting for productive factors in various branches of production, there is no way to reorient excessive demand from one branch into another branch of production. In cases where demand has increased for products of an industrial branch which has used up its capacities and the sources of supply of factors of production, practically the entire increase in demand will be absorbed through price rises. If there are multiple branches in the economy which have already entirely used up their capacity, but for whose products there exists additional demand, this will cause a greater rise in the general level of prices, with a very weak impact if any on additional employment.
4. *The impact of changes in the unit wage* During a period of economic growth and increasing effective demand, employers are ready to increase many wages. Accordingly, the money value of the unit wage does not appear to be out of step with changes in the prices of goods from the breakdown of consumption in working households. Keynes, however, assumes that workers will not seek an increase in their money wages simultaneously with every increase in effective demand. Thus the money value of the wage-unit changes, but in discontinuity, which results in putting up prices. Keynes calls these points of discontinuous rise in the costs of the factors of production the conditions of semi-inflation,

in distinction from a condition of true inflation, which appears when rising effective demand combines with full employment:

> When a further increase in the quantity of effective demand produces no further increase in output and entirely spends itself on an increase in the cost-unit fully proportionate to the increase in effective demand, we have reached a condition which might be appropriately designated as one of true inflation. Up to this point the effect of monetary expansion is entirely a question of degree, and there is no previous point at which we can draw a definite line and declare the conditions of inflation have set in. Every previous increase in the quantity of money is likely, in so far as it increases effective demand, to spend itself partly in increasing the cost-unit and partly in increasing output.[29]

5. *Different marginal products of the factors of production that enter into marginal cost* Because the money compensation differs for different factors of production which are an integral part of the marginal costs of production, due to different levels of the marginal value of the products of the factors, the elasticity of supply of these factors is also different. Keynes particularly stresses the importance of user cost:

> Perhaps the most important element in marginal cost which is likely to change in a different proportion from the wage-unit, and also to fluctuate within much wider limits, is marginal user cost. For marginal user cost may increase sharply when employment begins to improve, if (as will probably be the case) the increasing effective demand brings a rapid change in the prevailing expectation as to the date when the replacement of equipment will be necessary.[30]

Keynes, accordingly, shows us the direction in which we should focus our attention if we wish to explain phenomena and changes over the long-term, but he does not establish the relationship between the level of prices, the level of nominal wages, and user cost over the long-term. A more precise consideration of the impact of user cost on price levels approximates to the concept of unit cost. The unit cost represents the average value of costs which enter into marginal prime costs. The unit cost and the volume of production are key determinants of the level of prices.

2.16 Notes on the Business Cycle

Keynes explained fluctuations in the business cycle using his own method, where the independent variables are the propensity to consume, the marginal efficiency of capital and the interest rate. Keynes gave pride of place in explaining the business cycle to changes in the marginal efficiency of capital:

> But I suggest that the essential character of the trade cycle and, especially, the regularity of time-sequence and of duration which justifies us in calling it a cycle, is mainly due to the way in which the marginal efficiency of capital fluctuates. The trade cycle is best regarded,

[29] John Maynard Keynes, Op. cit., p. 151.

[30] John Maynard Keynes, Op. cit., p. 151.

> I think, as being occasioned by a cyclical change in the marginal efficiency of capital, though complicated and often aggravated by associated changes in the other significant short-period variables of the economic system.[31]

In the phase of late boom, excessively optimistic expectations tend to form regarding the future returns on capital assets and, accordingly, so do excessive rates of growth of investment. Because, during this phase of the business cycle, capital becomes abundant (there ceases to be a relative scarcity of capital as a major factor justifying investment), expectations of future returns change suddenly from optimistic to (overly) pessimistic, bringing about a sharp fall in the marginal efficiency of capital. As a result of this fall, there will be a fall in employment and a consequential fall in aggregate demand.

If the fall in the marginal efficiency of capital is steep and sudden, because of a sudden switch from optimistic to pessimistic expectations about future yields and returns, just cutting the interest rate is not enough to revive business activity. The reason for this lies in the sudden loss of confidence in the effectiveness of economic policy measures and the very pessimistic estimate of future yields. A certain passage of time is necessary for recovery, which Keynes said lasts between 3 and 5 years. This period corresponds with the average lifetime of equipment in which investment has already been made and whose marginal efficiency of capital is unsatisfactory. Inseparable from this position of Keynes is his further position that a necessary condition for getting out from depression is reviving the economy through new investment. The new investment, however, must be directed towards more effective techniques and technological solutions.

One of the most important factors of the length of the phases of the business cycle is the cost of surplus stocks. Thus, during a phase of excessive investment and inability to sell products, it is a necessary consequence that surplus stocks of finished products grow. The carrying costs of these inventories represent major elements of cost and it would be desirable, to the degree possible, to reduce the price of goods in stock so as to free the entrepreneurs from their stocks and the associated costs. However, selling stocks represents disinvestment, which has a negative effect on employment. A similar problem arises with stocks of materials and semi-finished product (variable inputs). Keynes divides the business cycle into the following four phases:

1. The first phase—the phase of falling business activity. The characteristics of this phase are a reduction of stocks of raw materials and unfinished goods (disinvestment) and an increase in stockpiles of finished goods (production for stockpiling).
2. The second phase—the phase of the lowest level of business activities. This phase is characterised by further reduction of investment in raw materials and semi-finished product, with a simultaneous reduction of production and, consequently, a reduction in the stockpiles of finished products.

[31] John Maynard Keynes, Op. cit., p. 155.

3. The third phase—the phase of the recovery. During the recovery phase, entrepreneurs increase investment in raw materials and semi-finished products, while still reducing their stockpiles of finished product.
4. The fourth phase—the phase of prosperity. The phase of prosperity is marked by an increase in investment in raw materials and semi-finished products, and with an increase in investment in stockpiles of finished products.

According to Keynes it is possible to avoid the cyclical fluctuations of business activities over the longer term by cutting the interest rate during the boom, rather than letting it rise, which is a possible remedy for excessive optimism:

> Thus the remedy for the boom is not a higher rate of interest but a lower rate of interest! For that may enable the so-called boom to last. The right remedy for the trade cycle is not to be found in abolishing booms and thus keeping us permanently in a semi-slump: but in abolishing slumps and thus keeping us permanently in a quasi-boom.[32]

Keynes's particularly negative approach to the use of high interest rates with a goal to restraining investment is clear from the following statement:

> Thus an increase in the rate of interest, as a remedy for the state of affairs arising out of a prolonged period of abnormally heavy new investment, belongs to the species a remedy which cures the disease by killing the patient.[33]

2.17 The Social Philosophy of the General Theory

The fundamental economic problems of any society are ensuring full employment and preventing an unjust distribution of income and wealth. For society to realise full employment, there is a need to continuously maintain a favourable relationship between the marginal efficiency of capital and the interest rate. The volume of investment depends on the interest rate and the marginal efficiency of capital. The problem consists in the fact that the incentive for an increase in savings is a higher interest rate, while at the same time a higher interest rate causes the marginal efficiency of capital to fall. Therefore, it is the main task of society to establish such relationships between the interest rate and the marginal efficiency of capital as to allow a sufficient level of saving and a sufficient quantity of investment, on the basis of which new jobs can be opened and full employment attained. This is why Keynes, as already stressed in the previous section, was a strong advocate of low interest rates and a high propensity to consume, which would facilitate sufficient demand and efficiency of investment.

---

[32] John Maynard Keynes, Op. cit., p. 159.

[33] John Maynard Keynes, Op. cit., p. 160.

Keynes also advocated expanding the role and competencies of the state in economic life, in order to allow the executive in a given country to secure stable conditions for the conduct of business activities:

> The State will have to exercise a guiding influence on the propensity to consume partly through its scheme of taxation, partly by fixing the rate of interest, and partly, perhaps, in other ways.[34]

Keynes advocated for an active role of the state, not in the sense of participating in ownership, but with a view to ensuring the necessary and sufficient conditions for the system to operate as a whole and, in that way, enable stable operations by private entrepreneurs:

> Whilst, therefore, the enlargement of the functions of government, involved in the task of adjusting to one another the propensity to consume and the inducements to invest, would seem to a 19th-century publicist or to a contemporary American financier to be a terrific encroachment on individualism, I defend it, on the contrary, both as the only practicable means of avoiding the destruction of existing economic forms in their entirety and as the condition of the successful functioning of individual initiative.[35]

With regard to the social distribution of income and assets, Keynes stressed that there exist both social and psychological justifications for significant inequality in the distribution of both, but that there were no justifications for inequality of the scale that existed in the period when he was writing his work. In fact, it was for precisely this reason that Keynes thought that the rentier form of capitalism had no future in the long term, any more than a society in which such a form of capitalism dominated:

> I see, therefore, the rentier aspect of capitalism as a transitional phase which will disappear when it is done its work. And with the disappearance of its rentier aspect much else in it besides will suffer a sea-change. It will be, moreover, a great advantage of the order of events which I'm advocating, that's the euthanasia of the rentier, of the functionless investor, will be nothing sudden, merely a gradual but prolonged continuance of what we have seen recently in Great Britain, and will need no revolution.[36]

It is clear from the views cited above that Keynes was fighting for a greater role for the state, which, through an effective economic policy, the basic levers of which were fiscal policy (the public debt) and monetary policy (the interest rate), was to ensure an adequate propensity to consume, a low interest rate, and a relatively high level of taxation on the owners of non-productive assets. In this way, the state would secure sufficient room for the individualist-entrepreneur to conduct innovative business activities.

The state, however, could not be allowed to suppress individualism in entrepreneurship through its measures, because that was the very foundation of the system:

---

[34] John Maynard Keynes, Op. cit., p. 187.

[35] John Maynard Keynes, Op. cit., pp. 188–189.

[36] John Maynard Keynes, Op. cit., p. 186.

> But, above all, individualism, if it can be purged of its defects and its abuses, is the best safeguard of personal liberty in the sense that, compared with any other system, it greatly widens the field for the exercise of personal choice. It is also the best safeguard of the variety of life which emerges precisely from this extended field of personal choice, and the loss of which is the greatest of all the losses of the homogenous or totalitarian state.[37]

The task of the state is, accordingly, to secure stable conditions in which market competition can unfold:

> The task of translating human nature must not be confused with the task of managing it. Though in the ideal commonwealth men may have been taught, inspired or bred to take no interest in the stakes, it may still be wise and prudent statesmanship to allow the game to be played, subject to rules and limitations, so long as the average man, or even a significant section of the community, is in fact strongly addicted to the money-making passion.[38]

Respect for private property is a precondition for respect for private initiative:

> There are valuable human activities which require the motive of money-making and the environment of private wealth-ownership for their full fruition. Moreover, dangerous human proclivities can be canalised into comparatively harmless channels by the existence of opportunities for money-making and private wealth, which, if they cannot be satisfied in this way, may find their outlet in cruelty, the reckless pursuit of personal power and authority, and other forms of self-aggrandisement. It is better that a man should tyrannise over his bank balance than over his fellow-citizens: and whilst the former is sometimes denounced as being but a means to the latter, sometimes at least it is an alternative.[39]

Consequently, the wealthy are not undesirable, but, rather, to the contrary, one should allow as large a number of people as possible to become wealthy, on the one hand, and allow that social group to increase its propensity to consume so as to allow the reproduction of capital and achieve full employment:

> Thus our argument leads towards the conclusion that in contemporary conditions the growth of wealth, so far from being dependent on the abstinence of the rich, as is commonly supposed, is more likely to be impeded by it. One of the chief social justifications of great inequality of wealth is, therefore, removed.[40]

2.18 Keynes's Theory of Capital, the Speed of Economic Growth and a Possible Answer to the "Inflation Trap"

Keynes's *General Theory of Employment, Interest and Money* beyond any doubt provoked a great change in the economic mode of thought, creating an academic basis for a great growth in the significance of macroeconomic management of the

---

[37] John Maynard Keynes, Op. cit., p. 188.

[38] John Maynard Keynes, Op. cit., p. 185.

[39] John Maynard Keynes, Op. cit., p. 185.

[40] John Maynard Keynes, Op. cit., p. 185.

business cycle and the scope for realising full employment. At the same time, the main weakness of Keynes's approach was the unsolved question of how to control inflation as economic growth from a phase with a low degree of capacity utilization (actual output is much less than potential output) becomes economic growth stimulated by a low interest rate in a phase of a higher or high degree of capacity utilization (actual output is close or equal to potential output).

In this context, it is useful, as a potential contribution to the analysis of this problem, to investigate or analyse whether user costs are a key element in long-term price level stability. Keynes wrote about user cost in the appendix to the 6th chapter of the General Theory, ascribing them great importance for the stability of economic growth. According to Keynes, the short-term price of a unit of production is formed on the basis of marginal prime costs, which are made up of the factor cost and the user costs. The long-term stable price of supply should equal the sum of the marginal prime cost, supplementary cost, the cost of risk and the cost of interest. Keynes points out that the fluctuation in the user cost is more significant and outweighs fluctuation in labour costs. As the degree of capacity utilization increases (growth of employment) user costs increase due to the law of diminishing returns.

On the basis of the preceding commentary on Keynes's General Theory, we may draw a clear conclusion that Keynes's preoccupation was thus with solving the problem of mass involuntary unemployment and determining effective measures for exiting economic depression in the short term, that is given a set of productive techniques and available resources. In such an environment, investors are very pessimistic, unwilling to invest and they need incentive. Keynes's introduction of targeted measures of expansionary monetary policy and equally expansionary fiscal policy, with the outright Central Bank purchase of new issues of government bonds was a key recommendation for the conduct of economic policy in the 1930s, with a view to reducing of private entrepreneurs' uncertainty over future revenues and on that basis to increasing the incentive to invest, based on an increased marginal efficiency of capital, derived from a policy of low interest rates. Moreover, Keynes's analysis, as contained in the General Theory, is primarily an analysis of the instruments of economic policy focused on attaining internal equilibrium, because of the nature of the problem and the time in which that problem arose (the Great Depression). The time when Keynes wrote the General Theory was not a time in which measures based on the liberalisation of flows of goods and, more particularly, of capital were dominant.

Thus, the primary focus of the analysis was on attaining the internal equilibrium which had to be attained over the short term. By contrast, in his analysis of the long-term, Keynes did not consider the impact of technological progress on changes to the structure of costs, even though in the appendix to his 6th chapter he did explain in somewhat more detail his theory of costs based on user costs. Moreover, in his analysis of capital accumulation, Keynes did stress that in the long-term unit cost would probably be under the significant influence of user cost, and to a lesser extent of changes in wages. In this way, he implicitly assumed changes in material costs and changes in depreciation, but did not work out that aspect, that is he did not work it out fully as a special chapter or part of a chapter in which he would be analysing the effects of investment in research and development on changes to the quality of techniques and technologies of production, or of changes on that basis on the structure of costs, or provide answers to the question of how to maintain price stability in the long term, while also ensuring that investment in physical capital continues to pay off, if the remedy for maintaining prosperity is low interest rates.

In his theory of the capital formation, as we have already discussed, Keynes's key assumption is that what ensures investment pays off and capital increases is its relative scarcity. Capital is sufficiently scarce so long as the marginal efficiency of capital is greater than the interest rate, so that, during the phase of economic boom and employment at or close to full employment, the remedy for maintaining that level of economic activity is a low interest rate. For aggregate demand to be sufficient to maintain employment at a level of full employment, the marginal propensity to consume must be kept at a relatively high level, that is at a level that allows stable share of personal consumption in the creation of GDP. Business sector investment is the condition sine qua non of maintaining a level of full employment, as a sufficiently attractive marginal efficiency of capital is maintained by growth in the money supply and suppressing the interest rate to a very low level. Under such circumstances, however, capital becomes ever more abundant and one of the bases in the analysis of the efficiency of investment is lost. So, growth in the money supply under conditions of full employment and capacity utilization equivalent to full capacity (the potential output) means that, according to Keynes's analysis, the entirety of the additional money supply is spent on rising unit costs (higher prices). This is the phase that Keynes called the stage of true inflation.

A mathematical response to the theoretical dilemma left by Keynes can be found in the work of Branko Horvat, a candidate for the Nobel Prize for economics in 1983.[41] Modelling economic growth and testing Marx's so-called law of the faster growth of sector one (sector one in Marx's theory is the production of capital goods), Horvat proved mathematically that the law was not valid. Horvat's mathematical proof that more intensive investment in the production of capital goods than in the production of consumer goods was not a necessary condition of accelerated economic growth can be used with success to clarify Keynes's theory of capital, in which the scarcity of capital is the key condition to ensure that investment pays off and that the economy can be maintained in a condition of boom witouth inflationary pressures.

Horvat developed his model using differential calculus in sixteen gradually developed formulas, and at the end of the proof established the following relation[42]:

$$r'(t) = [r(t)/km][\alpha(t) - (um - uo)/I(t)]$$

[41] That was the year the Nobel Prize for Economics was awarded to Gerard Debreu. Branko Horvat was on the shortlist, which contained only two names. That is, in the final round of voting, Debreu and Horvat were the only rivals.

[42] See Ref. [1], pp. 55–62.

where

r′(t) the marginal rate of growth of the economy;
r(t) the average rate of growth of the economy;
km the marginal capital coefficient;
α(t) the rate of growth of the sector producing capital goods (sector one);
um the marginal coefficient of user cost (the ratio of the difference between user cost in the current and in the previous period, to the change in production in the current period compared to the previous period);
uo the average coefficient of user cost (the ratio between user cost in the current period and the value of production in the current period);
I(t) the index of the increase of overall production at time (t) compared to the base time period (to).

Here it should be stressed that the marginal and average coefficients of use (Um and Uo) in Horvat's model correspond to the marginal and average share of Keynes's user cost in the additional (in the case of marginal) or total value of production (in the case of average share). Thus, Horvat's use represents the sum of the values of material costs and depreciation, and so is equal to Keynes's user cost.

From the above identity, it follows that a change in the general rate of growth of a given economy depends on the following variables: the average and marginal coefficients of user cost and the marginal capital coefficient. Horvat's model is based on the early Harrod-Domar model (an early post-Keynesian model of growth), and this is particularly relevant to the case where the marginal capital coefficient is equal to the average capital coefficient. So, the key variables of the model are the following:

1. *The average coefficient of user cost* (*uo*) is the ratio of the total user cost and total production, or gross domestic product. Total user cost represent the sum of all material costs in a given economy, and the total costs of depreciation in the economy:

$$uo = U(to)/P(to)$$

2. *The marginal coefficient of user costs* (*um*) is the ratio of the difference between user costs in the current and in a previous time period and the difference between the total value of production in the current and in a previous time period:

$$um = [U(t) - U(to)]/[P(t) - P(to)]$$

3. *The marginal capital coefficient* (*km*) is the ratio of the difference between the total value of available fixed and working capital in the current period and the value of available fixed and working capital in the preceding period, with the difference between GDP in the present and in the preceding time period:

$$km = [S(t) - S(to)]/[P(t) - P(to)]$$

In the preceding three expressions, U(t) and U(to) represent the total user cost in the current and the baseline period, S (t) and S (to) the total value of fixed and working capita (the value of physical capital) in a given economy for both the current and the baseline period, while P (t) and P (to) represent the values of total production (output) in the current and base year.

Change in the general rate of growth, i.e. acceleration of economic growth, depends on the expression in brackets on the right side of the equation. In other words, for the economy to achieve an accelerated growth rate, the expression in brackets must be positive, i.e.:

$$\boldsymbol{\alpha}\,(\mathbf{t}) \geq (\mathbf{um}-\mathbf{uo})/\,\mathbf{I}\,(\mathbf{t})$$

The coefficient **α(t)** represents the rate of growth for the sector producing capital goods (sector one). The relationship of these variables gives rise to the following regularities:

(1) If the marginal coefficient of user cost is greater than the average coefficient (a situation in which an economy is becoming less efficient—consuming more inputs in the current period per unit of output than in the previous period), the primary sector producing capital goods must grow faster, for the economy to grow faster as a whole than in the previous period. In Keynes's theory, exit from depression requires investment in capital goods because the investment multiplier and the employment multiplier act faster for these forms of investment. This means that in the recovery phase, investment in capital goods is still sufficiently scarce and still has a sufficient rate of return, on the one hand, and allows for a faster rate of employment growth, on the other, to ensure the first phase of recovery, the transformation of a stagnant or declining economy into a growing one.
(2) Nevertheless, the continuous accelerated growth of the sector manufacturing capital goods is not a necessary condition for economic growth, given that it follows from the equation given above that the sector producing capital goods does not have to grow (it can stagnate or even have a negative growth rate) for the economy as a whole to grow. This condition is met when the marginal coefficient of user cost is less than the average coefficient of user cost, so that even a zero value for the α(t) coefficient satisfies for the expression in brackets to be positive, because of the negative difference between the marginal and the average user cost coefficient. Thus, after the recovery phase facilitated by growth in investment in capital goods, there follows a boom stage in which investment in the production of capital goods does not have to continue to grow, can in fact continue to be scarce, without compromising the continuation of the boom. This continued boom unfolds with increased investment in the production of consumer goods, which were not a priority during the recovery phase, because of the relatively low marginal propensity to consume and low effective demand. Keynes himself stressed that during the boom phase, the solution to maintaining a sufficient scarcity of capital will be

increased investment in the production of consumer goods. It follows from Horvat's model that a faster rate of growth of production of consumer goods than for capital goods happens when changes in the cost structure are such that the marginal user cost is less than the average user cost, and this is possible thanks to the change from the law of decreasing returns to the law of increasing returns as a consequence of the previous phase of increasing the capital/labour ratio based on advanced technology.

(3) The economy as a whole, accordingly, grows even when the sector manufacturing capital goods is stagnating (the primary sector), i.e. when the rate of growth in that sector has fallen to a lower percentage than the percentage by which the coefficient of user cost has improved over the average coefficient of user cost. It is a necessary condition of economic growth, however, in this case that the production of consumer goods grow faster than the production of capital goods, i.e. that the production of consumer goods records a rate of growth that is greater than the rate of the reduction of growth of the production of capital goods.

Reducing the marginal coefficient of user cost in comparison to the average coefficient of user cost in an economy means that the economy as a whole, i.e. its manufacturing and service sectors, are performing successfully and, so, consuming fewer inputs per unit of additional production than previously. In periods where such a situation exists, resources previously invested in expanding productive capacity (previous accelerated growth of the production of capital goods) will have resulted in a greater or improved efficiency of production and so in greater productivity and lower average costs. Accelerated growth of the sector manufacturing consumer goods, consequently, is a consequence of more efficient production, realised on the basis of investment in increasing the availability of capital goods. If the marginal coefficient of user cost, after the period of adjustment to the new structure of the economy is over, is greater than the average cost, this means that productivity in the economy is falling, which in turn means that the resources previously invested in capital equipment cannot be paid off at the initially planned rates of return of investment. Economic growth in this phase, based on a higher rate of growth in the production of capital goods, does not represent a sound basis for sustainable economic growth. It is, in fact, a phase of excessive investment in which capital ceases to be scarce, so that not even a policy of low interest rates can maintain the marginal efficiency of capital over the long-term as a level which will secure the reproduction and accumulation of capital itself.

On the basis of our application of the Horvat model, through which he proved that Marx's so-called law of faster growth of the first sector does not hold, and adaptation to Keynes's terminology, we have shown that price stability can be maintained under conditions when the actual output is close or equal to the potential output (to the level of full employment), given improvement in the cost structure based on a reduction of user cost per unit of production, based on technological change, that is investment in higher quality capital equipment and the production of better quality inputs of production, which would increase the productivity of labour and transform the law of diminishing returns into production with growing or at least constant returns.

References

1. Horvat B (1987) Radna teorija cijena i neki drugi neriješeni problemi ekonomske nauke (Labor theory of prices and some other unresolved problems of economic science). Rad, Beograd
2. Keynes JM (1936) The general theory of employment, interest and money. Macmillan & Co. Available on: The ISN ETH Zurich webpage—www.isn.ethz.ch
3. The Royal Economic Society (2013) The collected writings of John Maynard Keynes—volume VII: the general theory of employment, interest and money. Cambridge University Press, Cambridge

Chapter 3
Impact of Financial Globalization on the Scope of Economic Theory and Policy

Abstract This chapter deals with the changes in the global financial and economic environment over the last two decades that have led to an unprecedented increase in the global capital flows based on foreign direct investment and portfolio investment. The chapter shows that the process of globalization of production based on vertical intra industry trade between the advanced and largest developing countries has caused the production costs of globally active multinational companies to decrease, on the one hand, and has led to an increase in the global systemic risk of internationally active investment funds. An enormous growth of the global capital flows has led to a significant increase in the importance of prices of assets in the measurement of inflation, as well as in analysing the effects of monetary and fiscal policy. Therefore the process of globalization has actually caused that almost all the macroeconomic models and theories, whether based on the neoclassical or the new Keynesian thought, have failed in predicting and suggesting measures of economic policy that have been necessary to prevent the global collapse. Since the theories are mostly based on the law of decreasing returns, and on the role of the private sector companies and households' goal functions in establishing the general equlibrium, not taking into account the importance of profit-motives of financial institutions and the importance of speculative demand for money for the analysis of investment cycle, they haven't been able to offer a consistent theoretical explanation of the major causes of the global crises. The author ends this chapter by pointing out that Keynes's theory of the demand for money with a special reference to the speculative demand for money has been an extraordinary theoretical achievement in the history of economic science, and is especially important for the analysis of current economic and financial problems in the world. In addition, the author calls for a new macroeconomic model based on two large economies one of which is a representative of the advanced countries and the other a representative of the developing countries.

Keywords Globalization · Foreign direct investment · Portfolio investment · Asset prices · Monetary policy · Speculative demand for money · Macroeconomic theory

F. Čaušević, *The Global Crisis of 2008 and Keynes's General Theory*,
SpringerBriefs in Economics, DOI 10.1007/978-3-319-11451-4_3

3.1 Changes in the Balance of Economic Power

Major changes took place in the relations of economic power between the developed and developing countries over the first 13 years of this century. One of the best indicators of the increasing economic power of developing countries is the data on their foreign exchange reserves. According to data from the Bank for International Settlements (BIS), the total foreign exchange reserves of the developed countries in 2000 were approximately 1.7 times greater those of the developing countries. Five years later (2005), for the first time since the Second World War, the developing countries' foreign exchange reserves had become greater than those of the developed countries. The BIS data for 2013 show that the developing countries' reserves are now almost twice those of the developed countries. Total foreign exchange reserves worldwide were approximately US$11.7 trillion in 2013, with developing countries accounting for more than US$7.86 trillion of that.[1]

China saw the greatest absolute and relative growth of its foreign exchange reserves. Between 2000 and 2013, they grew from US$184 billion to more than US $3.8 trillion. According to fundamental laws of macroeconomics, this growth in developing countries' foreign exchange reserves is only possible on the basis of a marked increase in their export capacity and a surplus in trade in goods and services. China's economic growth model has from the beginning of the 1990s been based on a policy of attracting foreign direct investment from the United States, Great Britain and the countries of Western Europe, a policy based on a predominantly export orientation on the part of those investors, which, in return, allowed major growth in the trade surplus and, as a consequence, in the country's foreign exchange reserves (Fig. 3.1).

In an article recently published in *Finance and Development*, Eswar Prasad[2] explains the reasons for this major growth in foreign exchange reserves in developing countries and their readiness to shoulder the high opportunity costs arising from the very low yield on US government securities, in which most Asian countries (in particular China, India and the countries of Southeast Asia) hold most of their allocated foreign-exchange reserves. He sees the explanation for these trends in a desire to invest in safe or at any rate the least risky financial assets in the current constellation of global finance. That is, the very negative experience of the Southeast Asian countries during the crisis in 1997 and the indicators of a rapid reduction in the foreign exchange reserves of some developing countries during the global crisis of 2008–2009 suggest that it requires very large amounts of foreign exchange reserves to ensure that developing countries can effectively intervene on foreign exchange markets. The Figs. 3.2 and 3.3 also illustrate this very significant change in the relations of economic power in the world, showing changes in the US,

[1] See the IMF webpage—Currency Composition of Official Foreign Exchange Reserves (COFER) http://www.imf.org/external/np/sta/cofer/eng/cofer.pdf (accessed 9 March, 2014).

[2] Reference [11]. http://www.imf.org/external/pubs/ft/fandd/2014/03/prasad.htm.

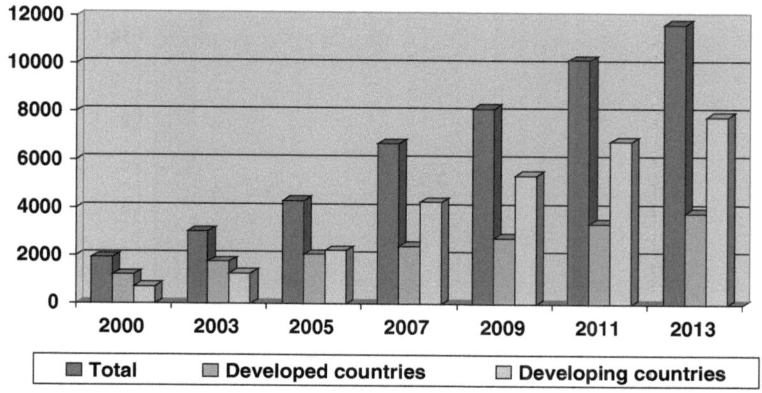

Fig. 3.1 Foreign exchange reserves of developing and developed countries: 2000–2013. *Source* The Bank for International Settlements

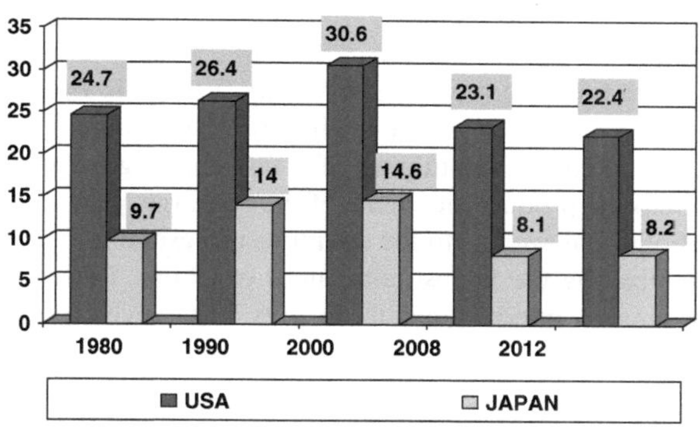

Fig. 3.2 Change in the US and Japanese share in world GDP: 1980–2012. *Source* the author's construction based on World Bank data http://data.worldbank.org/indicator/NY.GDP.MKTP.CD

Japanese and BRIC countries' shares in global GDP between 1980 and 2012 (US and Japan) (only the first twelve years of this century for the BRIC countries).

The opening up of China to international flows of capital, goods and services, which began with the coming to power of Deng Xiaoping in 1978 and gathered steam through the 80s and 90s, alongside weakening and eventual dissolution of the former Eastern bloc (the former socialist countries with a Soviet style of socialism), their subsequent transition to market economies, the increasingly intensive economic reforms aiming at accelerated and more dynamic private sector development in India, and the stimulatory role played by the developmental state in the countries of South-east Asia,[3] represented a major change in the international economic,

[3] Reference [17].

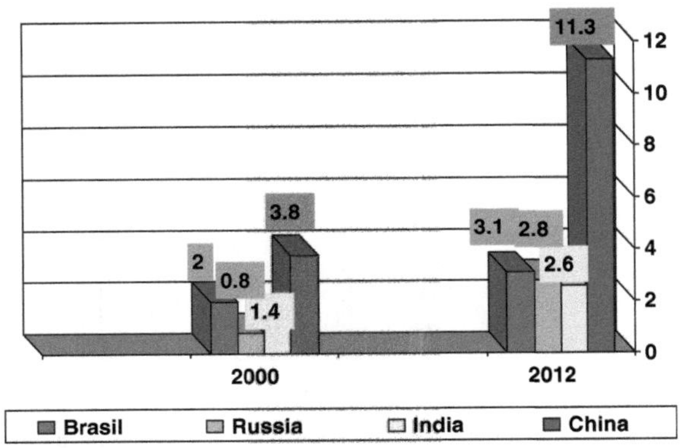

Fig. 3.3 Change in the share of the BRIC countries in world GDP: 2000–2012. *Source* the author's construction based on World Bank data http://data.worldbank.org/indicator/NY.GDP.MKTP.CD

political and geopolitical environment over the past 25 years, particularly compared to the first four decades after the Second World War. This opening of China to international flows of capital based on foreign direct investment from the US, the UK and the EU, resulted in the building of a fast-growing, export-oriented economy, integrated into the production chains of the countries providing this capital to China in the first place, as well as to the countries representing final demand for the goods being produced by the largest companies in China. Tables 3.1, 3.2 and 3.3 show major changes in international trade over just eleven years (2000–2011), which reflect the fact that China has become the world's largest exporter of goods and that trade in goods between the US and China in 2011 was equivalent to 80 % of the merchandise trade between the US and the European Union.

The sharp rise in interconnectedness and interdependency of the developed and developing countries over the past 25 years have created very different conditions for the geographic distribution of production and the composition of the costs of

Table 3.1 The US trade in goods with China

Year	Exports	Imports	Trade balance (in billions of USD)
2000	28.4	107.6	−79.2
2005	41.8	259.8	−218.0
2008	71.5	356.3	−284.8
2009	69.6	309.5	−253.9
2010	91.9	383.0	−291.1
2011	103.9	417.3	−313.4

Source The World Trade Organization

Table 3.2 The US trade in goods with the EU 27

Year	Exports	Imports	Trade balance (in billions of USD)
2000	168.4	234.7	−66.3
2005	187.4	319.6	−132.2
2008	275.2	376.7	−101.5
2009	221.1	286.7	−65.6
2010	240.2	326.3	−86.1
2011	269.1	375.5	−106.4

Source The World Trade Organization

Table 3.3 China's trade with the rest of the world (goods and services)

Year	Exports	Imports	Trade balance (in billions of USD)
2000			
Goods	249.2	225.1	+24.0
Services	30.4	36.0	−5.6
2005			
Goods	762.0	660.0	+102.0
Services	74.4	84.0	−9.6
2010			
Goods	1.577.8	1.396.2	+181.6
Services	162.2	193.3	−31.1
2012			
Goods	2.048.8	1.818.1	+230.7
Services	190.9	282.1	−91.2

Source The World Trade Organization

production, as well as for the sales of goods and services being produced by multinational companies, than those which were prevalent with regard to the development of markets for mass consumer goods and consumer durables during the period when the main economic theories were being developed. As indicated in the first two parts of this book, the most influential currents and schools of economic thought and economic models arising from those theories took shape during the 1960s, 1970s, 1980s and first half of the 1990s. Global geostrategic, political and international economic relations were, at the time when these theories were being developed, based on a division of the world into blocs and an almost absolute withdrawal of China and the former socialist countries (with the exception of the former Yugoslavia) from economic exchange with the West or the Far East.

Given such an environment, economic theories were developed in a context where it was possible to arrive at both internal and external equilibrium in trade in goods, services and capital flows, which related primarily to trade between developed market economies, the formation of labour costs primarily within the framework of those economies, and a far lesser degree of mobility of capital at the international level, compared to what we have seen over the past 30 years, again because of the very marked global political divisions and oppositions of the time. One result of this context was the inflationary pressures in the 1970s, which were the result of a combination of a demand inflation and cost push inflation, which ensured that the rational behaviour of market actors became the focus of economic model making, under conditions of rational expectation, a high accessiblity of information, and a predominant orientation on the analysis of internal equilibrium or a combination of internal and external equilibrium, as a consequence of mutually interrelated economic policy measures being applied by the developed economies (the US, Japan, the EEC or later EC countries).

The new classical economists claimed that thanks to rational expectations market actors quickly adapt to new circumstances, that economies function at the level of full employment, and that change in the monetary policy only has an impact on nominal variables. On the other hand, the new-Keynesians introduced sticky prices and then sticky information into their models to show that there was room for an active use of monetary and fiscal policy even under conditions of rational expectations, that is within the models of Phelps and his colleagues under conditions of autonomous expectations. Behavioural economists, and particularly economists working on behavioural finance (Schiller and Akerlof for example), showed that business cycle crises were inevitable, because of the domination of psychological factors over decision-making on financial markets and that therefore did exist a need for the government to manage business cycles through economic policy, so that there would not be a breakdown of the system in periods of steeply growing pessimism.[4]

The authentic economic thought of John Maynard Keynes and his recommendations for conducting anti-cyclical economic policy, as contained in the General Theory, and those of the advocates of post-Keynesian economic theory, particularly the branch that developed the monetary circuit theory holding that the money supply is primarily endogenously determined, experienced almost absolute satisfaction (justification) in 2008 and the years that followed, alongside the behavioural theory of finance. Even Robert Lucas himself, one of the greatest critics of Keynes's economic doctrine as contained in the General Theory, announced at the beginning of the crisis that there was a need to sharply increase the money supply, to save the system. As Robert Skidelsky said of Lucas, this statement of Lucas's meant effectively: "This concession to reality involves Lucas in theoretical breakdown."[5]

[4] Reference [1].

[5] Reference [12], p. 47.

3.2 The Changed Nature of Managing the Money Supply in the Context of Globally Integrated Finance

The changed nature of monetary management and the scope of monetary policy, due to the great strength of the financial markets and their influence on both short-term and, more particularly, long-term interest rates formation is perhaps best described by Alan Greenspan. In “Conundrum,” the 20th chapter of his book *The Age of Turbulence*, Greenspan explains his confusion and that of his colleagues on the FOMC at the events of 2004 and 2005 as follows:

> “What is going on?” I complained in June 2004 to Vincent Reinhart, director of the Division of Monetary Affairs at the Federal Reserve Board. I was perturbed because we had increased the federal funds rate, and not only had yields on ten-year treasury notes failed to rise, they’d actually declined.… Our hope was to raise mortgage rates to levels that would defuse the boom in housing, which by then was producing an unwelcome froth.… But by June, market pressures seemingly coming out of nowhere drove long-term rates back down.
> …
>
> I did not come up with an explanation for the 2004 episode, and I decided that it must be just another odd passing event not be repeated. I was mistaken. In February and March 2005, the anomaly cropped up again.[6]

The strength and power of financial markets in determining the rate of return and costs of financing investment through the yield on bonds and the required rate of return on equity on financial markets, which are predominantly based on “arm-length” financial systems (US, UK, Canada, Australia), are best reflected in the data on the value of institutional investors’ assets. According to the data of the Bank for International Settlements (BIS), at the beginning of this century, the assets of institutional investors in Great Britain amounted to 176 % of GDP, while in the United States they were 155 % of GDP. According to the most recent available data from the Fed, the assets of American institutional investors in June 2013 had reached the value equivalent to approximately 220 % of American GDP.[7] According to the same source, the total value of the net assets of American households in June 2013 amounted to US$74.8 trillion, of which the households themselves directly owned US$17.74 trillion, in the shares of companies, while US$18.88 trillion represented the assets of American citizens invested in pension funds, of whose assets the shares of companies make up something more than a third.

In the 5th chapter of *Banking on the Future*, a chapter dedicated to the analysis of the impact of changes in asset prices (financial assets and property) on monitoring inflationary trends and the efficiency of monetary policy, Howard Davis and David Green explained the problem of the (in)efficiency of monetary policy as conducted by the world’s major central banks, and particularly the Fed, which they

---

[6] Reference [4], pp. 377–378.

[7] Source: Board of Governors of the Federal Reserve System, *Financial Accounts of the United States—Flow of Funds, Balance Sheets, and Integrated Macro Economic Accounts*, 2nd quarter, 2013.

see as arising from neglect of trends in asset prices and the impact on financial stability:

> Perhaps the fiercest controversy in the world of central banking in the last few years has centred on the extent to which monetary policy should respond to changes in asset prices—whether equities, property, or, most particularly, housing. The arguments for and against doing so rumbled on for the early years of the century, in Basel, the IMF, and the scholarly journals and burst into the open in late 2007, as the severity of the credit crisis began to be understood. The issue lies at the heart of the critique of central banking in the Greenspan years, which has now become more fully articulated.[8]

Considering the quite urgent need for central banks to include changes in the price of assets in the focus of their interest when determining monetary policy measures, Davis and Green cite Andrew Crockett, then the general director of the BIS, and, according to them, the most powerful man in the group of G 10 governors at the beginning of this century in the BIS, who in 2003 defined the focus required of the monetary policy makers as follows:

> In a monetary regime in which a central bank's operational objective is expressed exclusively in terms of short-term inflation, there may be insufficient protection against the build-up of financial imbalances that lies at the root of much of the financial instability we observe. This could be so if the focus on short-term inflation control meant that the authorities did not tighten monetary policy sufficiently pre-emptively to lean against excessive credit expansion and asset price increases. In jargon, if the monetary policy reaction function does not incorporate financial imbalances, the monetary anchor may fail to deliver financial stability.[9]

On the other hand, in a text, "The Great Failure of Central Banking," published in 2007, in *Fortune* magazine, Stephen Roach, former chief economist at Morgan Stanley, expressed direct criticism of the excessively narrow and dangerous focus by the Fed on relying on the CPI (consumer price index) to follow inflation, as according to Roach it gave very wrong information about real changes in prices on the market in the United States and, accordingly, led to incorrect conduct of monetary policy, contributing to the creation of the "bubble" in property prices in the US, particularly between 1995 and 2005.

In his most recent book *The Map and the Territory*, published in 2013, Greenspan gave partial recognition to the major role that financial assets and property prices had in changing the business cycle in the US. In a chapter on asset prices, Greenspan states:

> The economic collapse of 2008 reinforced what previous experience had clearly shown me: Stock prices are not merely a leading indicator of business activity but a major contributor to changes in that activity. Stock prices gyrations have a profound effect not only on financial markets and financing but on the real economy as well. Capital gains and losses are key factors in the ups and downs of the business cycle. Most pronounced is their effect on consumer spending. Over the past six decades, the market value of all stocks held by American households and non-profit organisations directly or indirectly through equity

[8] See Ref. [2], p. 115.

[9] Reference [2], p. 120.

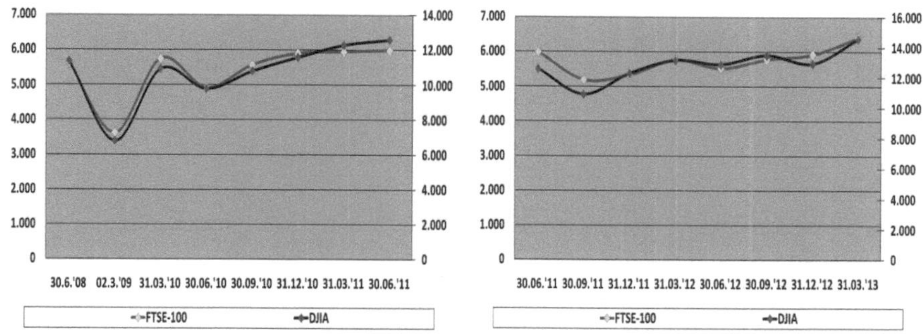

Fig. 3.4 Change in the Dow Jones Industrial Average and the Financial Times Stock Exchange indices: 30th of June, 2008–31st of December, 2013. *Source* the author's construction based on the Bloomberg's data

> holdings of pension and mutual funds, insurance companies, and other financial intermediaries has risen in value by nearly $20 trillion.[10]

According to the dominant financial theory, it should actually be easier to structure major investors' (institutional investors, investment banks) portfolios of financial assets under conditions of liberalised capital flows, insofar as the standard arguments for financial globalisation and liberalisation are that it broadens the field of choice with regard to financial assets and diversifes risk as a result of the greater potential to invest in different regions of the world and in different capital markets. Thus, starting from the basic propositions of Markowitz's portfolio theory, one achieves a minimisation of risk per unit of expected return by investing in securities whose returns are negatively correlated.[11] A negative correlation of returns on securities (shares) of companies in different parts of the world would entail that fast-growing developing countries would have to develop specific forms of production dominated by products which could compete with producers from the West, that is from developed countries. The way, however, in which developed countries integrate with fast-growing developing countries is on the basis of foreign direct investment and vertical intra-industry integration of companies from developing countries in the production chains of developed countries. This has resulted in a high degree of positive correlation between the return on shares from companies in fast-growing developing countries and that on shares from companies in developed countries.[12] This interrelationship between developed and developing countries meant that in the period from the second half of 2008 to the first half of 2009, the

---

[10] Reference [5], pp. 80–81.

[11] Reference [7], Part II and Part III.

[12] There is an excellent description of the way in which the production chains of multinational companies are connected with China, Japan, South Korea and the countries of south-east Asia in the following article: Yoshitomi M (2007) "Global Imbalances and East Asian Monetary Cooperation" (Ref. [19])

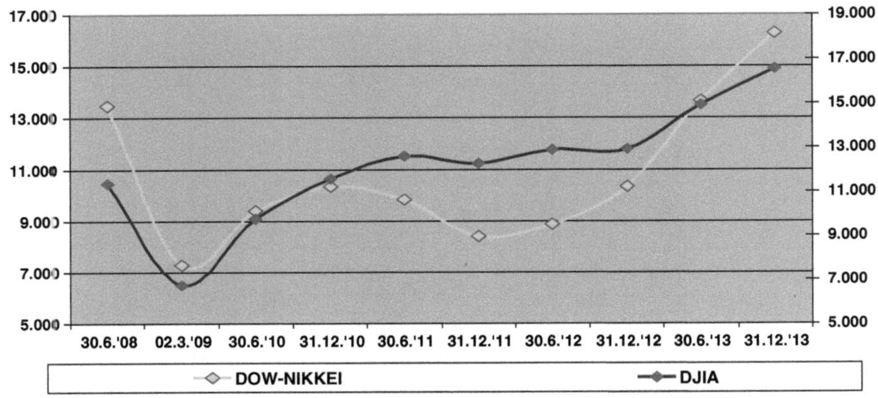

Fig. 3.5 Change in the Dow Jones Industrial Average and the Dow Nikkei 225: 30th of June, 2008–31st of December, 2013. *Source* the author's construction based on the Bloomberg's data

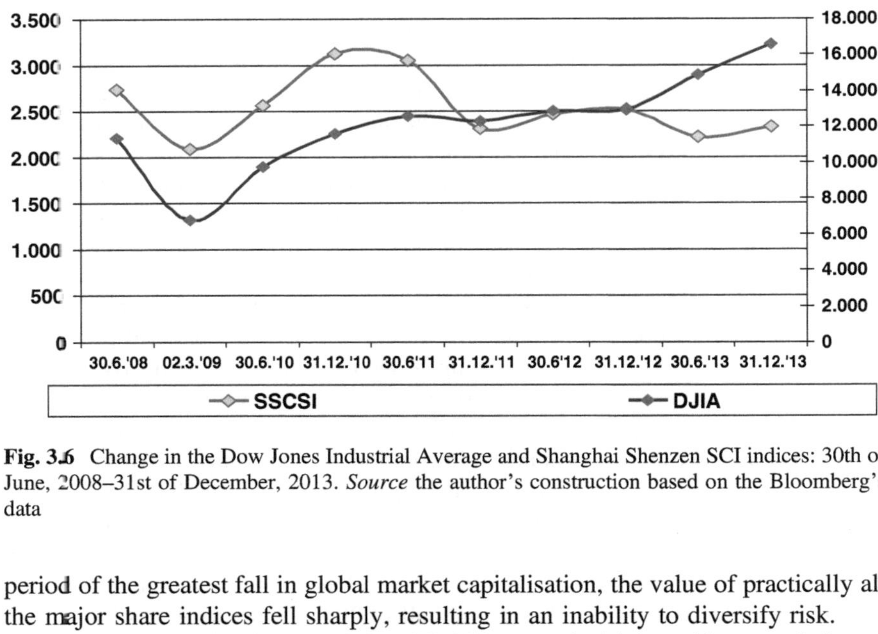

Fig. 3.6 Change in the Dow Jones Industrial Average and Shanghai Shenzen SCI indices: 30th of June, 2008–31st of December, 2013. *Source* the author's construction based on the Bloomberg's data

period of the greatest fall in global market capitalisation, the value of practically all the major share indices fell sharply, resulting in an inability to diversify risk.

The four graphs in Figs. 3.4, 3.5 and 3.6 illustrate the high positive correlation of yield and share price for some of the most prominent global share indices (the Dow Jones Industrial Average, the FTSE 100, the Dow Nikkei 225 and the Shanghai Shenzen CSI) in the period between the second half of 2008 and the end of 2010. During this period, capital markets in the US, Japan, the UK and China were "floating" in the same or very similar direction. The marked fall in the price of financial assets on markets in the US quickly "spilled over" onto the markets of other countries, including the Eurozone countries which were not included in these graphs. The expansionary monetary policy conducted in the US, and then in Great Britain, Japan and China very quickly had an impact on changing the value of financial assets, but its impact on employment growth and business sector investment entailed a time lag of 18–24 months. The lesson that the world of global finance learned (unfortunately only partially) is that the international economic links established in the real sector over the past 20 years, based on foreign direct investment, have led to the development of a global system of risk and the need for faster and more effective coordination of economic policy by the main world economies, with a view to preventing global recession slipping into global depression.

3.3 The Impact of Financial Liberalisation on the Effectiveness of Economic Policy

In works from 1973, Edward Shaw[13] and Ronald McKinnon[14] initiated the theoretical debate on the effects of financial liberalisation or, rather, the need to accept financial liberalisation measures as preconditions for improving the relationship between saving and investment and the acceleration of economic growth to be expected from that. Both authors argued that financially repressed economies, in which the interest rate is exogenously (administratively) determined and which have legally determined maximum interest rates for deposits (deposit rate ceilings), cannot achieve an optimal ratio between saving and investment and so cannot create the conditions for savings growth as a precondition to investment growth. Their key claims were that by removing the legal limits with regard to the interest rate and increasing deposit interest rates, after having implemented liberalisation, one could create a basis for growth in savings and productive investment, of investment projects which could withstand the test of higher interest rates and achieve higher returns. The distribution of lending in financially liberalised environments is, according to Shaw and McKinnon, more efficient, because investment is channelled into industries with a higher productivity and higher rates of return.

Thirteen years passed between the publication of these works, which corresponded with the decision definitely to take the US dollar and other main world currencies off the gold standard and put them on floating exchange rate regime, and the effective removal of Regulation Q. In the period between June 1980 and March 1986, all restrictions on interest rates on demand and term deposits were removed in the United States.[15] Over these 6 years, numerous financial innovations were introduced, starting with the introduction of swap-agreements (interest rate swaps),

---

[13] Reference [13].

[14] Reference [8].

[15] Reference [3].

NOW accounts, and floating rate notes (FRN). During the 1980s and 1990s, financial liberalisation became one of the fundamentals of European Community expansion and then of the formation of the European Union. Similarly, during the 1990s and 2000s, financial liberalisation was one of the integral elements of the transition package in all the countries in transition and in the fast-growing developing countries.

The accelerated growth of financial innovations other than the aforementioned bank accounts and securities with floating interest rates was primarily based on the issue and sharply growing importance of derivative financial instruments, starting with futures contracts, then options, swap arrangements, credit default swaps (CDS) and mortgage-backed securities (MBS), or collateralised debt obligations (CDO), and the asset-backed commercial papers (ABCP) mentioned in the first part of this book. Some of the most important former chairs of central banks across the world were and remain highly sceptical with regard to the use of these financial innovations introduced between the 1970s and the end of the first decade of this century. A good example of such a position regarding this form of financial innovation is the views of the former Fed Chair, Paul Volcker, who headed the institution from 1979 to 1987 and was chair of the Committee for Economic Recovery set up at the beginning of his first mandate by US President Barack Obama. Paul Volcker said, in an interview with the Wall Street Journal, that the American economy had been better and more stable without CDS-s, without securitisation, and without CDO-s.[16]

While this statement by the former Chair of the Fed may be interpreted in various ways, one of the most detailed studies of the impact of financial globalisation, published in April 2002 by Maurice Obsfeld and Alan Taylor, showed that the major capital flows during financial liberalisation and globalisation over the last three decades of the 20th century had been largely directed between rich and rich, and that nearly three quarters of financial flows resulting from financial liberalisation had been flows of "diversification finance." By diversification finance, Obstfeld and Taylor meant financial flows based on investing in financial products whose primary purpose was protection from risk and in market trades in such instruments.[17] In the study, they compared the effects of the first financial globalisation (1870–1914) and the second financial globalisation (1970–2000), ending with data for late 2000, but global financial flows in the first decade of the 21st century have essentially confirmed their conclusions.

According to the data of the BIS, lending by internationally active banks between 2002 and 2008 grew at an annual level twice the average between 1985 and 2002.[18] Their total claims grew from US$9 trillion in 2000 to US$23.07 trillion

[16] Reference [18].

[17] Reference [10].

[18] Reference [16], p. 6.

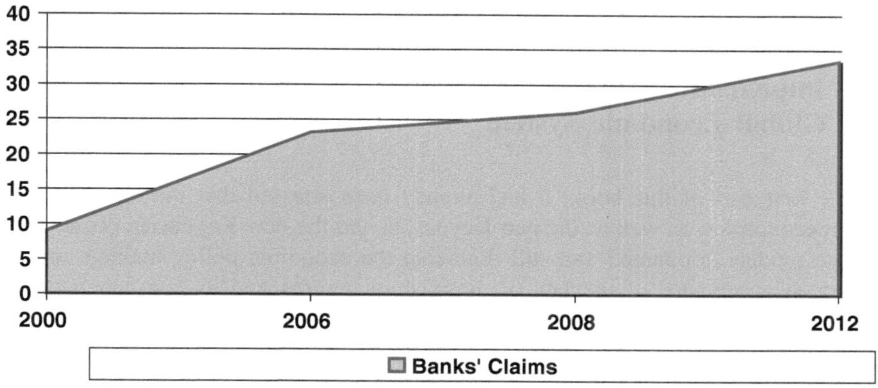

Fig. 3.7 Total claims of internationally active banks (in trillions of US dollars). *Source* The Bank for International Settlements

in 2006 and US$26.69 trillion at the end of June 2009.[19] Of this total claims of US $23.07 trillion in 2006, US$18.46 trillion or close to 80 % were claims between rich and rich (developed countries). The situation was similar with regard to late June 2009, when 77 % of the figure given above for total claims by internationally active banks was made up of claims internal to the group of wealthy countries. According to the latest available data, the total claims of internationally active banks at the end of the third quarter of 2013 amounted to US$34.5 trillion, while the concentration of these claims between developed countries remained at almost the same level as in the last 3 years of the first decade of this century (Fig. 3.7).

One of the greatest paradoxes of financial liberalisation as conducted in the United States and Great Britain during the 1980s, and then in the EC countries (later European Union) and, to some degree, in developing countries, is the fact that the two financially most sophisticated environments, the United States and Great Britain, became the greatest importers of capital. That is, from 1982 to 2013, both these economies had a deficit of savings to investment in every year of that period.[20] In other words, both the United States and Great Britain became net importers of capital, from the point at which they started to implement measures of full financial liberalisation.

[19] The Bank for International Settlements—Committee on the Global Financial System, Op.cit., p. 11.

[20] Data on the ratios of savings to investment as percentages of GDP for all the countries in the world may be found in the IMF World Economic Outlook for any given year that the interested reader or analyst wishes to analyse: http://www.imf.org/external/pubs/ft/weo/2012/01/weodata/index.aspx.

3.4 The Challenges Facing Economic Science and Economic Policy as a Result of the Measures Implemented During the Global Crisis in the Integrated Global Economic System

In the first part of this book, it had already been stressed that the new classical macroeconomics, as well as the neo-Keynesian and the new Keynesian economics, whose models dominated and still dominate the economic policy making, are all based on acceptance of the law of diminishing returns and its consequence, the inevitable growth of the marginal cost of production, which becomes ever more intensive as actual output comes closer to the level of potential output. Thus, the problem of inflation and maintaining employment at the level of full (or close to full) employment has been central for these models, on the very important assumption that marginal cost necessarily rises as the degree of capacity utilisation increases. Under conditions of rational expectations with sticky prices and sticky information (with Fisher, Mankiw and Reis), an expansionary monetary policy will, in the short run, according to the new Keynesians, affect growth in production and employment without changing prices, but in the medium to long run, prices will change as a result of demands for a higher price of labour and goods, as new information comes in from markets. Consequently, there will, in the medium and perhaps in the long run, when the economy reaches the level of full employment, be a problem of overheating, so that conducting an expansionary fiscal and expansionary monetary policy necessarily results in acceleration of inflation.

Very intense international flows of long-term capital in the form of foreign direct investment being directed from the developed to developing countries during the late 1980s and particularly during the 1990s and the first years of this century led to major changes in the cost structure of globally active, multinational manufacturing companies, however. The major differences in the price of labour between the developed and developing countries, like China, India, and the countries of Southeast Asia, created a basis for a very significant reduction in the production cost of consumer goods, with a major impact on the average level of prices and the rate of inflation in developed countries. Thus, the production of the consumer goods and components for the manufacture of final capital goods, which global companies, using the possibilities of foreign direct investment in the developing countries of the Far East and Southeast Asia, produced in those countries and sold on the markets of their countries of origin at far lower prices than when those same goods were being produced in the developed countries using a domestic workforce, contributed to a reduction in inflationary pressure on developed countries' markets, as well as on those in developing countries.

In other words, the international capital mobility arising from the financial liberalisation measures implemented in developing countries, particularly the most populous ones like China and India, predominantly in the area of long-term capital flows based on foreign direct investment, brought about a sharp reduction in the costs of production, compared to the same costs on the national markets of developed countries. Consequently, one of the fundamental assumptions of both the new classical model and the new Keynesian model of production in developed market environments, that is the assumption of growing marginal costs and the consequent preoccupation with inflationary pressures, ceased to be a key problem in the period from 1990 to 2010 in the globally connected major economies (both developed and developing).[21]

The globally connected major economies and the specific economic "alliances" and interests arising therefrom, the key segment of which is the specific G-2 axis between the United States and China, give rise to a need for new macroeconomic models. What needs to be stressed here is the challenge and need for macroeconomics to invest additional effort in grounding a macroeconomic model that we might call a new macroeconomics of the integrated global economy. Starting from the Obstfeld-Rogoff model, which, as we reminded ourselves in the first part, is based on two big developed open economies in which households have similar preferences, insofar as their incomes are at a similar level and their needs similarly differentiated, given that they are at the same level of development, it would be useful, if we are to analyse the effects of relative changes in the key instruments of economic policy (relative changes in the money supply, relative prices, the levels of production, and the impact on utility maximisation), to develop an economic model of trade between two big open economies in which one is a developed and the other a developing country. The developed economy invests long-term capital in the developing country and links that economy to its own markets, through foreign direct investment based primarily on vertical intra-industry trade. The preferences of the populations differ significantly, because of major differences in income level, and additional money supply has different effects on priorities for investment in these two economies. The largest part of the money supply growth goes, in the developed economy, to the investment portfolio (speculative demand for money), while in the developing country the lion's share goes on increased investment in capital goods.

Thus, redefining the open economy macroeconomics model from one for a world comprising two big developed open economies to one for a world comprising two big open economies of which one is developed and the other is a big developing economy entails the assumption that the primary flow of capital as a result of the process of financial liberalisation is directed from the developed country to the developing country through foreign direct investment, while flows in goods are

---

[21] According to McKinnon, the United States' interest does not in fact lie in putting constant pressure on the government of China to significantly appreciate the Yuan against the Dollar, with a view to eliminating trade imbalances. McKinnon thinks that it is in the American interest to control the level of prices, that is to avoid pressure on the prices of mass consumer goods imported from China as a result of appreciation of the Yuan. Were the Yuan to appreciate significantly over the short term, this could have a significant impact on destabilising price levels in the United States, which is not in this country's interest, because it wishes to maintain the leading role of the dollar in the world. See: Ronald McKinnon [9], "US Current Account Deficits and the Dollar Standard's Sustainability—a Monetary Approach," published in: Eric Heilleiner and Jonathan Kershner, edds. (2009), *The Future of the Dollar,* Cornell University Press, Ithaca and London, pp. 45–68.

interconnected and conditioned by the fact that as a result of the foreign direct investment the big developing economy is integrated into the production chains of the big developed one. Such interrelationship entails that one of the central assumptions of the Rogoff-Obstfeld model has to be abandoned, namely the assumption of a constant elasticity of substitution of goods (the CES function), as the CES function and the derived theta-coefficient have to be modified and adapted in the new model to the structure of the flows of trade in which there is on the demand side a vertical industrial specialisation, while on the supply side exchange of goods is based on intra-industry trade and imperfect markets.[22] Customers from developing countries tend to seek smaller quantities of goods at significantly lower prices, while customers on developed countries' markets seek larger quantities of more sophisticated products at higher prices, and accordingly, with a higher level of satisfaction of their needs.

Such a structure of trade between a big developed economy and a big developing economy that is primarily vertically integrated within the structure of the real sector entails that the portfolio investment flows (investment in financial assets) are determined by this type of production and trade relationship. If we were to represent in our model the real sector of both economies in terms of a single aggregate share, where the returns on that share would be determined by the progress of the business cycle, which is to say the profit potential of the real sector of both economies, then, in such a model, in which the economies are vertically integrated, the possibility of diversifying risk in portfolio investments becomes very reduced or in fact practically disabled. The global system is interrelated intersectorally, so that the returns on equity in the real sector of the major developed economy has a direct impact on returns on equity in the major developing economy. Consequently, the correlation coefficient of expected returns on the global capital market is positive, and, starting from the principle of effective investment and optimisation, as established in Markowitz's theory and later refined by James Tobin, it is not possible to optimise the structure of investment on financial markets. This would entail a continuous threat to the possibility of increasing capital, i.e. to household savings invested in institutional investors.

The financial liberalisation measures implemented in the United States, Great Britain and the states of the European Union during the second half of the 80s and the 90s thus resulted in very significantly reduced capacity on the part of democratically elected representatives and the specialised institutions of Western democratic societies to control the flows of money. Ever stronger demands for increased intensity of financial liberalisation in the late 1990s, particularly in the United States, resulted in the already mentioned measures by the Clinton administration in the last 2 years of Bill Clinton's second mandate. By repeal of the Glass-Steagall Act (1999) and introduction of the Commodity Modernisation Act (2000), the legal obstacles were removed to complete domination by the interests of private banking

[22] Reference [6].

groups, which have, for the most part, remained outside the control of democratically elected institutions.

The pace at which private banking group interests have been strengthening was additionally reinforced by lobbying on the part of the largest financial groups originating from the United States, Great Britain, France, Germany, Spain and Italy, to change the model for managing risk and capital. Thus, the capital adequacy ratio standards of 8 % established under Basel I created unnecessarily large reserves of capital, in the view of the largest internationally active banking groups, and so caused major opportunity costs and significantly reduced their global competitiveness and profitability. Under powerful lobbying by these institutions from 1999 to 2003, the Basel Committee for Banking Supervision adopted three consultative papers, opening the path to the adoption of new standards in international operations, called Basel II. Basel II was finally passed in June 2004 and the standards, along with the already mentioned financial liberalisation measures affecting OTC derivative markets, gave the largest banking groups in the world the right to set their own capital adequacy ratios on the basis of internal models for determining client rating and risk, that is the line of operations.[23] In effect, with Basel II or, more precisely, the earlier adopted Consultative Paper no. 2 (in January 2001) the megabanks were allowed to set a capital adequacy ratio at a level nearly four times less than that had been demanded by Basel I.

The above-mentioned measures allowed internationally active banks effectively to become their own regulators. Such situations, in which a person (in this case a legal person) is allowed to create their own rules and to act according to those rules (naturally, primarily in their own interest), represented the basic source of global financial disturbance. George Soros referred, in *The Crisis of Global Capitalism*,[24] to this phenomenon as a "reflexive situation." That is, situations where "rule takers" are at the same time "rule makers," so that they are lobbying for rules which they then adapt in application to their own interest, often at the expense of the public, or society and the global system of which they are part, are reflexive situations, according to Soros, from which major problems for the future of open societies are to be expected.

Thus, the measures of financial liberalisation adopted between 1999 and 2004 created a situation in which the money supply was predominantly endogenously determined, that is determined on the basis of the business policies and profit motives of banking groups which unified the operations of commercial and investment banking, as well as those of trading in financial derivatives on rapidly growing and, between 2000 and 2009, almost entirely deregulated over-the-counter markets. Given a US monetary policy that was, during the periods in which financial bubbles were being created, powerless (or uninterested) to step in, through determined measures to increase the interest rate, the enormous growth in lending

---

[23] For a detailed description of these activities and the impact of adopting CP1, CP2, CP3 and the Basel II standards, See Reference [15], pp. 104–135.

[24] Reference [14].

activity from 2002 to 2008, particularly on the interbank market, and given the multiple systems for ensuring through the issue and sale of financial derivatives that risk transferred de facto onto the public budget, a situation was created which is best described in theoretical terms in the works of the post-Keynesian economists who developed the monetary circuit theory. Between 2000 and 2008, the behaviour of private banking groups, on the one hand, and the behaviour of the most significant central banks of the world, on the other hand, corresponded almost exactly to the predictions of that current of Keynesian fundamentalists who believe that the central banks "close the circuit of money" and, in fact, are just adjusting to the consequences of the business decisions made by private financial groups, who dictate through their lending activities the issue of money.

References

1. Akerlof G, Robert S (2009) Animal spirits: how human psychology drives the economy, and why it matters for capitalist societies. Princeton University Press, Princeton, NJ
2. Davis H, David G (2010) Banking on the future—the fall and rise of central banking. Princeton University Press, Princeton and Oxford
3. Gilbert AL (1986) requiem for regulation Q: what it did and why it passed away, Federal Reserve Bank of St Louis, February
4. Greenspan A (2007) The age of turbulence—adventures in a new world. Allen Lane, Penguin Book, London
5. Greenspan A (2013) The map and the territory—risk, Human Nature, and the Future of Forecasting, Allen Lane
6. Kovač E, Krešimir Ž (2014) International competition in vertically integrated markets with innovation and imitation: trade policy versus free trade, Economica. London School of Economics, London
7. Markowitz HM (1991) Portfolio selection: efficient diversification of investment. Blackwell, Massachusetts and Oxford
8. McKinnon R (1973) Money and capital in economic development. The Brookings Institution, Washington
9. McKinnon R (2009) US current account deficits and the dollar standard's sustainability—a monetary approach. In: Eric H, Jonathan K (eds) The future of the dollar. Cornell University Press, Ithaca and London, pp 45–68
10. Obstfeld M, Alan AT (2002) Globalization and capital markets. National Bureau of Economic Research, April
11. Prasad E (2014) The Dollar Reigns Supreme, by Default. Finance and Development, March, vol 51, No 1
12. Skidelsky R (2010) The return of the master. Penguin Books, London
13. Shaw ES (1973) Financial deepening in economic development. Oxford University Press, Oxford
14. Soros G (1998) The crisis of global capitalism—open society endangered. PublicAffairs, New York
15. Tarullo DK (2008) Banking on Basel—the future of international financial regulation. Peterson Institute for International Economics, Washington, DC (August)
16. The Bank for International Settlements (2010) Committee on the Global Financial System "Long-term issues in international banking," *CGFS Papers No 41,* July 2010

17. The International Bank for Reconstruction and Development/The World Bank (1993) The east asian miracle—economic growth and public policy. Oxford University Press, Oxford
18. Volcker P (2009) Financial Innovations Disastrous. http://www.newsmax.com/StreetTalk/paul-volcker-financial-innovationdisastrous/2009/12/16/id/343421/. Accessed Dec 16
19. Yoshitomi M (2007) "Global Imbalances and East Asian Monetary Cooperation". In: Duck-Koo C, Barry E (eds) Toward an east asian exchange rate regime, Brookings Institution Press, Washington, DC., Chapter 2

Conclusions

Measures taken by the central banks and finance ministries of the most developed countries, and in particular by the Fed and US Treasury in 2008 and the 5 years that followed, represented in effect a return to Keynes's original teachings as contained in the General Theory. The remedy for preventing the Global Crisis of 2008 becoming a global depression, which is to say the recipe for the struggle against steep falling economic activity and steeply rising unemployment, was a highly expansionary monetary policy combined with rapid fiscal expansion. Keynes's recommendations, given in the final chapter of the General Theory and cited in the initial pages of this book, as to the need to socialise investments, but only insofar as the state must ensure through its fiscal and monetary policy measures that growth of aggregate demand is sufficient to produce exit from the bottom of the business cycle and simulate investment by private economic agents, were taken on board by the United States, Great Britain and Japan, and somewhat later by the countries of the European Union. So, monetary and fiscal measures came to play a decisive role in preventing a steep fall in aggregate demand and an equally steep fall in the confidence of private investors as to the prospects for making a profit in the near future. Additional confirmation of how directly Keynes's recommendations for exiting a major economic crisis had been taken over was provided by the measures recommended by traditionally the most conservative of the international financial institutions, at least when it comes to the conduct of fiscal policy—the International Monetary Fund. After a speech made by the former head of the IMF at the Banco de Espana in December 2008, a speech on the urgency and necessity for embarking upon a very expansive fiscal policy on a twofold basis—both in terms of allowing public debt to rise and in terms of cutting taxes, Keynes was clearly and quite definitely back in the heart of economic policy and theory. A further indication of this return of interest in Keynes and Keynesian recommendations for economic policy came in the 9 October, 2008, issue of the The Economist, whose cover page highlighted a text on *Saving the System* and a special report on the global crisis entitled *The World Economy*. Even though the conclusions to the report stressed how very wrong it would be to conclude that the free action of the market was exclusively to blame for the most severe economic crisis since 1929, explicit emphasis was nonetheless put on the urgent need for swift action by the central banks and ministries of finance of the developed countries.

F. Čaušević, *The Global Crisis of 2008 and Keynes's General Theory*,
SpringerBriefs in Economics, DOI 10.1007/978-3-319-11451-4

The rejection of Keynes's teachings during the 1980s by the economic policy makers in the United States and Great Britain, based on the arguments of the new classical macroeconomics, had resulted in the suppression, rejection or just plain ignoring of some of the most important sections presented in the General Theory. The second part of this book has been dedicated to presenting his basic positions as set out in the General Theory, chapter by chapter, not least because one sometimes gets the impression that a good many of the "interpreters" of the General Theory, in both West and East, have failed to read what is unquestionably the most significant work of economic science of the past century with sufficient attention (if at all). Krugman has compared it to a meal, whereby the first and last parts of the book represent "a tasty starter and a very fine dessert," but points out that if the intention is to understand the General Theory then the reader can hardly avoid the major chapters in the central part of the book, which are not particularly easy to understand or read, but nonetheless contain the essence and basis on which the entire theory stands. The key variables of Keynes's system of thought, in this work, are psychological in nature—the propensity to consume, the liquidity preference, and the marginal efficiency of investment, which depends upon expectations of future return. While the key actors in the economic system—consumers, which is to say households and private companies in the business sector, form their expectations on the basis of the information available, on the one hand, financial markets are dominated, on the other, by mass psychology, which is not founded on the rationality of actors, but on the attempt to adapt one's own behaviour so as to "guess" what the majority think will be the majority's choice and on the basis of that form one's own behaviour to be in line with what the behaviour of majority is expected to be, and not with any form of rational judgement based on an unbiased analysis of the relevant information. This is why the dominant characteristic of private sector behaviour is—herd behaviour, which is subject to frequent (and sometimes sudden) changes of expectations and mood.

Such changes in the state of expectations lead to sudden and sharp fluctuations in investment, given that investment is to a large extent a function of returns expected in the years to come, and expected returns depend on the psychological propensity to consume, which tends to fall as incomes grow. If the increase in income is accompanied by growth in the concentration of wealth, and so by growth in inequality in society, as a consequence of mistaken tax policy, whose regressivity fosters a concentration of wealth in social groups of rentiers, the marginal propensity to consume amongst the most wealthy strata of society will diminish steadily and a problem arise of a mismatch between savings and investment. Savings outrun investment and aggregate demand is insufficient to absorb the level of production developed in preceding periods, leading to a growth in stocks, falling investor confidence regarding future yields, growth in pessimism, the appearance of large-scale involuntary unemployment and social pressures. Such social pressures become very dangerous for the reproductive prospects of an economic system based on private ownership and the accumulation of capital and government intervention in the investment cycle becomes necessary. It is the government that "steps in," increasing demand, converting the sharp growth in pessimistic expectations into a moderate optimism. The rise in capital expenditures, financed by the issue of government bonds, bought by the central bank, induces an increase in private investment, while moderate inflation at unchanged nominal money wages brings about a fall in real wages and opens up room for private sector profits to grow, along with investment and employment.

Keynes's analysis was primarily based on psychological factors and uncertainty over future prospects for profit. It undoubtedly represented a major event in economic science. Keynes's theory of money and the interest rate, based on the psychological liquidity preference, on the one hand, and of the demand for money, based on the transactions motive, the precautionary motive, and most particularly, the motive for speculation, represent a contribution to economic science which has not been superseded even today. Even though, at the time when Keynes wrote his General Theory, the overall development of financial markets and the integration of national financial markets into global markets were both at a considerably lower level than they would after the second half of the 1970s and in particular after the build-up during the last two decades, his theory and analysis of the ways in which demand for money affects the real sector through the financial markets has not lost any of its significance. In fact, it has become even more significant than it was at the time when it was written. Keynes's analysis of the role and significance of speculative demand for money in his analysis of changes to the interest rates and the monetary policy transmission mechanism is now one of the central pillars of economic theory and forms the groundwork on which must rest any analysis of the creation of money by central banks and the significance of the role of private financial transactors' speculative demand for money at the level of the interest rates and the transmission of changes in the expected return of financial assets onto investment activity in the real sector of the economy.

In the 15th chapter of the General Theory the psychological and business incentives for the analysis of demand for money are explained in more detail. Keynes lays out clearly that the transaction demand for money and the demand for money for precautionary purposes represent a relatively stable ratio between income, or the scale of economic activity, and the demand for money, which arises from these motives. Keynes dedicated special attention to speculative demand for money in his analysis of the transmission mechanism onto the business cycle. The liquidity preference is the key variable in Keynes's theory of the interest rate, insofar as it is through liquidity preference that Keynes defines the interest rate as the price which private transactors require to forego liquidity, and not as the price for postponing consumption. Keynes, however, also gives great significance to the central bank and its impact on the level of interest rates on money markets. On the other hand, in the analysis of the significance of speculative impulses, Keynes stressed that the expectations of financial investors and their assessment of the level of equilibrium between the interest rate and the rate of return on financial assets have a very significant impact on the level of long-term interest rates, the consumption of capital and the marginal efficiency of investment.

In his analysis, Keynes wields what remains today an unquestioned authority, insofar as, through his introduction of speculative demand for money and the analysis of the behaviour of financial transactors on capital markets, he laid the foundations for the incorporation of changes in the price of financial assets into the analysis of macroeconomic equilibrium. By insisting that a focus on open market operations and on buying and selling short-term government securities did not allow central banks to react rapidly enough to changes in the state of expectations on the capital markets during periods of crisis, Keynes drew explicit attention to the need to expand the their influence in the direction of direct purchasing of long-term government bonds. The changes to the structure of the balance sheets of the Fed, primarily, but also the Bank of England, the Bank of Japan and the European Central Bank over the past six years, show that the Fed and its most important allies have been applying Keynes's recommendations quite directly with regard to the need for rapid reaction in periods of crisis and more significant change to the structure of central banks' balance sheets, in order to head off sharp growth in pessimism by preventing a fall in the prices of long-term government securities and a cut in long-term interest rates.

Keynes's theory of the scarcity of capital, his analysis of the change in the structure of costs, and his direct view that a policy of very low interest rates is required to maintain full employment, which in turn would require monetary expansion, represent those parts of the General Theory which are most susceptible to criticism, given that Keynes never gave an answer as to how to prevent inflation under conditions of full employment from accelerating towards ever greater intensity. Keynes's theory of inflation was primarily based on the cost inflation, insofar as he stated that "true" inflation would come about at the level of full employment, when any additional money supply would be largely or even entirely absorbed by a rise in user costs and product prices, but that it would have practically no impact on increasing output. Given that he was analysing the economy at a given state of technology and that he did not devote a particular chapter to the analysis of technological progress or the structure of costs, Keynes's theory of employment rising to the level of full employment, both in theoretical and economic and political terms, left open the problem of how to maintain full employment without moderate inflation becoming rapid inflation and so economic stagnation, followed by the creation of a crisis arising on this very basis. This internal contradiction in the General Theory is what opened up room for the intellectual offensive of the new classical economists and monetarists during the 1970s, or rather more specifically in the second half of the 1970 and 1980s. It was the thesis of these new classical macroeconomics that monetary and fiscal policy could not affect real variables, given rational expectations, and their thesis presupposed the constant capacity of economic actors to maintain the economy in equilibrium at the level of full employment through the system of market exchanges. The new Keynesians accepted rational expectations and introduced sticky prices for labour and goods into the analysis, attempting to prove that economic policy could affect real variables, even under conditions of rational expectations. In contrast to the original Keynesian interpretation of how to get out of a crisis, which was essentially to keep nominal money wages unchanged but allow the prices of goods to grow, with a consequent reduction in real money wages, the new Keynesians, by introducing their sticky prices and rational expectations, which are equally contrary to the spirit of Keynes's original authentic teaching, significantly deviated from the foundations on which the General Theory itself rests.

MIX
Papier aus verantwortungsvollen Quellen
Paper from responsible sources
FSC® C105338

If you have any concerns about our products,
you can contact us on
ProductSafety@springernature.com

In case Publisher is established outside the EU,
the EU authorized representative is:
Springer Nature Customer Service Center GmbH
Europaplatz 3, 69115 Heidelberg, Germany

Printed by Libri Plureos GmbH
in Hamburg, Germany

SpringerBriefs in Well-Being and Quality of Life Research

More information about this series at http://www.springer.com/series/10150

Youngwha Kee · Yunji Kim
Rhonda Phillips
Editors

Learning and Community Approaches for Promoting Well-Being

Springer

Editors
Youngwha Kee
Soongsil University
Seoul
Korea, Republic of (South Korea)

Rhonda Phillips
Purdue University
West Lafayette, IN
USA

Yunji Kim
Cornell University
Ithaca, NY
USA

ISSN 2211-7644 ISSN 2211-7652 (electronic)
SpringerBriefs in Well-Being and Quality of Life Research
ISBN 978-3-319-12438-4 ISBN 978-3-319-12439-1 (eBook)
DOI 10.1007/978-3-319-12439-1

Library of Congress Control Number: 2014955609

Springer Cham Heidelberg New York Dordrecht London

Printed on acid-free paper

Springer International Publishing AG Switzerland is part of Springer Science+Business Media (www.springer.com)

Preface

This volume began when a group of scholars, the Community Well-being Research Network, convened in 2012 at Seoul National University. With representatives from throughout Asia, Europe, and the US, ideas began to coalesce around community well-being as applied to local governments and learning in communities. This collection of chapters represents the outgrowth of the convening, along with another volume in the *Springer Briefs* series, *Community Well-being and Community Development: Conceptions and Applications*, edited by Seung Jong Lee, Yunji Kim, and Rhonda Phillips. It is our intent to spur interest in community well-being with both conceptual and applied work.

The first two chapters present around the idea of governance—that is, active and participatory approaches by local government including meaningful participation by residents and other stakeholders progressing towards desired outcomes. The idea of governance versus government has become quite popular of late, due to the appeal of having more transparent and effective decision-making within local governments. Further, the notion of including people in the process of governance can build community capital as well as increase likelihood of achieving outcomes for enhancing community well-being.

Youngwha Kee, Yunji Kim, and Rhonda Phillips, the editors of this volume, present concepts in "Modeling Community Well-Being: A Multi-dimensional Approach". Beginning with concepts and definitions that range across related terms, the authors present underlying assumptions influencing development of a model for defining community well-being. Discussion of differences between individual and collective well-being drive the basis of the model. This chapter is presented in the context of local government, with considerations for incorporating community well-being goals into governance for achieving desirable outcomes.

Youngwha Kee and Okchae Joo present intriguing ideas about communities of practice in "Community Learning and Self-generated Communities of Practice: The Case of Jangheung School". The authors explore public policy approaches to development guided by resident participation. Considerations and implications for communities of practice are presented, along with stages of development and activities for community learning.

Wen Juan Lu and Longgen Chen present an applied case study with "Promoting Community Integration of New Migrant Workers in China". This chapter focuses on tying the concept of community well-being, which is relatively new to Chinese scholars and policymakers, to facilitating new migrant workers into society. Issues around migrant workers include how to transition rural worker skills to other settings successfully, how to facilitate new migrant workers' integration into cities effectively, how to guarantee their equal civil rights and privileges, and how to promote their survival ability and employability in cities. These and related issues demand attention and the context of community well-being provides the platform for governance approaches.

This volume concludes with "Community Well-Being Through Intergenerational Cooperation" by Ann-Kristin Boström, Seung Jong Lee, and Youngwha Kee. The authors point out that the population of the world has grown old while the generation gap between younger people and older people has widened in industrialized countries. Intergenerational initiatives and projects have been developed in various industrial countries in order for younger and older people to meet more frequently than is often otherwise the case. These models of intergenerational learning hold much promise for enhancing community well-being and encouraging learning in communities while at the same time addressing social issues related to aging. This chapter presents an applied project from Sweden, the Granddad Project, that partners elders with elementary schools.

We sincerely hope that you will find this volume both useful and inspiring in your own work around the issues of governance and learning. Community well-being is quickly gaining ground as a basis or lens through which to understand and explore ideas, models, programs, and policies for encouraging more desirable processes and outcomes in our communities worldwide.

Youngwha Kee
Yunji Kim
Rhonda Phillips

Acknowledgments

This work was supported by the National Research Foundation of Korea grant funded by the Korean Government (NRF-2013S1A3A2054622).

Contents

About the Contributors

Ann-Kristin Boström works as a researcher at Encell. Her research area includes international and comparative education, intergenerational learning, and social capital. Boström is active in several international networks and is also responsible for the evaluation of the European project, ECIL. The aim of this project is to pilot a collection of intergenerational life-long learning training approaches developed in the UK with a number of partners across Europe. Since she received her Ph.D. in International Education, Boström has worked as a director of education for the Swedish Ministry of Education and Research and as a special advisor for the Government Offices of Sweden. Here, she has mainly worked with research questions and she has also represented Sweden in the European Council, within the Organization for Economic Cooperation and Development (OECD) and the European Union (EU). She has been responsible for two national reports in connection with two OECD-studies: "Migrant Education" and "Evaluation and Assessment in Education."

Longgen Chen is a professor and dean of the School of Adult and Continuing Education, Zhejiang University of Technology, China. His research interests include adult education, higher adult education, and community education. He has chaired or participated in more than 39 national, provincial, and ministerial projects. He is the author of *China's Higher Adult Education in the New Century* and numerous articles in the *Journal of Higher Education in China.*

Youngwha Kee is a professor of the Department of Lifelong Education at Soongsil University and the president of the National Institute of Lifelong Education. She currently serves as the director of the Korea Institute of Local Development Education. Previously, Dr. Kee was president of the Association of Adult and Continuing Education of Korea and researcher of Korea Association for Community Education. She has served on several advisory committees in relation to educational policies and has been deeply involved with community education among multicultural families and education for the disadvantaged. She serves on the editorial board of the International Journal of Continuing Education and Lifelong Learning (Hong Kong) and the Lifelong Education Magazine (Taiwan),

Lifelong Education (China). Her research interests include older adult learning, community education, civic education, community development, and governance.

Yunji Kim is a doctoral student in the Department of City and Regional Planning at Cornell University. She received her master's degree from the Graduate School of Public Administration, Seoul National University in 2011. Her master's thesis (Title: "Development and Application of a Community Well-being Index for Korea") focused on defining community well-being as a guiding principle for local governments and suggested a new community well-being index for Korean metropolitan districts. Her current research interests include the relationship between community well-being and local government services; citizen participation; and community development. Her recent articles include, "The Development and Application of a Community Well-being Index in Korean Metropolitan Cities," "An Analysis of the Relative Importance of Components in Measuring Community Well-being: Perspectives of Citizens, Public Officials, and Experts," and "Sing, Dance, and Be Merry: a Strategy for Successful Urban Development?"

Okchae Joo is a public official in the Ministry of Public Security and Public Administration. Her research interest includes lifelong education and local self-governance. She holds a master's degree in Public Administration and a doctoral degree in Lifelong Education. Her master's thesis is titled, "A Study on the Inspection Administration of Local Government" and her doctoral thesis was titled, "Building Process of Community of Practice in Village Community."

Wenjuan Lv holds a doctoral degree in Adult and Continuing Education and is a lecturer at the School of Educational Science and Technology, Zhejiang University of Technology, China. Her research interests include community education, older adult education, and vocational education. She has participated in several research projects at the provincial and ministerial level and is the author of *Successful Aging of Chinese Older Adult Learners* and the co-editor of *Transforming Development of Continuing Education in Higher Institutions*.

Rhonda Phillips, Ph.D., AICP, is Dean of the Honors College at Purdue University and a professor in the Agricultural Economics Department. A former senior sustainability scientist in the Global Institute of Sustainability for Arizona State University, she served as professor in School of Community Resources and Development. Community well-being and development comprise the focus of Rhonda's research and outreach activities including community-based education and research initiatives for enhancing quality of life. Honors include serving as the 2006 UK Ulster Policy Fellow Fulbright Scholar and a 2012 Fulbright Senior Specialist to Panama. Rhonda is author or editor of 18 books, including *Community Development Indicators Measuring Systems*, and *Introduction to Community Development*, and serves as editor for the Springer series, *Community Quality of Life and Well-Being*. Rhonda is president of the International Society for Quality-of-Life Studies, www.isqols.org.

Modeling Community Well-Being: A Multi-dimensional Approach

Youngwha Kee, Yunji Kim and Rhonda Phillips

The last decades of the 20th and 21st century have been a busy period in civilization: the invention of the Internet, decoding the human genome, and subsequent explosion of technological advances from cell phones to biomedical procedures are only some examples. Yet, the search for a better life—for something more—seems to be persistent in both academia, politics, and popular culture. If the world is so much better as some claim, then why are we seeing greater demand and interest in well-being and happiness? For example, although calorie intake among the poor is increasing, this indicator does not capture the full range of impact. Questions about quality, and "what *kinds* of calories are these?" is important to answer in regards to quality of life status. Pagani and Huot (2007) use data from Canada to show that children living in poverty are more likely to have an unhealthy diet, leading to obesity. Beaulac et al. (2009) and Larson et al. (2009) pointed out that in low-income neighborhoods there are significantly less numbers of grocery stores or farmers' markets where people can access fresh fruits and vegetables.

We see this demand as a pushback response to the Gross Domestic Product (GDP)—centric paradigm that focuses predominately on economic, quantitative accumulation rather than a more balanced goal. However, we also see limits to the current focus on happiness and individual well-being in academia as well as political realm. In response, we suggest that community well-being is a more appropriate concept to connect governments—and governance—to citizen lives. Yet, community well-being has its critics as well. For example, Scott (2012) has pointed out that this term has been used by those on the left as well as right on the political spectrum due to the ambiguity around its meaning, and thus can mislead

Y. Kee (✉)
Soongsil University, Seoul, Korea, Republic of (South Korea)
e-mail: key@ssu.ac.kr

Y. Kim
Cornell University, Ithaca, NY, USA
e-mail: yk634@cornell.edu

R. Phillips
Purdue University, West Lafayette, IN, USA
e-mail: rphillips@purdue.edu

Y. Kee et al. (eds.), *Learning and Community Approaches for Promoting Well-Being*,
SpringerBriefs in Well-Being and Quality of Life Research,
DOI 10.1007/978-3-319-12439-1_1

citizens. Lee and Kim (2014) have also emphasized the danger surrounding this ambiguity and this chapter explores some of these concerns. In detail, we situate community well-being in the welfare state framework as the *raison d'être* of local governments and governance, presenting an asset and capital-based model of community well-being. We hope this model will serve as a map for interdisciplinary studies that are crucial for the realization of community well-being.

Why Community Well-Being? Function of Local Governments

Community well-being is a key concept in the 21st century welfare state discourse (Seo et al. 2012). Since the prior century, the theory of the welfare state has received much attention as the value and goal of nation states and successful cases of welfare states have been widely circulated. GDP and meeting material needs became the main criteria for assessing the state of progress or policies in nation states. However, in the 1970s criticisms against the GDP were voiced and furthermore, efforts to measure progress on an individual, collective, and national level were made. GDP uses material production and consumption as the essence of human life and was not able to answer criticisms that human life is not defined by material needs only. This led to concepts other than GDP being used to discuss social issues.

Happiness is one example. Happiness became a popular term in the 1970s and in particular, the tiny country of Bhutan attempted to replace the GDP with a Gross National Happiness (GNH) Index. However, happiness was focused on an individual's subjective emotions and thus limited in its use as a national and collective criterion. As a response, there were efforts to replace happiness with the concept of quality of life and well-being. Continued demand for a new concept can be identified in various disciplines (Smits and Hoekstra 2011) and these demands were in line with the social indicator movements led by the United States and Scandinavian states (Ilić et al. 2010). The need for a concept that encompasses the non-economic, non-material aspects, such as culture and context became clear. National governments made various efforts to discover something more than simple GDP as policy visions. Britain's Prime Minister David Cameron spoke of general well-being as a national policy vision, while former U.S. President Bill Clinton's election victory is an example that shows economics is not the only important element in deciding the outcome of elections (Ormerod 2012). In addition, French President Nicholas Sarkozy focused on social welfare rather than simple GDP to evaluate the nation's standing. These countries are key actors in international organizations and thus their experiences are being widely circulated and shared with other countries. Community well-being is deeply connected to a government's capacity and ability to realize the social welfare state.

If a nation state's goal is to realize a welfare state, then the question of who is the actor in this process is important. In *Development as Freedom*, Sen (1999) discusses this question using the term "agent." He defined progress as the expansion of capabilities on the basis of individual freedom and emphasized that public policies can strengthen these capabilities. The agent's evaluation is based on citizens' goals. Given this, even in a small, centralized nation state, it will be difficult for the national government to directly act as an agent. Instead, the task of coordinating the several activities of welfare realization will be more realistic. It is difficult for national governments to directly connect with citizens, and thus local governments will become the key agents. Local governments may pursue the welfare state on a national level but local governments will be the main agents and the goal of a welfare state is inappropriate for local governments.

The reason for local governments acting as the agents of the welfare state is that the welfare state and community well-being share commonalities that can be separated by the level at which this is being realized. Well-being is the point at which the national government connects to local governments because both community well-being and welfare states contain the concept of *well*. The term well signifies good, best, or ideally acceptable. The only difference is that one sees *well* as a *fare* (i.e. state) while the other sees *well* as *being*. Despite this difference, the state of well and the being of well can be seen as synonymous. *Fare* is a more macro condition whereas *being* is an individual aspect of this condition. If the national government pursues welfare and local governments pursue well-being, the end product is actually the same: a collective body of well citizens.

Thus we see that community well-being can be the value and goal of local governments in the realization of a welfare state. We have also argued in previous studies that community well-being is the value and goal of local governments (Seo et al. 2012; Kim and Lee 2013). Our arguments are based on the following assumptions. Local governments affect the national government. Local governments deliver the services that local residents need and can affect the level of citizen satisfaction with respect to national governments. Second, the issues faced on a local level are multidimensional. The local community is where individuals' daily lives are practiced. Third, local issues require the involvement of local governments. The local community is where individuals' needs and demands are voiced and they expect local governments to respond to these voices.

In sum, there needs to be an appropriate value goal for local governments, and this chapter aims to model community well-being as the appropriate value goal. The community well-being model is based on an analysis of related concepts and highlighting the uniqueness of community well-being. It is also connected to community development as local government is typically very concerned with this dimension. Community development can be defined as both a process and an outcome for communities:

> A process: developing and enhancing the ability of a community to act collectively and an outcome: (1) taking collective action and (2) the result of that action for improvement in a community in any or all realms: physical, environmental, cultural, social, political, economic, etc. (Phillips and Pittman 2014: 6).

Community development builds on the concept of assets or capital in an area across all realms as just defined. Building this capital and capacity entails active governance, with both local governments and their residents involved and participating in these processes. Governance essentially implies an active state of governing where residents and other stakeholders work with local government leaders to accomplish goals. Further, it implies that it is participatory, which has been found to be essential in successful local government models (Phillips and Pittman 2014). Within community development, we can see the direct relationship to community well-being, as both a goal/outcome and process of progressing towards desired states of well-being.

Comparing and Contrasting the Concepts of "Beyond GDP": The Uniqueness of Community Well-Being

Comparing and contrasting concepts of quality of life, well-being, happiness, and sustainability show that not all agree on their definition (Lee and Kim 2014). Some previous studies have attempted to tease out the difference between and among these concepts, while others have treated them synonymously. Easterlin (2001) studied the relationship between happiness and income at the national level and treated subjective well-being, well-being, satisfaction, utility, and welfare as synonymous. Booth (2012) also shows that a study by the Institute of Economic Affairs in the U.K. that analyzed more than 100 countries used well-being and happiness as the same concept. Well-being and quality of life both focus on an individual's psychological evaluation while subjective well-being is treated as an element of quality of life and thus it is difficult to precisely distinguish these terms. Meanwhile, happiness is similar to well-being in that they both focus on a subjective level of satisfaction but in general, well-being includes more than subjective evaluations, such as objective environmental and economic elements. Well-being can be seen as more objective than happiness since there is a certain criterion for levels of well-being while the criterion for happiness is entirely subjective and differs for every individual.

Lee and Kim (2014) compare and contrast happiness, quality of life, individual well-being, and community well-being using the following aspects: individual or collective, domain of study, quantity or quality, objective or subjective, static or dynamic, approach, goal, and value judgment. In sum, they see community well-being as the most comprehensive concept. For example, happiness is seen as being limited to the psychological domain, having an affective (i.e. emotions) approach, and limited to the subjective while community well-being includes cultural, economic, environmental, social, physical, and political domains and including both subjective and objective aspects. Quality of life is seen as being limited to qualitative aspects while community well-being encompasses both qualitative and quantitative. The term well-being is seen as an individual concept that is inappropriate for guiding public policies or local government decisions that affect the collective group. Community development is included to show the similarities of dimensions (Table 1).

Table 1 Comparison of community well-being and related concepts

Community well-being	Happiness	Quality of life	(Individual) well-being	Community development
Individual/collective (collective)	Individual	Individual	Individual	Collective
Domain Cultural, economic, environmental, social, physical, political	Psychological	Economic, social	Economic, social, physical, psychological	Cultural, economic, environmental, social, physical, political
Quantity/quality (both)	Quantity	Quality	Both	Both
Objective/subjective (both)	Subjective	Objective	Both	Both
Static/dynamic (dynamic)	Static	Static	Dynamic	Dynamic
Approach (asset)	Emotion	Capital	Asset	Asset
Goal (production; accumulation; flourishing)	Induce positive emotion	Fulfillment of deficiency	Production; accumulation; flourishing	Production; accumulation; flourishing; fulfillment of deficiency
Value judgment (value-driven)	Value-neutral	Value-neutral	Value-driven	Either

Source Lee and Kim (2014)

Modeling Community Well-Being

We use the definition of community well-being to analyze its characteristics and ultimately suggest a model of community well-being. To define community well-being, we compared previous literature on community well-being and modeled community well-being for in depth studies. We focus on the group of community for community well-being. In this process, given that individuals are the building blocks of a community, we also include individual elements. In particular, we see community well-being as an asset based approach. In contrast, quality of life aims to fill up deficiencies to achieve a zero base while well-being aims to go beyond a zero base of accumulation, flourishing, and production.

While there have been previous attempts to define and model these concepts, Adler (2013) points out that differentiation amongst these concepts has been weak and lack normative definitions. In fact, with a weak understanding of well-being among philosophers, the empirical studies of well-being by psychologists and economists are severely limited (p. 6).

The Multidimensional Approach

We use a multidimensional asset approach to modeling community well-being. This approach was chosen with an explicit attention to the fact that individuals have different levels of assets at their disposal. This is in line with the capability approach of Sen (2005) where the consumption focused models proposed by economists were critiqued because they ignore the fact that individuals have different levels of assets and more consumption from an unequal starting point cannot be a sustainable solution.

This study used an asset and capital approach to model community well-being. White (2008) discusses two approaches to community well-being: the first approach views community well-being as the sum or mathematical mean of individual well-being; the second approach sees community well-being as a unique feature of the collective. These approaches have implications for the indicator development process. In detail, the first approach focuses on measuring individual well-being and then calculating the total level of well-being, while the second approach searches for a unique measure of collective well-being that is not entirely derived from individual well-being measures. His community well-being measurement model consists of identical domains as that of individual well-being, such as material, subjective, relational elements. However, he offers different indicators within these domains. The material elements are related to welfare or quality of life, while relational elements are related to the individual or social relationships, and subjective elements include subjective values, perceptions, and experiences.

NWMO (2009) treats community well-being as a complex concept that combines an abstract idea with human behavior. Just as does community development, community well-being reflects the interests of individuals within a community, and also reflects the collective interest of the group. Thus, the term is unique to each group and also to each individual. NWMO defines community well-being by defining community and well-being separately and then combining them, which is similar to the first approach of White (2008). According to their definition, well-being includes social, economic, psychological, and cultural elements and include an individual's health and safety, is connected to the satisfaction level and quality of life within a group. A community is defined by geographical proximity or interests.

Assumptions of the Community Well-Being Model

Community well-being needs to examine the basic characteristics of well-being first. Community is the level at which we examine well-being and thus we focus on the characteristics of well-being in this section and use them as the following four main assumptions to model community well-being.

We assume that, community well-being shares the unequal characteristic of well-being. That is, this study uses a asset approach to community well-being which inherently contains inequality. Assets are already unequally dispersed throughout a nation and across the globe. Thus, there are unequal levels of community well-being. This study aims to achieve a balance and harmony despite this given structure, and thus uses an asset and secondary capital approach. This is in line with the freedom that Sen emphasizes: the freedom to realize one's capabilities. The government has a responsibility to provide the opportunities to realize these capabilities. This is why community well-being needs to be studied in close relation to local governments. Utilitarianism is also based on these assumptions and is an effort to maximize efficiency in the consumption of such goods. Local governments can balance the inequality of opportunities in its role of delivering services. The task of balancing the unequal distribution of assets at the stage of transforming assets to capital and well-being is given.

Community well-being is dependent. Well-being is dependent because it depends on the process of transforming assets to capital. Well-being has a strong subjective characteristic and is influenced by individual feelings or level of satisfaction. At the same time, this implies that there needs to be an agent that can rationally utilize individual assets in the best way possible and transform them into capital. Local governments can play this role of coordinator.

Community well-being is social. In other words, well-being presupposes a collective group. Social characteristics are based on collective groups, and thus we include both subjective indicators and objective indicators to measure community well-being. Subjective indicators concern the individual feelings, perceptions, and psychology while objective indicators go beyond individual levels. This study does not propose an ideal set of community well-being indicators, but rather a realistic, practical set of indicators.

Community well-being is made up of elements that are interdependent. Well-being can be seen as having several elements that affect and are affected by each other. For example, health, education, and income can be examined together. We see relationships wherein people with high education levels have higher incomes and are more likely to have access to better healthcare and are able to maximize their opportunities to prevent illnesses. In sum, the elements of well-being have intricate connections.

Modeling the Relationship Between Assets and Capital

The community well-being model in this study is based on the following assumptions about the relationship between assets and capital. We use this approach because the assets that each individual has is inherently unequal. We use asset and resource synonymously while capital is the broader, processed format of an asset.

First, community well-being is made up of assets and capital. Assets are static and when assets are processed they become a dynamic secondary asset. Hicks (1939) mentioned the asset capital of developed countries that garnered much interest in capital. The capital of human life was treated as the welfare of assets and the possibility of fulfilling this was assumed. However, this position was changed to accommodate capital and especially the human being himself or herself becomes an asset and capital. Welfare was defined as consumption, that is consumption of materials but in fact, human well-being is more than consumption—that is consumption of assets, and this approach ignored the emotional aspects. In sum, the aggregate of consumption is not the realization of welfare. This turn of attention of assets to capital began with human capital and human capital was seen as education levels and this framework expanded to include natural capital, intellectual capital, and social capital. Welfare is possible when there is capital and so we connect capital to welfare and capital can be made when there is enough assets. Thus we see assets and capital as primary asset and secondary asset. Assets are the raw format before processing, while the processing of assets produces capital. Capital is what enables humans to function and therefore it is a secondary asset.

Second, capital is produced when individuals use primary assets and process them. Capital is again used to accomplish another goal and can be seen as a secondary asset.

Third, as capital is a secondary asset, local government intervention is possible. By local government intervention, we mean that local governments can manage and utilize the primary assets so that individuals can use a new asset. This process of producing capital will differ according to a local government's capacity.

Fourth, assets are collective while capital is individualistic. Assets are limited within a community, but capital is made from assets and thus becomes individualized depending on each individual's capacity. Assets are used by individuals, transformed into capital, and thus becomes an individual's possession.

Fifth, primary assets will be depleted while secondary assets can be artificially produced and can be renewed. Figure 1 shows these relationships between asset and capital.

The relationship between assets and capital can be summarized in Fig. 1, and in this model we identify social, human, and natural assets. Natural assets are the environmental assets within the geographical limits. Human assets are the artificially created elements.

Fig. 1 Characteristics of asset and capital

Asset		Capital
Static	↔	Dynamic
Collective	↔	Individual
Material	↔	Psychological
Depleted	↔	Rechargeable
Direct	↔	Indirect
Consumption	↔	Renewal

Elements of Capital

The assets discussed above are processed into capital that individuals can use. Capital is the processed form of these assets and we identify the following six types:

1. Economic capital consists of built capital and financial assets.
2. Natural capital consists of the physical and ecological environments.
3. Human capital consists of labor, education, health, and housing.
4. Social capital consists of trust, commitment, community bonding, and participation.
5. Cultural capital consists of leisure, sharing, and mutual help; it also centers on societal accumulation and as such is an important aspect of community quality of life and well-being (Phillips and Shockley 2009).
6. Infrastructure capital consists of physical and social aspects (Fig. 2).

Based on these assumptions, the national government can be connected to local governments and the effective practice and response to resident needs becomes the opportunity to participate. Community well-being is discussed in this framework.

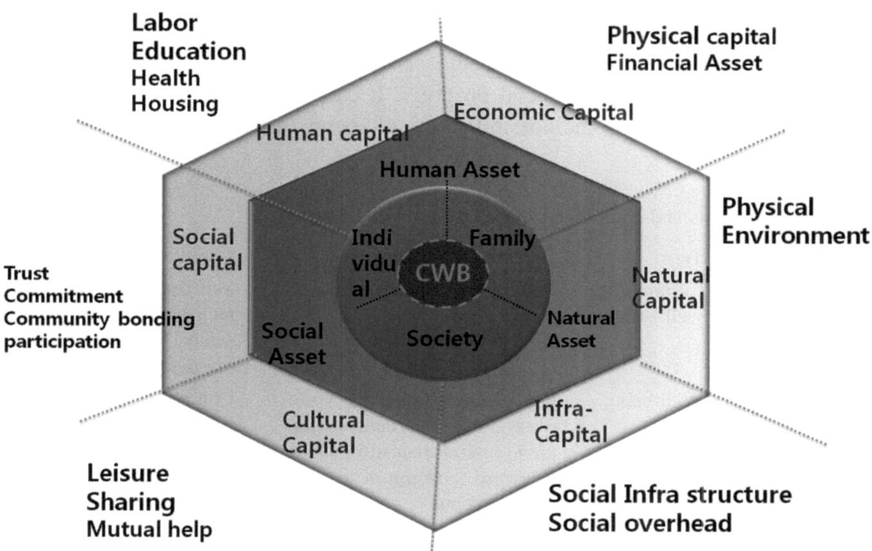

Fig. 2 Modeling community well-being

Defining Community Well-Being

Residents' expectations of government services emphasized specific concepts, but the singular focus on certain concepts cannot effectively realize well-being or enhance quality of life. This is why we need a more comprehensive concept. Based on this need, this study recognizes that the several elements of well-being influence each other and focus on the relationships among these elements. There is no generally accepted definition of community well-being (Forjaz et al. 2011). Community well-being will differ depending on the culture and society (Cox et al. 2010) (Table 2).

Community well-being signifies the most ideal condition of community life and is a concept that signifies the ideal of people living together harmoniously in a sustainable, thriving community (Rural Assistance Information Network 2004). According to Wiseman and Brasher (2008), community well-being is the necessary social, economic, environmental, cultural, political conditions for satisfying the needs of the members of a community and is the ultimate goal of such a process and strategy. These conditions are the necessary items for realizing individual well-being and capacity. Thus he emphasizes an ecological understanding of well-being and to go beyond the individual, collective well-being to human and environmental well-being. That is, community well-being enhances mental health, mental health is a necessary condition of realizing community well-being and this is not an individual issue but a social, psychological, mental well-being. The necessary socio-economic conditions for an individual's mental health include freedom from violence and prejudice, equal rights and sense of community, economic participation and safety, as well as social progress. It cannot be measured by economic growth alone but needs to include indicators of social and environmental problems as well.

Table 2 Definitions and conditions of community well-being

Author	Term	Definition and keyword	Conditions of well-being
Hay (2003)	Community well-being	Fulfillment of individual and collective desires Keyword: fulfillment	Self determination, reciprocity, equality, safety, civic responsibility, democracy
Ramsey and Smit (2002)	Well-being	Psychological, material, social environment that enables people to realize their full potential Keyword: creation of environments	Conditions of World Health Organization's healthy communities
Ribova (2000)	Well-being	Emotional, cultural, social demands of the individuals and the community Keyword: requirements	Economic, social structures
Scottish Development Center (2003)	Community well-being	Flourishing continuously in a healthy way Keyword: thriving	Positive values, local infrastructure, service, opportunities, safety

Source Lee et al. (2013)

Although there are previous studies that treat community well-being as the simple sum of individual well-being, this study differentiates the two concepts and define community well-being as a unique concept. The simple sum of individual well-being is strongly subjective and does not reflect the community characteristics. Treating individual and community characteristics leads to confusion in understanding underlying theory and its application. In terms of policy analysis, an implied causal relationship is that the sum of individual well-being affects individual well-being. However, as discussed before, when governments focus on individual well-being as a policy goal it is very inefficient and unrealistic. Thus, we explicitly differentiate individual well-being and community well-being (Fig. 3).

Also, community well-being can be differentiated as a concept of state or a concept of process (Kee et al. 2013). Community well-being as a state focuses on the present and emphasizes current thoughts and behaviors of citizens who affect local policies periodically through elections. Also, flexibility refers to the relationship among government, residents, and local community that are constantly affecting each other. This leads to the possibility of innovative policy making. On the other hand, community well-being as a process has high sustainability and can help achieving capabilities through service delivery and evaluation based on a lifetime process.

The uniqueness of community well-being is identified by comparison with related concepts and extraction of key elements. The following reasons account for why community well-being can be the key value of welfare states and local government administration. Community well-being is not restricted to the realm of individual control as is the case with quality of life, well-being, and sustainability. These related concepts can be controlled by individuals but community well-being cannot and must be coordinated and managed by another actor (see Table 3). Community development on the other hand, is inherently of the collective and is more akin to community well-being in aspects, particularly that of process and concepts of assets and capital.

We propose community well-being as a necessary condition for maximizing happiness, quality of life, and well-being. This is connected to the previous point in that community well-being cannot be controlled by an individual. Kee et al. (2013) discuss the relationship between individual need and impacts on these at the community level. We support that individual happiness or well-being is affected by

Fig. 3 Hypothetical relationship between individual well-being and community well-being. *Source* Lee and Kim (2014)

Table 3 Characteristics of community well-being in comparison to related concepts

Community well-being	Happiness	Well-being	Quality of life	Sustainability	Community development
Subjectivity (low)	High	High	Medium	Low	Medium
Related to economic elements (low)	High	Medium	High	Medium	Medium
Relativity (comparative needs) (low)	High	Low	Medium	Not applicable	High
Internal (low)	High	Low	High	Low	Medium
Usability of assets (high)	Medium	High	High	High	High
Capital (high)	High	Medium	Medium	High	High
Individual as actor (low)	High	High	High	Low	Medium

community well-being and we can see community well-being as the necessary precondition for happiness, quality of life, and well-being.

Community well-being has a relationship of necessary condition and sufficient condition with related concepts. Community well-being needs to be distinguished from other concepts to be discussed in this way. It is difficult to discuss community well-being without distinguishing it as a unique concept. Community well-being needs to be separated because happiness, quality of life, and well-being become customers of community well-being. In other words, community well-being affects happiness, quality of life, and well-being so it is important to understand what their characteristics are and how they are affected. Community well-being is discovered by local governments and used for local government actions but it is important how the individual components, the residents, feel. These feelings are reflected in elections and this is the focus of politicians and public administration is again connected to politics. This is why community well-being also considers the individual, subjective perspective of happiness, quality of life, and well-being. Community development, both as a practice and a discipline, can be seen as a complementary concept that has at its core some of the same principles as community well-being, predominately to improve people's lives.

Table 3. shows these comparisons with related concepts of community well-being. Community well-being is used in various ways by different researchers for different purposes. This study aims to use community well-being as a key value of public administration and thus defined and modeled community well-being through a literature review. We compare these concepts on seven criteria, assessing the level of high, medium, low. These levels are not absolute but only relative.

Community well-being is not only the physical space, but also the community's changing population and interest that focuses on the context and is the necessary social, economic, environmental, cultural, political elements for an individual and community to reach their capabilities (Kee et al. 2013). This definition assumes that the individual, community, and local government checks and decides on the necessary conditions and includes the community member, community organization, and community members' well-being in a dynamic network. By meeting the life cycle needs of the members it contributes to the process of realizing capabilities and community well-being becomes the goal of local governments. Thus community well-being is enhanced by distributing, delivering, producing the goods and services that can be the conditions for local members, local government, and local organizations. In addition, it contributes to an individual's realization of capabilities by meeting the life cycle needs of an individual. That is, community well-being is the state and process of individuals and communities to flourish and reach its capabilities through the necessary social, economic, natural, cultural, and political conditions.

Conclusion

This study has presented community well-being as a consideration for local governments to help foster better states for their residents. We modeled community well-being as a guiding principle of local governments and governance by examining its origin and use. We examined the background of community well-being and well-being as common factor on the national and local level. We identified the characteristics of community well-being and used assumptions of local government roles to define the concept and build a model. The modeling approach is focused on the assets and the capital of communities. We looked at the related concepts of community well-being including happiness, quality of life, and community development and their previous studies and compared them in a comprehensive way to build a multidimensional model. Expert consultation and expert modeling consultation and forums were used and this model will be used in future studies.

Community well-being model positions local governments and has a close connection to the role of the state. It defines the role of state for building a welfare state and it can be connected to the happiness and happiness realization role of states. Happiness and well-being are not completely unrelated. These two concepts are used interchangeably in reality. The government needs to act to enhance citizen happiness and they do this through policy making and when these policies are connected to daily lives they involve local governments. Therefore, this study has suggested community well-being as directing the role of local governments in connection to the welfare state and modeled community well-being to explore the role of local governments.

As such, community well-being is the goal and value goal of local governments and this dictates the role of local governments. Community well-being requires studies on service delivery mechanism as local resident well-being is realized through efficient service delivery processes and it also requires indicator studies for residents to evaluate local governments based on levels of satisfaction. Local governments can serve as tools for realizing resident well-being while also "producing" well-being. Local governments can be thought of as the delivery actors of resident well-being, and can address distribution of well-being asset to the disadvantaged. Well-being has the characteristic of inequality and resident well-being is not distributed fairly in terms of needs. Thus individual well-being needs to be coordinated in order to ensure individual happiness and this is a primary role of local governments. If not, there will be social costs. This is why community well-being can assess the well-being of local communities and then help with evaluation and prescription of how local governments can enhance community well-being in the process of service delivery. We modeled community well-being as the first step towards developing community well-being indicators. Community well-being appeared as a new agenda in the process of moving from the central government to the local level. This is because the local level is where individual lives are lived out and the possibility of enhancing the quality of life is present. In fact, as life satisfaction has become important the local level became more vital because this is where individuals spend time on a daily basis and this has led to an increase in interest in local government involvement, or governance. This is transformed into local government services and thus local government services become even more important. This goes beyond the resident's freedom and self government and services related to the problem of unemployment, aging, pollution, safety, education, and culture are important for the local governments as appropriate actors for these problems. In other words, the various problems due to changes in environment that relate to resident life and happiness can be addressed very directly at the local government level. Obviously, there is a role and need for national governments to help address pressing issues and challenges; rather, we are implying that local government and good governance can address issues at the level of where the residents experience it most directly.

Acknowledgments This work was supported by the National Research Foundation of Korea grant funded by the Korean government (NRF-2010-330-B00259).

References

Adler, M. D. (2013). Happiness surveys and public policy: what's the use?. *Duke Law Journal, 62*, 1508–1601.

Beaulac, J., Kristjansson, E., & Cummins, S. (2009). A systematic review of food deserts, 1966–2007. *Preventing Chronic Disease, 6*(3), A105.

Booth, P. (Ed.). (2012). *... and the pursuit of happiness-well-being and the role of government.* London: The Institute of Economic Affairs.

Cox, D., Frere, M., West, S., & Wiseman, J. (2010). Developing and using local community well-being indicators: Learning from the experience of Community Indicators Victoria. *The Australian Journal of Social Issues, 45*(1), 71–88.

Easterlin, R. A. (2001). Income and happiness: Towards a unified theory. *The Economic Journal, 111*(473), 465–484.

Forjaz, M. J., Prieto-Flores, M.-E., Ayala, A., Rodriguez-Blazquez, C., Fernandez-Mayoralas, G., Rojo-Perez, F., et al. (2011). Measurement properties of the community well-being index in older adults. *Quality of Life Research, 20*(5), 733–743.

Hay, D. I. (2003). *Well-being: A conceptual framework and three literature reviews.* Vancouver: Social Planning and Research Council of B. C.

Hicks, J. R. (1939). The foundations of welfare economics. *The Economic Journal*, 696–712.
Ilić, I., Milić, I., & Aranđelović, M. (2010). Assessing quality of life: Current approaches. *Acta Medica Medianae, 49*(4).
Kee, Y., Kim, Y., & Namsuk, K. (2013). Developing community well-being indicators for Korea: Learning from international cases. *Journal of Regional Studies and Development 22*(1) (Korean).
Kim, Y., & Lee, S. (2013). The development and application of a community well-being index in Korean metropolitan cities. *Social Indicators Research*, 1–26. doi: 10.1007/s11205-013-0527-0.
Larson, N. I., Story, M. T., & Nelson, M. C. (2009). Neighborhood environments: Disparities in access to healthy foods in the US. *American Journal of Preventive Medicine, 36*(1), 74–81. e10.
Lee, S. J., & Kim, Y. (2014). Searching for the meaning of community well-being. In S. J. Lee, Y. Kim, & R. Phillips (Eds.), *Community well-being and community development: Conceptions and applications*. Berlin: Springer.
Lee, S. J., Kee, Y., Kim, Y., & Kim, N. (2013). "Comparing community wellbeing index evaluations of experts and public officials: Using the analytical hierarchy process method." *Korean Public Administration Review, 47*(1), 295–320.
NWMO (Nuclear waste Management Organization). (2009). *Applying community well-being: Lessons and experience of Canadian practitioners*. NWMO Report SR-2009-02.
Ormerod, P. (2012). The folly of well-being in public policy. In P. Booth (Ed.), ... *and the pursuit of happiness-well-being and the role of government* (pp. 40–43). London: The Institute of Economic Affairs.
Pagani, L. S., & Huot, C. (2007). Why are children living in poverty getting fatter? *Paediatrics and child health, 12*(8), 698.
Phillips, R., & Pittman, R. (2014). *Introduction to community development* (2nd ed.). London: Routledge.
Phillips, R., & Shockley, G. (2009). Linking cultural capital conceptions to asset-based community development. In G. P. Green & A. Goetting (Eds.), *Mobilizing communities: Asset building as a community development strategy*. Philadelphia: Temple University Press.
Ramsey, D., & Smit, B. (2002). Rural community well-being: Models and application to changes in the tobacco-belt in Ontario, Canada. *Geoforum, 33*(3), 367–384.
Ribova, L. (2000). Individual and community well-being. The Arctic. Retrieved from http://www.thearctic.is/PDF/Individual%20and%20Community%20well.pdf.
Rural Assistance Information Network. (2004). *Rural Assistance Information Network*. Retrieved from http://www.rain.net.au/community_well-being.htm.
Scott, K. (2012). *Measuring Wellbeing: Towards Sustainability?*: Routledge.
Scottish Development Centre for Mental Health (SDC). (2003). *Building community well-being an exploration of themes and issues*. Project report to the Scottish Executive Edinburgh: The Stationery Office Scottish Development Centre for Mental Health (SDC).
Sen, A. (1999). *Development as freedom*. Oxford: Oxford University Press.
Sen, A. (2005). Humaan rights and capabilities. *Journal of Human Development, 6*(2), 151–166.
Seo, J. H., Kim, H. J., & Lee, S. J. (2012). A study of community well-being as local government's value principle. *Korean Public Administration Review, 46*(1), 33–56.
Smits, J. P., & Hoekstra, R. (2011). *Measuring sustainable development and societal progress: Overview and conceptual approach (D. o. M.-e. s. a. dissemination, Trans.)*. Netherlands: Statistics Netherlands.
White, S. C. (2008). But what is Well-being? A framework for analysis in social and development policy and practice. *Conference on regeneration and well-being: Research into practice* (pp. 3–7), University of Bradford.
Wiseman, J., & Brasher, K. (2008). Community wellbeing in an unwell world: Trends, challenges, and possibilities. *Journal of Public Health Policy, 29*(3), 353–366.

Community Learning and Self-generated Communities of Practice: The Case of Jangheung School

Youngwha Kee and Okchae Joo

Introduction

The origin of Korea's Community of Practice (CoP) goes back to the 14th century Seowon (educational institution), Hyangyak (village codes), Gye (mutual financing societies), Doorye (famers' cooperative groups), and Poomasi (labor sharing) of the Joseon Dynasty. Through farmers' cooperative groups and exchanges of labor, a variety of local communities were formed. Village codes and mutual financing societies allowed associations to be created for various purposes and regulations were set up to promote mutual interests. Seowon was a nationwide private institution established to foster outstanding scholars during the Joseon Dynasty. It soon became the foundation of learning communities within each village. A learning CoP was created through exchanges of knowledge and it promoted the self-development of its residents and community. Seowon not only allowed residents to deeply understand their nation and the local community life but enabled them to participate, providing many implications in the process of voluntary participation and democratic movements (Kee 2011).

In Korea, the Local Assembly was formed in 1952 and was suspended temporarily under the military takeover of 1961. In 1995, with the first election of local government mayors, the official era of localization began. The history of these

A Korean version of this paper was published in the Journal of Local Government Studies vol. 24 issue 4 in 2012.

Y. Kee (✉)
Soongsil University, Seoul, Korea, Republic of (South Korea)
e-mail: key@ssu.ac.kr

O. Joo
Ministry of Public Security and Public Administration, Jongno-Gu, Republic of Korea
e-mail: ruby115@korea.kr

Y. Kee et al. (eds.), *Learning and Community Approaches for Promoting Well-Being*,
SpringerBriefs in Well-Being and Quality of Life Research,
DOI 10.1007/978-3-319-12439-1_2

events have unfolded in such a way that community development and the learning of local residents became inseparable, for learning was the mechanism that led to a sense of ownership and the power to solve various community problems. The various community activities encouraged residents to participate voluntarily and experience discussions and negotiation processes, which became the paths of intervention towards enhancing the community's wellbeing (Lee 2011). In addition, there have also been demands of change from the top-down policy to a system that is led by residents and the spontaneity of the local community. As such, efforts to seek individual progress and local community changes through CoP are being made. As communities get bigger and more complex, the role of CoP that could change fragmented relationships into communal relationships has become crucial. In other words, with reciprocity as the basis, members need to establish CoPs that will allow residents to cooperate in achieving mutual goals. Overcoming the rift between knowledge and practice is what Wenger et al. was referring to by CoP where local residents are the main agents. Here, through voluntary participation and sharing learned topics, insight, and experience, mutual interaction and knowledge creation becomes possible (Wenger et al. 2002).

However, there has been little interest in the spontaneous CoP activities by local residents acting as main agents, which serves as a foundation for individuals and community development. Furthermore, civic movements based upon regions and led by institutions have been discussed but thorough examination of the formation, composition, and maturation process of CoP that developed through voluntary participations of the residents has not been carried out.

Accordingly, this study searched for cases of spontaneous CoP established by local residents as a result of education integrated into the communities. In particular, we were interested in a public policy approach to CoP. We chose Jangheung Hakdang, which is located in Jeollanamdo province, Korea as our case because it is focused on learning and based on voluntary actions. Since its voluntary establishment by local residents in 1994, Jangheung Hakdang has been changing community members' frames of thinking and perspectives and striving to search for a new paradigm for local development. This CoP is well-known for its autonomous operation and is still being maintained today. The idea of Jangheung Hakdang all started from a retired high-ranking government official returning to his home village and Jangheung Hakdang is now serving as a place for democratic participation and academic pursuits.

The purpose of this research is to analyze the process of local residents voluntarily organizing groups to change their own lives and the local communities around them and as a result creating a CoP. The study aims to find implications to assist the successful operations of spontaneous CoP. The research question is how was this spontaneous CoP of local communities formed. The findings of this research, involving the role of local governments, democratic participation of residents, and transition process of spontaneous CoP can be applied to finding effective strategies to encourage local participation. Moreover, it is expected that this research can provide a model of CoP development that is based on increased participation of residents.

Theoretical Background

Community of Practice

Previous Literature

CoP is a group of people who share mutual interests, a series of issues, and specific topics, and through constant interaction, collectively gain more knowledge and professionalism of a certain field (Wenger et al. 2004). That is, a learner in the community may acquire social practice through an informal situated learning called 'Legitimate Peripheral Participation'. Hereafter, CoP gradually develops into a community concept where members carry out joint duties for the creation of knowledge. According to Wenger (1998), a CoP has a fluid characteristic. In order for a CoP to change, the increased participation of its members is crucial. For a beginner, strained relations exist in the process of participating in a CoP. This is due to the existence of conflicts around allocations of learning opportunities, the continuation and maintenance of CoP, and replacement and renewal as well. After repeated changes and creations, a CoP is newly constituted in the process of a beginner overcoming conflicts and strains, and becoming a complete participant of a community. This is the relational context that Lave and Wenger (1991), and Wenger (1998) was referring to, in which situated learning in a CoP takes place. Thus, a CoP is not stagnant, but possesses a dynamic attribute that constantly changes according to the increase of legitimate peripheral participation.

Previous research on CoP can be classified into two kinds. One type of research sees CoP as a crucial element for organizational vitalization and improvement of the power of execution. These studies focus on the strategies and effects of CoP. Some examples of these studies are Jeong (2008), Jang et al. (2010), Kim (2008b), Ham (2008), etc. Jeong (2008) conducted a comparative analysis on factors that influence the vitalization of public organizations and profit-making organizations. Jang et al. (2010) analyzed the effects of CoP participation types regarding organizational commitment of members and Kim (2009) analyzed the methods in which a CoP can be developed for the success of knowledge management system of government organizations. Kim et al. (2009) conducted an empirical analysis on the relationship between the nature of CoP of small and medium enterprises and citizen's action of organizations. Ham (2008) analyzed the importance of performance factors of public sector CoP.

Another type of CoP research is on the creations of knowledge and the effects of knowledge management through CoP in terms of creating, sharing, and distributing knowledge. These include the researches of Ku and Song (2011), Jang et al. (2009), Lim and Kim (2009), Kim (2008a), Baek and Kim (2008), Lee and Jung (2008), Choi and Yoo (2003), Bang and Lee (2000). Ku and Song (2011) conducted a study on the relationship between CoP participation willingness and commitment. Jang et al. (2009) analyzed the effects of CoP formulation on the vitalization of organizational learning. Lim and Kim (2009) carried out an analysis study of the effects of organizational culture in public agency on CoP activities, the types of organizational learning, and organizational performances. Through the case of a network-based community for teachers, Indi-School, Kim (2008a) analyzed the developmental process of network-based CoP established for the purpose of knowledge sharing and improvement of professionalism of teachers. Moreover, Baek and Kim (2008) analyzed the influence factors of CoP knowledge sharing and Lee and Jeong (2008) conducted an empirical analysis on the impacts of CoP on community spirit, organizational commitment, and knowledge management activities. Choi and Yoo (2003) analyzed a real case of developmental process of CoP as a knowledge creation and sharing strategy. Bang and Lee (2000) suggested a management strategy of CoP towards a more successful knowledge management.

In short, most of the preceding researches of CoP are closely related to the effects of CoP on improvement of organizational performances, CoP vitalization measures from an organizational performance approach, and strategies for CoP vitalization from a knowledge management approach. Yet, these researches set the relationship between the components of CoP and other components as a passive factors and analyze the relationships instead of seeing members of organizations as independent and autonomous beings, and accordingly analyzing the independent interaction processes or learning practice processes of each member.

Components of CoP

CoP is defined in many perspectives, but there are common components of CoP. According to Wenger et al. (2004), CoP consists of ‘the domain’ of knowledge where issues are defined, ‘the community’ of people interested in a certain domain, and ‘the practice’ that develops and shares for the purpose of one’s ability being guaranteed in a certain field. Wenger (2007) viewed CoP as consisting of ‘reciprocal participation’, ‘joint domain’, and ‘joint practice’. Though the terms differ, the components are the same (Yang 2011).

First of all, ‘the domain’ is the scope of knowledge or a series of key issues relevant to the common issues, interests, and topics of CoP members. According to Wenger et al. (2004), a desirable and effective domain does not relate to abstract areas of interest. Rather, it is related to complex and long-range issues and requires constant learning of its members, and consists of primary issues and matters that are commonly experienced by members. That is, it is not merely fixed by duties but is a range of knowledge that must be explored and developed (Wenger et al. 2002). A well-prescribed domain unites members and provides a mechanism to legitimize the purpose and the value of the CoP to the outside (Wenger et al. 2002). In other words, it ascribes meaning to the existence and actions of members (Wenger et al. 2002).

Secondly, another component of CoP is the community itself. According to Wenger et al. (2004), a ‘community’ is a group of people who interact, learn together, form relationships, and in the process a sense of belonging and reciprocal commitment becomes a bond that ties the people together. This is a necessary and crucial component of an effective knowledge structure. This component allows a CoP to become a place that contributes not only to personal intellectual competence, but to the formation of social capital (Kim and Kim 2005; Lesser and Prusak 1999).

Thirdly, 'the practice' is a chain of methods socially defined in a specific domain. According to Wenger et al. (2004: 21), practice is the way tasks are carried out in a certain area, and this practice is socially defined. It signifies a shared standard set and a common approach that forms the basics such as behavior, communication, problem-solving, performance, responsibility, etc. The resources of a community includes examples and stories, theories, regulations, frames, models, principles, tools, professional knowledge, theses, lessons from experience, best practices, assessment guidelines, and also explicit and implicit sides of a community's knowledge. In other words, practice is the intangible and tangible knowledge that was developed, shared, and accumulated with the joint domain placed in the center of the mind of the members. It reflects the history of a CoP.

Besides the three components of domain, community, and practice, depending on the analyzing fields of CoP and the purpose of a CoP, other components have been identified (Wenger et al. 2004). This research is based upon the previous finding that when all the suggested components operate successfully, a CoP can become the ideal knowledge-based structure, that is, a social structure that carries the responsibility of developing and sharing knowledge (Wenger et al. 2002).

Types of CoP

There are various types of CoP depending upon the scale, duration, compactness, homogeneity, boundary, spontaneity, artificial organization, visibility, etc. (Lee 2008). Andriessen et al. (2004: 15–16) suggest five types of CoP according to the standards of its purpose, whether its boundary is open or closed, composition, interaction, and scale.

The first is a daily CoP where people with experiences and new people jointly participate. It focuses on the discussion of daily business experience through periodic face-to-face contacts. The second is a CoP of official experts. As a group consisting of selected experts with local or systematic representativeness, the main focus is on the exchange and development of knowledge. The third type is an informal network CoP. This type is formed to discuss common issues and interests of people who approach each other informally and freely. The fourth CoP is a problem-solution CoP. Workers in certain fields interact on and offline to obtain solutions to specific problems that arise in the process of job performance. The fifth is a potential network CoP. It is a group formed for the common interests of people in differing organizations and fields such as committees or associations of government ministries. Here, CoP is introduced in an organizational leading method, which is the main interest of this research, and the next paragraph examines related classification types of Nickols (2000).

Nickols assorts CoP into a self-organizing CoP that is created by the members themselves and a sponsored CoP which is designed and sponsored by an organization. The main focus of a self-organizing CoP is the pursuit of individuals' common interests. Yet, these types of CoP also have value from an organizational aspect since many best practice cases or important cases are self-organizing CoP types. Meanwhile, productions of scalable outcomes that are beneficial to organizations are expected from a sponsored CoP. In these types of CoP, the necessary resources and official roles or responsibilities are granted by organizations.

The Stages of CoP Formation

CoP is viewed as an organism that is consistently experiencing ecological changes such as birth, maturation, and dissolution. The formation and development of CoP will be discussed through a life cycle model and function-oriented model.

Life Cycle Model

The life cycle model describes the development of CoP in comparison to the developmental cycle of an organism. In other words, a CoP is assumed to be equal to organisms, growing and becoming extinct, and its typical types of activities, challenges, and changes of activities that appear throughout the maturation process are captured.

Wenger (1998: 2–3) and Wenger et al. (2002: 93–111) describe the expansion of a CoP in five stages: potential, coalescing, maturing, stewardship, and transformation stage. The potential stage is where potential members who feel the necessity of association and collaboration prepare and plan. The second stage is the coalescing stage where a CoP is established; members select specific topics that accords with their knowledge domains, plan and develop activities, and begin the process of carrying out such plans. Bettoni et al. (2002) combine this coalescing stage with the potential stage, naming it a formation stage. The main activities of a formation stage are to unearth common footing between the existing network and potential members, and devise an opportunity for the creation of new values. The third stage is the maturing stage where the CoP establishes its role in the relationship with the organization, actively shares the knowledge, and creates new knowledge based upon the shared knowledge. In the fourth stage, the stewardship stage, the activities of a CoP are maintained, but the interests of members are decreased and activities become stagnant. Bettoni et al. (2002) viewed the maturing stage and stewardship stage as an integration stage where the main activity of a CoP is to introduce and create new knowledge and ideas. The last stage is the transformation stage where activities are terminated or new changes are sought out, for a CoP can no longer fulfill its role as the center of knowledge activities.

Like Wenger (1998) and Wenger et al. (2002), McDermott (2000) explains the unfolding process of a CoP through a life cycle perspective. According to McDermott, a CoP goes through planning, establishment, and closure stage. Each stage corresponds with the potential, coalescing, maturing, stewardship, and transformation stage of Wenger et al. (2002) and its contents are similar. Wenger et al. (2002) combined the model of McDermott and proposed the stages of CoP. According to them, the visibility and energy of a CoP generally increase from the potential stage to maturing stage, and then begin to decrease in the maintenance stage, and rapidly decrease in the conversion stage.

Function-Oriented Model

Gongla and Rizzuto (2001) viewed the life cycle model as insufficient to portray all the various unfolding aspects of a CoP that appear in reality. Gongla and Rizzuto regard the fundamental functions that characterize the activities of CoP as connection, context creation and memory, approach and learning, collaboration, innovation and production. The conditions where these functions appear are suggested through potential, composition, participation, vitality, and adaptation stage. The potential stage is the stage where a CoP begins to formulate. In the second, the composition stage, a CoP is formed with its general operations established. With its vital members as the center, experiences and knowledge are shared and these activities begin to be documented. Thirdly, in the participation stage, plans from the previous stage are carried out, and all the members contribute to the sharing and extension of knowledge, and thus maintaining and managing the knowledge system. The fourth stage, the vitality stage, with cooperation as the foundation, knowledge creation activities are actively developed and this becomes further linked to problem-solving of members, organizational solutions to pending issues, and expansions of businesses. Lastly, in the adaptation stage, new knowledge is consistently being created and the environment is redefined and expanded; 'innovation' and 'production' become the two key functions in this stage.

Table 1 shows the developmental process suggested in common by both models and the primary activities of each stage.

Methodology

Case Selection

The purpose of this study is to investigate the formation process of CoP where the activities are unfolded. Thus, in order to examine the developmental process of a CoP developing amid the intentions and actions of members, instead of measuring the performance of CoP, this research selected a case where researchers are able to directly observe the interactions between the community and the individual members.

Table 1 The primary activities of each developmental stage of CoP

Developmental stage	Primary activities
McDermott (2000), Planning stage Gongla and Rizzuto (2001), Potential stage Wenger et al. (2002), Potential stage	Unearthing potential network and issues Contacting and connecting to potential members Setting main purpose and goals of CoP
McDermott (2000), Establishment stage Gongla and Rizzuto (2001), Composition stage Wenger et al. (2002), Fusion stage	Selecting a desirable leader Confirmation of the domains of knowledge and the scope of roles Organizing basic framework such as arrangement of managerial regulations
McDermott (2000), Growth stage Gongla and Rizzuto (2001), Participation stage Wenger et al. (2002), Maturation stage	Selection or development of specific topics Development of unique methods, tools, terms Sharing of experiences and knowledge through conversations or discussions Documentation, recording, or management of activities and shared knowledge Management of new or withdrawing members and the scope of activities
McDermott (2000), Growth persistent stage Gongla and Rizzuto (2001), Vitality stage Wenger et al. (2002), Maturation management stage	Development and sharing of phronesis of the highest degree Arrangement of optimum application plan for the organization regarding results of the activities Application of results of the activities to solve individual and organizational problems Measurement of activity performance and visualizing its application Maintenance or renewal of the most advanced phronesis Member management as well as association with and sponsoring other CoP Formation of relations with external Organizations

Source Choi Eunhee and Kwon Daebong (2007). Developmental aspects of local education offices CoP (revised)

A CoP operates in a complex and intrinsic context and it is difficult to grasp what is truly happening from limited anecdotal observances. Therefore, an open case approach where the purposes and intentions of introducing a CoP, the operation methods, and the culture of the CoP can be directly observed was necessary. Jangheung Hakdang, which was established in 1994 by the residents themselves and is still being operated independently, was chosen because it fulfilled these requirements. Not only did Jangheung Hakdang reject every sponsorship offered by the district and is currently operating solely on voluntary dues of members and donations, but also the dedication and interest of Jangheung residents has been sustained. In fact, Jangheung Hakdang has received much attention from local governments and is becoming a benchmark for community development initiatives.

Data Collection and Analysis

Data were collected and constantly triangulated through in-depth interviews, observations, and document analysis. Since this research intended to investigate the developmental progress aspects of CoP activities, data collection was carried out from 2010 to 2012 at regular intervals. We began data collection with a tentative design, and when a valued topics newly emerged in the process, the next stage of data collection was revised to accommodate further exploration of these topics.

The characteristics of the collected data and the collection process are as follows. First, through the managing director and general affairs manager of Jangheung Hakdang, the contour of the organization, purpose and activity outline, the performances are assessed. In addition, all related documents such as promotional pamphlets, articles of the association containing the regulations of Jangheung Hakdang, a certified copy of the register as a corporate body were collected.

Secondly, data were collected by observations. Participant observations were conducted twice on July 16th, 2010 and August 18th, 2012 along with the study meetings. From the welcoming of lecturers that occur the day before the study meeting to separate meetings held after the study meeting by the participants themselves, the complete and overall progress was observed through participation observation. On the night before the study meeting, primary members including the head of Jangheung Hakdang meet with the lecturers and have dinner together. During this occasion, news of the Jangheung area and various talks from the perspectives of the lecturers naturally come into view. On the day of the study meeting after arriving at 6:30 a.m., the participating residents were observed through eating meals together and conversing. Jangheung Hakdang makes nametags with each name of the participants written in Chinese characters. Throughout the study meeting, the materials that were distributed in advance were referenced, notes on significant aspects were taken down, and photos were taken. After the study meeting, the primary members gathered for lunch and feedbacks on the study meeting were provided. During this, two people were interviewed and tape-recorded but due to the noise of the dining hall, we failed to transcribe these recordings. Furthermore, main cultural heritage sites within the jurisdiction, five industrial settings, and two sites that were discussed as pending issues in the presentation debates were surveyed.

Thirdly, in-depth interviews were used as the primary resource of the case. The method of in-depth interviews was adopted through considering the suitability to the research topic and access possibility to the research field. During the case observation period, semi-structured in-depth interviews of seven people were conducted. Apart from the official interviews, conversations frequently occurred throughout the observation process and brief interviews and conversations took place on the phone and through messages and e-mails.

Fourthly, the 17 years of Jangheung Hakdang's annual journals containing the data of study meetings, workshops, and presentation debates and the related lecture data were collected as well. Lastly, external assessments and press articles on Jangheung Hakdang were used as references for secondary analysis.

The unit of analysis for this research is the transitional process of the CoP, Jangheung Hakdang. In order to achieve this, materials and data were examined thoroughly to identify and interpret their meanings. Likewise, until the research results were fully comprehended, constant collections and analyses of data took place, and the information gleaned from the data collections were selected, classified, and systematized. In addition, efforts were put into evince the qualitative traits by making frequent contacts with the participants. Afterwards, the analysis and interpretations underwent the process of materialization in the course of writing the research paper and forming the question at issue. During the writing process and among the diverse aspects of the case, the focus was on education theses, especially lifelong education. These portions combined with terms that appear in the case and that are used academically were used in conjunction to explain and interpret the results.

Accountability and Ethical Considerations

This research made the utmost effort to secure objectivity and accountability through the reliability criteria of Lincoln and Guba (1985): truth value, applicability, consistency, and neutrality. To accomplish this, first of all, raw data that did not undergo artificial processes were collected. Written sources, participant observations, in-depth interviews, field notes, and additional notes were used multilaterally to compose the data and supplement the interpretations. Secondly, reviewing process through peer debriefing followed. In the process of the research, the research was motivated to remain honest and the researchers made efforts to secure reliability of the research in the discussion process as well. The third was the use of member checking by the research participants themselves. By confirming the contents and the results of the research with the participants, the research aimed to ensure validity. Furthermore, seeking prior consent regarding information offering was included in the process and principles of secrecy and anonymity were complied with. By keeping an adequate distance between the participants and the researchers, relationships were maintained at an appropriate level to ensure that the focus of the research was in complete form and to prevent the exposure of unexpected information and sensitive issues which may upset the participants. In terms of participant interviews, decision to participate was made independently according to one's individual choice. In the case where the informer did not want disclosure, the interview would not be recorded, and in case where it was recorded, after the participants viewed the recorded materials, the contents were expunged upon the request by the participants.

Table 2 Characteristics of the research participants

Number	Name	Sex	Age (as of late 2011)	Position and role
1	Ko	Male	50s	Member
2	Yoon	Male	70s	Founding member
3	Kim	Male	60s	Founding member
4	Choi	Female	40s	Member
5	An	Female	50s	Member
6	Park	Female	40s	Member
7	Wi	Male	80s	Founding member

Research Participants

In the process of selecting research participants, to ensure that there would be a variety of participants in the research group, participants were divided into junior members and founding members according to their membership terms. Among the members, the criteria selected for persons who are able to recount their own stories with open minds, can simultaneously participate actively in the activities, and are willing to adapt to new changes. Table 2 contains brief information of the participants.

The Development Process of the Spontaneous CoP, Jangheung Hakdang

Jangheung-gun is located in the southwestern end of Jeollanam-do, which belongs in the southwestern part of the Korean Peninsula. Vertically extended, it assumes the central district of the area with the Tamjin River flowing in the southern sea and it borders five administrative districts. The development process of the CoP, Jangheung Hakdang, in Jangheung area is assorted into four stages. The first stage was from the year 1991 when the small scale Ae Jang Society was established, to 1994 when Jangheung Hakdang was first founded. The second stage is the period from 1994 to 1995 when the basic infrastructure, such as leader selection and arrangement of managerial regulations, was established. The third stage is when the leading founder of the village school operated the school as the head. The fourth stage is the period when learning was established as part of the culture of the Jangheung people through consistent learning. Though the process is divided by periods, each stage overlaps with one another. According to the four stages of the CoP development process, the formative years are assorted into the germinating period (1991–1994), system establishment period (1994–1995), participatory growth period (1995–2005), and sustainable maturity period (2006–2011) and the following section will describe these in detail.

Germination Period (1991–1994)

The efforts of a high-ranking government official served as the momentum for the birth of Jangheung Hakdang. The person behind this is Son, a former Minister of Transport. After retiring, he returned to the Jangheung area. With the resurrection of the local assembly in 1991, the voluntary participation of residents in 1995 instigated the four-position local elections and local self-government was reintroduced. The future of the local area in accordance with the changing times was beginning to become the main concern of people.

After coming back to his hometown in 1990, Son managed a rice processing complex left by his deceased father. In January of 1991, with a restaurant owner and 20 young and middle-aged people who were concerned with the future of the local area, he established a small village school called Ae Jang Society.

> We had a gathering named Ae Jang Society. The meeting was held with people from various age groups, and from a public official to farmers and doctors, it involved a wide array of people from all walks of life. (Kim)

'Ae Jang Society' means a gathering for people who love Jangheung. In the evening of 2nd and 16th of every month, people gathered in a local restaurant and opened the small village school. After one member made a presentation about a certain topic, the members would exchange their own opinions and life stories while having a meal together. Son often debated frankly while drinking with the young members.

> The small village school was the gathering of people who were present when Mr. Son initially started it. There were many discussions and debates including life stories. (Yoon)

As Son built the Ae Jang Society, he completely settled in Jangheung. In January 1993, the third year since the establishment of Ae Jang Society, the members of Ae Jang Society gathered and discussed the things they could do for the local residents of Jangheung.

> From 1993, we started discussing what it would be like to do such a thing. In 1994, we founded Jangheung Hakdang it has been operating ever since. (Kim)

After long discussion, the members agreed to provide the residents a place for study meetings and studies which would renew the awareness of the changing times and make Jangheung a place filled with vitality. This became the root of Jangheung Hakdang.

> We felt that the residents needed learning. The residents needed to grasp the situations they are in and the surrounding conditions to carry out one's role properly. The residents themselves wanted to learn through hands-on activities… (Wi)

The research participants mentioned that the establishment of Jangheung Hakdang was founded solely out of the attachment and affection for one's hometown and unwavering convictions for local development.

Institution Building Period (1994–1995)

Seeking a Paradigm for Local Development

In January 1993, the small village school, Ae Jang Society, made a decision to establish Jangheung Hakdang which would target every resident of the area. The following year, on November 30, 1994, Son and 30 other initiators officially founded Jangheung Hakdang with 'Ae Jang Society' at the center. The object of this establishment was to change residents' perceptions and frames of thinking and seek a new paradigm for the development of the area. Yet, some had pessimistic views about the establishment itself. This is the reason why community learning seemed strange and unfamiliar to the residents.

> So we decided to do this nationwide, build village schools, and compose establishment promotion committee, widening the scope of members. However, there were people opposing it and doubting its success. (Kim)

Soon, on December 1, 1994, Jangheung Hakdang completed its registration as a non-profit corporate body at Gwangju District Court with its main office located in 153–157 Giyang-ri, Jangheung-eup, Jangheung-gun, Jeollanam-do. Accordingly, the board of directors became not only responsible for the vitalization of the village school but also its bankruptcy as well.

The Purpose and Program Planning of Jangheung Hakdang

The purpose of promoting Jangheung Hakdang is to contribute to fostering healthy and wholesome democratic citizens and making a vibrant Jangheung through providing a place for study meetings and workshops. This would change the perspectives of the people regarding the changing world and the environment, and foster responding abilities that allow the people to find tasks and carry out those tasks. To achieve such purpose, there was the need to document learning contents from regular study meetings, field-centered workshops, and presentation debates on pending issues and publish those to share with the members and residents. It was decided that the regular study meetings would take place on the 2nd and 16th (twice a month) of every month and its location would be the readily accessible lounge room of the Jangheung District Office.

The time of the study meetings carried on from 7 a.m. to 8:30 (90 min). It was a rule to have the program in the order of breakfast (20 min), appreciation of poetry and music (10 min), introductions of lecturers and greeting from the head of the school (5 min), lecture (50 min), and a question and answer time (5 min). The selections of lecturers were made within the Jangheung jurisdiction by the heads of institutions and guest speakers or figures that left their hometowns but guest speakers would be selected as politics, community, economics, and culture education specialists by the recommendations of the members.

> We have to get out to the fields early in the morning. So, when they said we would start learning from 7 a.m., I thought it didn't make any sense. Of course, everything was hectic and it was hard. Sometimes, I would just fall asleep at the study meetings. I would roughly understand the words of the speakers but I could not concentrate and my mind was elsewhere. (Kim)

It was decided that the workshops of visiting fields inside and outside Jangheung or areas that were of members' interests would be held four times a year (March, June, September, December). For workshops, visits to educational spots, agriculture, forestry, fisheries, cultural sites, and construction sites were made. Presentation debates were hosted twice a year (May, October). The materials gathered from the study meetings, workshops, and Presentation debates were organized and published yearly.

Articles of the Association and Organization of Operating System

The operating system of Jangheung Hakdang at the time of establishment showed the organizational system operating separately under each department. The operating system of Jangheung Hakdang consists of the head of the school and an honorary head which assumes the role of a consultative body. According to the regulations and articles of the association, the head had a wide range of power and authority, representing the village school and handling everything from projects to activities while the authority of the board of directors was relatively weak. In this context, opinions were brought forward to allow the head of the school to only be the representative, the chairman of the general meeting, have the right to refer bills, and the authority to become the chairman of the board of directors. The first head of Jangheung Hakdang was the former secretary Son, who assumed the leading role in establishing Jangheung Hakdang. The honorary head, a consultant regarding overall affairs of the school, was elected among the former heads of the school through deliberations by the board of directors. Three vice heads of the school, consisting of vice heads of financial affairs, study meetings, workshops, aided the head and in cases of accidents, they became an acting head. A vice head of financial affairs takes charge of general affairs department while a study meeting vice head is in charge of the department of women and young men, and a workshop vice head controls the workshop department and publication department. In the department of general affairs, with a full-time general director and general affairs manager as the center, accounting and conventional administrative tasks are carried out. The study meeting department conducts and carries forward study meetings and presentation debates while striving to spur the participations of women and young people. Yet, the activities of the department of women and young people were insufficient and their participation rate was poor. Through reorganizations of the operating system, a vice head of the department of women was established and it was decided that the department of women, which was formerly under the control of the study meeting vice head, would be under the vice head of the department of women.

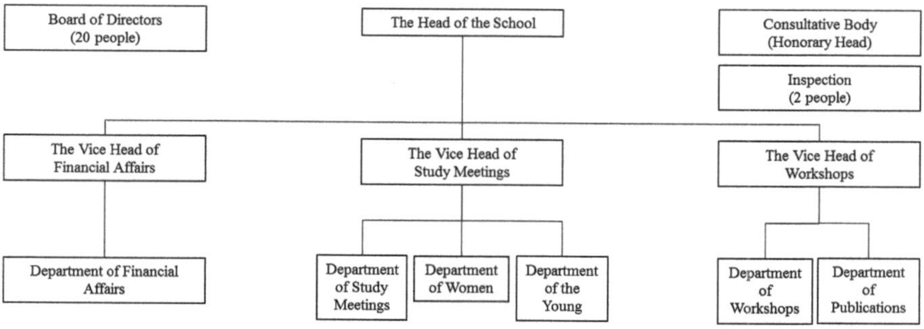

Fig. 1 Jangheung Hakdang organizational chart

> I was once a chairman of the department of women before becoming a vice head for the department of women. I never missed a study meeting and was active in mainly promoting the participation of others. The department of financial affairs should be headed by a person capable of handling financial matters and cooperation is always present regarding workshops. I think that everything is same here as in the society. The purpose is to promote vitalization. (Park)

The workshop department holds and promotes the field workshops. In addition, it assumes the role of publishing the yearly 'Jangheung' Journal from the materials gathered from the study meetings, workshops, and presentation debates. The legislative organ consists of an operating board, executive board, and regular general meeting. Issues and matters gathered and organized from the monthly meetings are decided by the operating board which oversees the overall operation. The executive board enforces the contents that are determined by the operating board and carries forward projects. Additional activities are supplemented through the regular general meeting held every January and the direction for the year is set then. With this, the structure, operating regulations and system of Jangheung Hakdang, was arranged to organize the general and basis system, and organs that were established for efficiency can be seen through Fig. 1.

Determining Future Goals of Jangheung Hakdang

After the general structure of association articles and the operating system had been established, specific guidelines and the knowledge domains that would become the goals of Jangheung Hakdang were enacted and became the four doctrines of Jangheung Hakdang. The domain, acting as the center and the range of common interests, issues, and topics of the community members, allows its members to make judgments regarding the importance and properness of knowledge based on common understanding and imbues significance to the CoP's reason for being and the actions of members (Wenger et al. 2002). When taking a look at the earlier mentioned four doctrines, the first doctrine is academic pursuit. It is the act of intensifying and improving one's knowledge through a vast array of information, various experiences, and field experiences. The second is new creation. After thoroughly grasping the old knowledge and making it completely one's own, based on that, a new creation of knowledge occurs (Cho 2008). This is based on constructive criticism, rational introspection, breaking away from stereotypes, and creative thinking. The third doctrine is golden mean. The concept of 'fittingness' is of the inclines standing straight up, not the meaning of center. The rightness is about filling the voids and insufficiency to become sound and whole. It is the practical awareness and methodology where excessiveness is drawn off and inadequateness is supplemented to make its subject stand tall and fitting (Son et al. 2010). Golden mean is represented through neutrality of political religion, eliminations of requests for favors, and realizations of democratic citizens. The last doctrine is self-reliance. The foundation of Jangheung Hakdang depends greatly on the participation and trust of residents, and the methods of gaining independence from local government or local institutions that may wield influence over the organization. To achieve this, Jangheung Hakdang repudiated the offers of financial assistance by Jangheung-gun and is running its operations solely on voluntary dues of residents and donations. This demonstrates the will of Jangheung Hakdang to maintain its morality and autonomy while remaining free from the local government and institutions, and adhere to independent and autonomous activities. This includes the values of participation with social responsibility, voluntary support, and rational operation.

Participation and Growth Period

Topic Selection and Development of CoP Learning

With the four doctrines (academic pursuit, new creation, golden mean, and self-reliance) as the foundation, the board of directors chose the subjects of study meetings, workshops, and presentation debates.

> The topic selection is carried out by the board of directors. I have attended several topic selection meetings. Exchanges of opinions occur and the item that seems most suitable to the development of individuals or the community is chosen. (An)

In the first stages, there was little information, knowledge, or awareness of lifelong learning so the focus was on the fact that local self-government was beginning. Accordingly, pending local issues were chosen as topics and the head of a local institutions within the Jangheung area was invited to be the lecturer. Gradually, the learning domains would be expanded further and further.

Study Meetings

As of late 2011, 390 study meetings have been held. Lecturers have been the heads of the school 17 times, the heads of institutions 96 times, and guest speakers 277 times. Topics of study meetings were chosen from topics of Jangheung's pending issues, appropriate in regard to the times, and these lectures were followed by discussions among lecturers and members. This was to enhance the intellectual abilities of individuals and contribute to creating healthy democratic citizens, ultimately leading to a vibrant, developing Jangheung.

At Jangheung Hakdang, before the commencement of a study meeting, a poetry reading by a member takes place. Then time for music and art appreciation follows. This has symbolic meaning to maintain the tradition of literary circles that Jangheung has. Through such study meetings, members as participation subjects of the CoP gain practical knowledge that can be applied to present or future situations in real life and learn and share the knowledge that will allow them to responsibly carry out a practical role. In other words, members are able to better grasp the effects of domestic and overseas conditions, which change rapidly in the information-oriented, internationalized, and globalized society, on individuals and the community. Members are to learn and put into action by positioning a variety of information, related to real life, into subject matters that will support rational decision-making. This is based on the actual lives of the people rather than theories or principles. Learning does not end in itself but connects to the practice process that solves the problems of daily lives. Furthermore, on the basis of information, knowledge, and professionalism gathered from learning, the interests towards the lives of the local residents are deepened and becomes linked and expanded to the activities that will solve the tasks of the community. The learning information of previous study meetings can be separated as follows by fields: social science (202 times, 51.8 %), techno science (101 times, 25.9 %), liberal arts such as new year prospects (33 times, 8.5 %), arts and culture (25 times, 6.4 %), humanities (19 times, 4.9 %), natural science (6 times, 1.5 %), history (3 times 0.8 %), and mythology (1 time, 0.2 %). Among these, the field of social science took up the majority. This reflects the emphasis on lifelong education, the social system, and political rights of the residents, along with the engagement of local self-government.

Field Centered Workshops

Field centered workshops are operated four times a year (March, June, September, December). After selecting topics that arose as problems during the study meetings or are in the interests of the members, the related fields are visited. A workshop is assorted into four categories: education, agriculture and fisheries, culture, and construction. Through the workshops, members explore historical and cultural heritage sites within and out of the jurisdiction. This encourages members to reflect and reorganize their thoughts on society and life. Yet, since the attendance rates for the workshops are poor, being around 10 %, efforts are expended to increase the attendance rate.

> Besides the study meetings, there are things called the workshops. Four times a year, we visit the industrial fields, successful cases in the agricultural and fisheries field, construction fields, education fields, etc.... (Yoon)

The workshops have been held 66 times in 121 different locations. The number of workshops held according to the fields are 16 times in education and industrial, construction, 14 times in agricultural and fisheries, 17 times in cultural heritage sites, two times in companies within the jurisdiction, and once in a local assembly.

Presentation Debates

In order to suggest visions for local administration and for the sake of community development, presentation debates are held twice a year in May and October. Along with education and environment issues and election of local representative, problems of the community and pending issues are unraveled in the form of a seminar or a symposium. Knowledge related to the presentation debate is gleaned and systematized through the study meeting and workshop that are held before the presentation debate. In a presentation debate, both specialists and residents participate to identify problems related to the issues and propose appropriate solutions. The presentation debates are meaningful in that by representing the public opinion and deploying active discussions, the direction for local administration is set forth and the ability to provide constructive criticism and create alternatives is cultivated.

> The purpose of gathering through discussions and debates is to accept yesterday as yesterday and by setting that as the base, we are able to analyze today objectively. Then, we figure out how tomorrow will approach us and how we will receive tomorrow. It is very valuable in that we are preparing ourselves for the future. (Wi)

However, in regard to the debates of candidates for the local election, there were times when only some of the candidates were present due to the differences of opinions. Presentation debates have been held 31 times in total. The fields of the presentation debates are divided into the following: ecological environment two times, local festivals four times, local education two times, local development nine times, public office election seven times, suggestions for military administration four times, and Jangheung Hakdang, gukak (Korean classical music), and health once.

Publication and Application of "Village School" Journal

From the year 1995 to 2011, Jangheung Hakdang has been documenting and objectifying the activities and shared knowledge of its core projects, which are study meetings, workshops, and presentation debates. Moreover, these are published yearly, distributed to the members and residents, and shared.

> By publishing the records of study meetings, workshops, and presentation debates, though it was not done by the members, we are able to inform the figures nationwide who left the Jangheung area or the people who used to work in Jangheung but left. For this reason, we have been publishing the journal every year. (Kim)

In this journal, every detail of the study meetings, workshops, and presentation debates are organized and printed without any adjustments. As of late 2011, the activity records indicate that 390 study meetings, 66 workshops, 31 presentation debates, and the activities of the school have been printed in 17 issues.

Management of Members and Financial Affairs

Though the financial situation of Jangheung Hakdang indicated its total assets at 10 million won at the time of establishment, the amount increased to 50 million won 3 years later. The investments came from donations, contributions, dues, and other revenues. Anyone who had the same objects and purposes like those Jangheung had the qualification to be a member. The members were divided into the general members who are the individual members, group members such as corporate bodies and organizations, and honorary members such as figures who left the hometown or residents of other areas. The finance of Jangheung Hakdang was covered with the dues of members and contributions of the residents. The fee, in the case of an individual member, was 50,000 won for membership and 30,000 won for annual fee. Corporate bodies were to pay 200,000 won for membership and 100,000 won for annual dues. The number of members of Jangheung Hakdang, which started out small with 30 members, increased to 302 members by 2007. As of late 2011, there were about 356 members, making Jangheung Hakdang an influential and vital CoP. Among the members, there are 310 individual members, seven corporate bodies, and 39 honorary members. About 50 members are actively participating and due to the large number of long-term residents of Jangheung, the sense of closeness between members is strong. The researcher of this research signed up for membership in 2010. Initially, the head of the school, Son and full-time members Ko and Choi operated the overall operation of the organization but soon, the horizontal communications between members made possible for the consistent establishment of the operation direction to be continued. However, as the members aged, flexibility of the organization decreased and the problem of young members being unable to easily associate with older members became apparent. Also, in terms of vitalization of the school, securing active participations of figures related to education, prosecution, and court came to the fore.

Maturation and Sustained Growth Period (2006–2011)

Disjunction Between Daily Lives and Academic Atmosphere

The daily lives of Jangheung residents were repetitions of quiet days without any radical changes. However, in any situation, until things are set in a regular track, difficulties are bound to arise. As such, there were many complications in the process of Jangheung Hakdang taking root. First, it was not an easy task to operate such a 'village school' for people of a farming village who may even be insensitive to the changes of times and the environment. Most of the local residents were engaged in agriculture and fisheries, and the image of them running busily from daybreak to make ends meet seemed distant from learning. Until then, 'lifelong education' was an unacquainted domain, with residents feeling it as being far from them and the issue of forming consensus as to the need for learning were difficulties faced in the process of urging residents to participate in lifelong education.

> The countryside is a relaxed place where you can get away with walking around backwards. In the countryside, most gatherings occur in the night and it might work if it were carried out at night after daily tasks are finished but would the meetings taking place in the mornings work? There were many opinions such as this. People said it would be hard... Yet, waking up at daybreak has its meaning. It is not the number of people who attends that matters but the fact that such a meeting is held at daybreak. So, it was pushed forward. It is true that the participation rate of the young is still low. (Kim)

Due to situations such as these, there were many limitations including the low participation rate of residents and the gap between daily lives and learning. Additionally, the task of collecting membership fees and dues from the participants was complicated and upon seeing this, some were astonished and claimed that Jangheung Hakdang would not last long.

> As the school started, there many who said 'What would that matter to old people like us...' They would possibly 'consider' going even if it was free, but paying with their own money... (Yoon)

Yet, regarding concerns such as these, Son remained unshaken in his convictions. Whether it is an individual or an organization, if the period of conflicts and confusion is successfully conquered, that becomes nourishment, bringing forth a more mature form. The influences of convictions and competence of a leader on the development of individuals and community were confirmed.

Misunderstandings of Residents and Entry Barriers

There were rumors among the residents on Jangheung Hakdang assuming the role of a political patronage of a certain figure. However, such suspicions were completely wiped out as time elapsed.

> Well, not much was expected or wanted from it... The members of Jangheung Hakdang must have a certain purpose for doing this. You know, like a suspicion that someone among the members wants to run for the National Assembly. (Kim)

> At first, there were people looking on at Jangheung Hakdang from a biased viewpoint, claiming that someone is running for the local election or the National Assembly but with the organization depoliticizing in every aspect, misunderstanding of the residents was resolved. (Yoon)

Additionally, there was a tendency among the residents to think of Jangheung Hakdang as the gathering of privileged classes such as former government officials.

Assessment of Jangheung Hakdang's Performance Results

In December 2004, after looking back to the last decade, the catch phrase 'Beyond a Decade, towards a Century' was established and posted. For this, 'How can Jangheung Hakdang be more successful and prosperous?' was selected as the presentation debate topic on April 7, 2005. 17 people including the head of the school, vice heads, members of the board of directors, members, and residents participated in the presentation debate. First, regarding the organizational (executive) domain, strategies instigating wide participations of people from all social standings and participations of young people who will substitute the aging population of members and reform of the head-centered operation methods were discussed. With these discussions as the base in terms of strategies to spur participation, at the regular general meeting held in February 2006, an address book of figures in a wide variety of domains was obtained and the figures were notified of the establishment purpose of Jangheung Hakdang. Furthermore, consistent efforts would be made for the participations of young people, which would substitute the aging population of the members. Son, who had substantially planned and operated Jangheung Hakdang for 11 years since 1994, judged the school to have reached a stabilized period and declared to leave the head position. After much argument, Yoon, the former director of Jangheung-gun Cultural Center, became the successor and operated Jangheung Hakdang for 2 years (2006–2007). During this period, due to the fact that Yoon was not originally from Jangheung, the participation rate of residents was relatively low. This indicates that even though Jangheung is known for its absence of exclusiveness, the local feelings, in regard to the head originally coming from another town, seem to have influenced to some degree. In 2008, the full-time governor Kim undertook the role of the head of the school and is currently assuming the role. In regard to member management area, appropriateness of fees, qualifications of members and clearance matter arose as key issues. The needs to clear away the members who have only registered without participating and to secure female members were discussed. Moreover, in financial affairs area, ordinary earnings such as fees and donations account for 55 % of all earnings and incomes other than the ordinary earnings such as business profits and contributions take up 45 %. Of the ordinary earnings, the pure earnings from the fees account for 18 %

and of incomes other than the ordinary earnings, contributions account for 40 %. Problems arising from such financial state were brought forward. To secure finances, talks were focused on measures such as the increase of fees, hunt for outstanding members and receiving financial assistance from the district office.

> There are many financial difficulties. At the end of the year when we settle accounts, there are the basic costs that are spent by the method of a deficit operation. You know, things like maintenance cost for managing the office, the minimum personnel expenses, food expenses paid to the restaurant for study meetings. We try to minimize these basic expenses, in addition to the publication expense but it is really hard. (Kim)

Nevertheless, participation with the mindset that dues are necessary for academic pursuit is required. Receiving financial assistance from the district office, since it comes from the collected taxes of the residents, goes against the fundamental founding purpose of operating the school with the strength of the school alone and for this reason, the school has been operating on its own funds to this day.

> Jangheung Hakdang is carrying out the local government's duty of educating the residents which provides sufficient grounds for justifying financial assistance from the district office. However, Jangheung Hakdang has always refused. It is really something to be thankful for. (An)

Nonetheless, in order to operate Jangheung Hakdang, it costs 40 million won to 50 million won a year, making members' dues insufficient to maintain the operation. The organization is in a state of depending on part of the contributions to make up for the shortage.

Ties and Sponsorships of Other CoP

Jangheung Hakdang is seeking opportunities for a new value creation through the ties and sponsorships of other CoP. Main organizations in mutual relationships with Jangheung Hakdang are Save the Gullies of Korea Gwangju, Jeollanam-do Headquarters, Korea Water Resources Corporation (KOWACO), 21 Century Namdo-Forum, Gwangju, Jeollanam-do Development Institute, etc. First, Save the Gullies of Korea Gwangju, Jeollanam-do Headquarters joined with Human Resources Fosterage Academy held the 'Solution Search to Save the Gullies of Korea (November 11, 2009)' discussion debate discussing the development of the participatory restoration of gullies model. Moreover, a forum on water was held jointly by Korea Water Resources Corporation (KOWACO) and Jangheung Hakdang with the sponsorship of Jangheung-gun.

> The forum on water was planned by Korea Water Resources Corporation (KOWACO) Jeollanam-do Southwest Conservancy, which supplies the water of Jangheung Dam to cities and districts including the city of Mokpo, to inform the people of the importance of water which surrounds us. Since this forum took interest in linking water resources to the community economy and was prepared well, I would like to assess that this debate had a great academic significance in regard to the Jeongnamjin Water Festival. (Wi)

This forum continues to be held (07/30/2008, 09/05/2009, 07/28/2010, and 07/29/2011). Furthermore, from March 2003, through mutual exchange visits with 21 Century Namdo-Forum, interchanges of information are occurring and connection with Gwangju, Jeollanam-do Development Institute has been made since July 1997.

Conclusion and Implications

The purpose of this research was to expose forming process of the CoP to provide implications that can be applied to the successful operation of the spontaneous CoP in local areas and promote community well-being. For this, the forming process of the CoP through the interactions between the members of Jangheung Hakdang and the environment became the focus and the factors that maintain and develop the CoP, which manifest in the developmental process, were explored.

The results of the research demonstrate the meaning of the establishment of Jangheung Hakdang in that the vision was realized under the leadership of a home-bound high-ranking government official and this CoP promotes the community well-being as well as community development. The forming process of Jangheung Hakdang, based on the characteristics and the background of the area, was identified as having four phases: germination period, system establishment period, participatory growth period, and sustainable maturity period. In the germination period, according to the changes of the local self-government circumstances, the main agent of Jangheung Hakdang who is the retired government official, made contacts and connections with potential members and the primary intentions of the CoP were discussed in this stage. In the system establishment period, a paradigm for the community development was sought out, the head of the school was selected, articles of association and regulations, which are the frame of an organization, were arranged, and the four doctrines of academic pursuit—new creation, golden mean, and self-reliance that are the future coordinates and specific guidelines of behavior of Jangheung Hakdang—were enacted. In the participatory growth period when action learning activities were actively carried out, participations of members and residents became diverse and inclinations started to change. Through discussions and debates such as study meetings, field workshops, and presentation debates, experience and knowledge were shared and 'Jangheung Hakdang Journal', that contains the activities and shared knowledge, was published every year to be of practical use to the residents and members. In the sustainable maturity period, problems that arose amid the operation of Jangheung Hakdang, such as the gap between daily life and academic atmosphere, misunderstandings of residents and barriers to entry, were checked, applied to the organization and visualized. In addition, creations of relationships with external organizations were executed by holding forums of topics, such as saving the gullies and water festival, through connections and sponsorships regarding other CoP. Likewise, Jangheung

Hakdang emerged as the large reception room of Jangheung area where the discussions of the subjects necessary in the latter part of the information age today occur.

Such results of the research exposed the resonance effect of the government official's retire to the country and his leadership on the residents, based on the local characteristic of the unique Confucian hereditary customs that Jangheung Hakdang has. Additionally, the research revealed that the active support and dedication, system reorganization for the school operation, participation of residents, and networks with related institutions exerted crucial influences over the 18 years of Jangheung Hakdang's continuation.

Based on the results of the research analysis, in order for a CoP to successfully take root and develop, the following implications can be suggested. First of all, to elicit dynamics and the participation from the majority of the residents, the role of a devoted leader of the CoP is important. For instance, common learning is realized through the overall lives of the residents by interlocking with the diverse domains of the region. For this reason, the leader must be aware and understanding of the current states of residents and the community, not to mention that the leader must lead a moral and ethical life and receive respect and trust of the residents. Secondly, voluntary support and service minds of members for financial stability of the CoP is necessary. In order to carry out the purpose of the desirable organization and make it shine brightly, the active attitudes and devotions of participants with concerns and love for the community are required. What is more, the role of a financial supporter is needed to make sure that the administrative organ is well-operated. Thirdly, the gap between generations needs to be eased and the participation rate of women should be enhanced. Unlike the urban area, in the case of the rural areas, the participation rates of women and the young are poor. To solve this, learning programs and time slots need to be composed flexibly to suit the young and women participants, and there is the need to improve learning accessibility through locally assigned study tour, not fixed study locations.

References

Andriessen, J. H., Huis, E. M., & Soekijad, M. (2004). Communities of practice for knowledge sharing. In J. H. E. Andriessen & B. Fahlbruchb (Eds.), *How to manage experience sharing: From organizational surprises to organizational knowledge*. Oxford: Elsevier.

Baek, Y. J., & Kim, E. S. (2008). Influence factors of knowledge share in the community of practice: modifying effects of structural and relational properties. *Knowledge Management Research, 9*(2), 63–86.

Bang, Y. S., & Lee, M. S. (2000). CoP operation strategy for successful knowledge management: examples of s company's knowledge management building. *Knowledge Management Research, 1*(1), 127–137.

Bettoni, M., Endress, G. A., & Schneider, R. S. (2002). Communities of practice. University of Fachhochschule beider Basel. (On-line) Available at: http://www.fhbb.ch/weknow/macro/A2002/kmp.

Choi, M. N., & Yoo, Y. M. (2003). Example study of development process of CoP as knowledge creation and sharing strategy. *Education Information Broadcasting Research, 9*(4), 177–208.

Gongla, P., & Rizzuto, C. R. (2001). Evolving communities of practice: IBM global service experience. (On-line) Available at: http://www.research.ibm/journal/sj/404/gonglaut.html.

Ham, J. (2008). *Research on relative importance of performance determinants in joint sector CoP*. Paper of masters degree at Konkuk University.

Jang, W. S., Kim, J. Y., Lee, D. H., & Lee, J. H. (2010). Research on the relationship between types of CoP participation and organizational commitment. *Industry Education Research, 21*, 1–19.

Jang, W. S., Kim, M. Y., & Yoon, J. H. (2009). Effects on organization learning by the community of practice. *Career Education Research, 28*(3), 209–228.

Jeong, Y. S. (2008). *Research on vitalization factors of CoP from public organization and nonprofit social service organization*. Paper of masters degree at Yonsei University, pp. 52–59.

Kee, Y. (2011). Community well-being. Seoul National University Knowledge Policy Forum materials for presentation, pp. 5–11.

Kim, J. W., & Kim Y. H. (2005). Example study of knowledge management and IT knowledge share: examples from communication company teams. In *Knowledge Management Scholarship Symposium*, vol. 14(1), pp. 259–268.

Kim, D. H. (2008a). Development process of network-based community of practice for knowledge share between teachers and specialty improvement: In-the-school cases study. *Educational Technology Research, 24*(2), 1–30.

Kim, G. (2008b). Development direction of CoP for the successful of knowledge management in administration organization: cases comparison of private sectors. *The Korean Public Management Administration, 22*(4), 329–355; *16*(3), 1–29.

Kim, G. P., Chae M. S., & Yeom D. S. (2009). Mediation role for small and medium enterprises CoP characteristics. Human Resource Administration Research.

Ku, J. M., & Song, H. D. (2011). The relationship between CoP participatory intention and immersion based on knowledge share intention. *Korea HRD Research, 6*(1), 75–95.

Lave, J., & Wenger, E. (1991). *Situated learning: Legitimate peripheral participation*. Cambridge: Cambridge University Press.

Lee, J. E. (2008). *Structural factors of the community of practice and the relationship between individuals and systematicity*. Paper of masters degree at Seoul National University, pp. 79–102.

Lee, S. J. (2011). Roles of self-autonomy commission. Study materials of Gwacheon-si self-autonomy commission, pp. 10–15.

Lee, E. S., & Jeong, J. S. (2008). Relation research among community spirit in CoP, immersion and knowledge management activities. *Educational Technology Research, 24*(4), 319–350.

Lesser, E. L., & Prusak, L. (1999). Communities of practice, social capital and organizational knowledge. *Information System Review: KSMIS, 1*(1), 3–10.

Lim, S. Y., & Kim, S. T. (2009). Analysis of efficacy on the achievement of organization through CoP activities from organizational culture's mediator effects, and on organization learning. *Career Education Research, 28*(4), 55–76.

Linclon, Y. S., & Guba, E. G. (1985). *Naturalistic inquiry*. Newbury Park: Sage Publications.

McDermott, R. (2000). Community development as a natural step: Five stages of community development. *KM Review, 3*(5), 1–10.

Nickols, F. (2000). Communities of practice. (On-line) Available at: http://www.home.att.net/~discon.KM/CoPS.htm.

Son, S. N., Gang, Y. H., Kim, Y. S., Nam, G. H., & Jang, D. J. (2010). *Asking the way to education in Asian classics*. Paju: Korea Research Information Service.

Wenger, E. (1998). *Communities of practice: Earning, meaning, and identity*. Cambridge: Cambridge University Press.

Wenger, E. (2007). Community of practice: Social ecology of knowledge sharing (M. H. Son & E. G Bae, Trans). Seoul: Hak-Ji Company.

Wenger, E., Mcedermott, R., & Snyder, W. M. (2002). *Cultivating communities of practice*. Massachusetts: Harvard Business School Press.

Wenger, E., Mcedermott, R., & Snyder, W. M. (2004). *CoP revolution* (S. K. Hwnag, Trans.). Seoul: Mulpu-le.

Yang, H. G. (2011). Compositional study of CoP's learning analysis framework. *Korea HRD Research, 6*(4), 41–59.

Promoting Community Integration of New Migrant Workers in China

Wen Juan Lu and Longgen Chen

Introduction

Community well-being is an umbrella concept, encompassing a wide range of concepts such as quality of life, happiness, sustainability and etc. Many scholars have tried to define the concept from different perspectives ranging from economic, social, environmental and cultural angles to governance goals and priorities. Haworth and Hart (2007) pointed out that community well-being is focused on understanding the contribution of a community in maintaining itself and fulfilling the various needs of local residents. Merriam and Kee (2014) defined it as a function of many factors working in concert to promote an optimal quality of life for all members of a community, young and old. Considering the great significance and influence of migrant workers' issues on society, there is now widespread concern and attention. Issues include: how to transfer rural labors successfully, how to facilitate new migrant workers' integration into cities effectively, how to guarantee their equal civil rights and privileges, how to promote their survival ability, and employability in cities.

Until now, community well-being has been a new concept for most Chinese scholars and local officials. This chapter aims to build a link between migrant workers and community well-being, and illustrate how to facilitate migrant workers' successful integration into communities through community education and finally achieve community well-being from the perspectives of new migrant workers.

W.J. Lu (✉) · L. Chen
Zhejiang University of Technology, Hangzhou, China
e-mail: lvwenjuan79@hotmail.com

L. Chen
e-mail: clg@zjut.edu.cn

Y. Kee et al. (eds.), *Learning and Community Approaches for Promoting Well-Being*,
SpringerBriefs in Well-Being and Quality of Life Research,
DOI 10.1007/978-3-319-12439-1_3

Migrant Workers

According to China's National Bureau of Statistics, by the end of 2011 China had a rural population of 0.94 billion, which occupied 70.15 % of the total population. In recent decades, along with the social-economic development and acceleration of urbanization in China, large number of surplus rural labor in countryside poured into cities and towns to make a living. Accordingly, a social group with Chinese features came into being—migrant workers. Based on the latest statistics, there were more than 0.25 billion migrant workers in China in 2012, and the number would increase steadily in the years to come.

The State Council Research Group (2006) defined "migrant worker as a special concept which took shape in the process of China's social and economic transformation, and it mainly referred to the staff whose household was registered as peasants owning contracted land, but they were mostly engaged in non-agricultural industries and depended on wage income as main financial resource" and "those who were born in 1980s–1990s were called new migrant workers, who received better education but presented different characteristics compared with old migrant workers". The new migrant workers were confronted with more difficulties when they tried to integrate into the rural life.

The survey data showed that among the new migrant workers, 85.23 % of them got middle school diploma or lower and only 14.77 % of them got the college bachelor degree. Limited by their poor education and insufficient vocational skills, most migrant workers worked in the fields such as manufacture (39.1 %), construction, service, accommodation and catering industry, wholesale retail industry, and transport postal industry. Undoubtedly, migrant workers have played a significant role in all walks of life in cities and made great contributions to the national economic construction, but unfortunately their self-quality failed to meet the demands of working in cities and in part hindered the economic development on the whole, and some time their efforts were not accepted and recognized in cities. As a matter of fact, at present most migrant workers were in inferior situation, they were lack of security due to unstable and low-income jobs, they experienced social exclusion, and furthermore they failed to enjoy the same rights as other city residents in terms of social welfare, education, medical treatment and other aspects. Xue (2011, p. 53) pointed out that "subjectively, new migrant workers did not agree with the rural life, but objectively they could not melt into the city life successfully, therefore the group was lack of sense of belonging in cities and greatly frustrated." Wen (2004, p. 43) declared that "under certain circumstance, such group could be changed into social alien force, and would even take revenge on the society via various inappropriate even illegal behaviors, which would finally lead to social chaos."

Integrating into Communities

Born in 1980s–1990s, new migrant workers are around 20–30 years old. Compared with old migrant workers, this new generation has a series of different features, such as higher levels of education, specific life goals, and strong desires to live in modern cities. However, migrant workers have encountered various problems in participating in city life and finding their role.

Ambiguity of Identity: Urban or Rural Residents

Differing from the old generation, most new migrant workers are landless with the acceleration of urbanization in rural areas in recent years. On the one hand, their household would be still registered as peasants, but they had no land, no farming knowledge and experiences, and were reluctant to work as peasants and lack of professional identity as peasants. Therefore, they were only symbolic peasants, not real ones.

On the other hand, the new migrant workers dreamed of city life, moved to non-agricultural industries, but the failure to integration into cities made them confused about their identity. According to a survey on young migrant workers in Chengdu city in 2006, 37.3 % still regarded themselves as peasants, 25.4 % gave the opposite answer, and 37.3 % were not sure about the answer. Another survey in 2007 showed that 87 % new migrant workers were ambiguous about their identity and failed to give positive answer (Xu 2007). As for the traditional migrant workers, they could still acquire sense of belonging in countryside even they were frustrated in cities, while the new migrant workers were confronted with the inferior situation of self-identity, neither classify themselves as urban residents nor peasants.

Marginalized Situation as Non-city-citizens

The new migrant workers have worked to pursue sense of belonging in cities instead of making a living; they desire to acquire citizenship in cities, and seek equal rights instead of endurance; and they have been more full of passion and vitality, and eager to promote overall quality via utilizing public resources in cities to help realize self-worth (Zeng 2012).

But subject to a hereditary dual household-registration system in China, known as "hukou", which divided citizens into urban and rural residents, the migrant workers met unprecedented difficulty in incorporating into urban life. They found it hard to achieve identity transformation, their rights and interests were damaged, and they were marginalized and excluded from political, economic, cultural, and social security system in cities. In other words, the system discriminates against migrants in employment, housing, health care and education and reinforced a widespread tendency to treat them as second-class citizens. There exists a big gap between their strong desire to integrate into urban life and the high requirement to acquire citizenship in cities. The poor integration sometimes arouses emotional conflict and in some cases, social chaos.

Weak Employability

New migrant workers are 16–30 years old now, and most of them obtained a middle school or high school diploma, and a small portion graduated from college or university. "Compared with the old migrant workers, the new generation received better education, were able to acquire information via books or internet, and were more likely to accept new things", which prepared knowledge basis for their future job in cities (Zeng 2012, p. 158).

But in past years, China's economic growth mode was experiencing dramatic change, and the government had been sponsoring a shift from labor-intensive to knowledge-intensive jobs and enterprises. The rapid and constant developments of knowledge-intensive enterprises needed the accumulation and innovation of knowledge and technology, which depended on the labor force who have received professional vocational training and were equipped with intermediate and higher-level skills. Generally speaking, the more education a worker received, the higher skills mastered, and the ability to obtain a well-paid job increased.

"Obtaining a stable job and higher income were the two important conditions for migrant workers to melt into cities, while several factors such as poor education, lack of skills restricted the realization of it (Xue 2011, p. 52)." Apparently, the current level of migrant workers' overall abilities did not meet the high demands of formal employment so many are engaged in the informal employment sectors such as manufacture, construction, service, accommodation and catering industry. Informal employment had many traits: wide field, flexible pattern, low cost and low entering condition, which could relax the employment predicament of vulnerable groups in cities, but the low income and instability could not afford the high living expenses in cities, such as children's education, accommodation, medical treatment and so on.

Community Education: A Path for New Migrant Workers

In recent years, Chinese government advocated vigorously about providing various educational and training opportunities for new migrant workers to enhance their employability and improve self-quality. Little progress has resulted, however, mainly caused by the gap between inflexible training and special requirements of migrant workers.

Traditional Education for New Migrant Workers

China's industrialization and urbanization necessitates the needs for training new migrant workers. Traditional education has not been able to satisfy their educational needs and desires. First of all, due to low income levels, training expenses may be too costly. Additionally, in spite of training needs, most migrant workers have to make money to support their family, and cannot attend long hours in off-job training, therefore they preferred a more economic and flexible training mode which would not affect their work. Another characteristic of new migrant workers is that they live more dispersed, which makes it hard to organize. As a result, it is difficult for traditional education to provide the desired impacts on educating the migrant workers (Wang 2008).

Transfer Training for New Migrant Workers

In the recent past, the Chinese government realizes that migrant workers transfer training is an important way to address problems of rural migrant workers and accordingly issued a series of macro policies and measures.

In 2003, the Ministry of Agriculture and the Ministry of Education co-enacted "*2003–2010 Training Plan on Rural Migrant Workers*". In 2004, the Ministry of Agriculture and the Ministry of Education co-executed "*Sunshine Project—Rural Labor Transfer Training*". In 2006, the State Council released "*the Opinions on Solving the Problems of Rural Migrant Workers*". In 2010, the Ministry of Education continued to vigorously implement "*Rural Labor Transfer Training Program*" from the following aspects: (1) first, continue to promote migrant workers training through urban vocational schools, adult schools and community schools; (2) second, to assign some rural labor transfer training schools or migrant workers training schools and offer the migrant workers with an extensive range of skill training, certificate training and degree education through holding day schools, evening schools and weekend schools; and (3) third, to make full use of community educational resources, develop urban civilization projects and provide skill training programs targeting to migrant workers. It was evident that for migrant workers' training and education, the Chinese government has expressed great expectations for community education (Chang 2009).

Community Education for New Migrant Workers

Community education had obvious advantages in terms of migrant workers' education. Differing from school education, community education provides a wide range of education to meet community members' educational demands of different levels based on community strength and resources. Community education's characteristics could be summarized as "service, popularity, diversity, openness and flexibility", which could satisfy the migrant workers' special needs as a whole and make it possible to carry out migrant workers' education based on community education (Wang 2006, p. 8).

Students would be free from traffic troubles by attending community schools in their neighborhood, with teachers there composed of volunteers or retired experts (saving costs). Practical educational contents in community education could largely resolve migrant workers' urgent needs by providing employment guidance and skills training. Humanization of educational aim in community education provides a view closer to the spirit of humanistic care, such as holding seminars regularly on mental health, neighborhood adjustment, elderly mental health care, parenting knowledge and legal knowledge, which could help migrant workers find the feeling of psychological reliance and home in the community. Last, community education can provide residents with learning resources whenever and wherever possible, which enables migrant workers to study and work at the same time (Chen 2010).

In a word, the benefits produced by community educational programs for migrant workers include the following: achieve collective identity in communities, improve quality of life, stabilize their economic and social condition over the long-term via updating vocational skills, improve their overall self-quality, acquire self-confidence, enhance coping ability in cities, and equip them with the ability to guarantee their own rights. On the other hand, migrant workers could contribute to community well-being in substantive ways through working in different sectors, volunteering, civic engagement and etc.

Development Obstacles

Considering its features as low-cost, openness, wide-participation and flexibility, community education has been gradually recognized as a best choice to promote social harmony and enhance overall quality of the labor force (Chen 2005). Nevertheless, in the implementation process, there existed various hardships and obstacles, such as lack of sufficient attention, shortage of educational funds, difficulties in educational management, inappropriateness in training contents and method, weak sense of participation, and lags in educational service.

Lack of Sufficient Attention

Despite progress achieved in the practice of both urban and rural community education, much more is needed. Currently, under the guidance of central government's policies, local governments have placed it on their agendas to emphasize professional skill training for migrant worker and have taken actions accordingly. It should be pointed out that migrant worker-oriented training programs have been on a scattered and short-term basis in most cases and remain fragmented in terms of management and implementation. As for providing training opportunities for migrant workers, some local governments and departments only spoke about it without taking real actions. Only by fully realizing the significant role community education plays in promoting migrant workers training from the policy level can it be possible to achieve the harmony between the migrant workers' quality improvement and communities' economic-cultural development, and form a new pattern of community education based on the city and countryside integration (Xi 2007).

Short of Educational Funds

According to national policies, community education should embody the principle of combining government and employers' input with individual payment. But the fact remains that many migrant workers are in poor economic conditions and can scarcely afford training costs. Furthermore, many enterprises and employing units lack of enthusiasm to train migrant workers. On the whole, educational funds for migrant workers need to be guaranteed at the policy level to ensure adequate provision.

Difficulties in Educational Management

Theoretically, community education is a bottom-up and spontaneous educational activity sponsored by the community residents. However, due to the weak self-awareness of community residents at present, it is hard for communities to achieve self-organization. And especially when offering training programs for migrant workers—characterized by big fluidity, lack of identity as a unit person, indefinite studying time and weak restriction—the communities' regulatory capabilities and organizational management were greatly challenged.

Inappropriateness in Training Contents and Method

According to the current survey on migrant workers' education and training, there exists the problem that educational programs and curriculum design have failed to coincide with the actual employment needs of migrant workers. Some training contents do not reflect practice, some contents were identical and lack relevance, and most training contents focused on manufacturing sectors without modern science and technology application guidance, which could not effectively improve migrant workers' employment capabilities. In addition, most training contents targeted to migrant workers emphasizes skills training and ignores how to improve survival skills and urban adaptabilities.

Weak Sense of Participation

Students are the subjects of learning; the key to realize the expected effects of migrant workers education lies in whether these special learners' learning passions could be stimulated. Survey showed that most migrant workers had not realized the importance of learning and training, and the effect of education on their own development. As a result, some of them even complained about it and took training as the responsibility of the government and only participated in a passive way. And meanwhile, limited by their poor educational background, many migrant workers experienced a difficult time learning and could not keep up with the training process. Due to this lack of confidence and perseverance, some dropped out half-way through programs.

Lag in Educational Service

Modern teaching management has been transforming from norm-oriented institutionalized management mode to service-oriented humanized management mode. It was only in recent years that the concern with training and educating migrant workers garnered increased attention, while at the same time, corresponding education services were far from keeping up with the training practice.

Limited by inadequate materials and marketing for community education and their narrow information channel, migrant workers could not get acquainted with the related education policies and failed to be aware of their rights to education. And furthermore, due to time limitations and other objective restrictions, migrant workers could not choose training flexibly and missed many learning opportunities. To guarantee the quality and effects of community education for migrant workers, supporting service systems coinciding with migrant workers' educational desires are needed, as well as making education closely related with employment and daily life, and embodying the principles of practical use and high efficiency.

Development Mechanism

Based on the above problems, the current study suggests setting up an overall developmental mechanism for ensuring the quality and impacts of community education for migrant workers.

Guarantee Mechanism for Educational Funds

Community education has the attributes of public welfare and services, therefore government input should be the main resource of educational funds. As a developing country, China is faced with the long-standing problem of a lack of adequate educational funds. To solve the fund shortage for community education, the nation's investment in migrant workers training needs expanding.

In addition, corresponding policies should be made based on the local actual situation, especially in economically developed regions, it would be workable to develop community education for migrant workers via a wide range of measures, such as inputting government's funds, absorbing charitable contribution and establishing educational operation organization for floating population. The employers should take the responsibility to train migrant workers and manage to attract more enterprise investment in community education for migrant workers through subsidy, duty-free, earnings and other forms.

Administrative and Management Mechanism

To strengthen educational management mechanisms for migrant workers, the key is to clearly define the responsibilities of government and communities. First of all, the government should play a leading role in rural labor transfer training, and therefore the communities should build management mechanisms for coordinating various social administrative organizations.

For communities, it is suggested to establish a leadership system with overall coordination and work promotion mechanisms for clear division of responsibilities, positioning training for migrant workers as important and as a new growth point in community education. Further, all available resources should be mobilized by formulating policies and system innovation. For governments in labor input areas, its role has been to provide strong organizational guarantee for training immigrant workers. On the one hand, the government should perfect the open and systematic management mode featured by the government's macro-control and the local's scattered implementation. In addition, the government should also mobilize the enthusiasm to offer educational programs at the grassroots level and explore bottom-up autonomy enjoyed by the community's residents. On the other hand, the government should adopt the workable funds management mode, establish strict monitoring measures, strengthen the guidance and supervision in educational programs and ensure the quality and efficiency of education.

Curriculum Development Mechanism

The curriculum in community education should be value-oriented and give priority to residents' demands instead of academic certificates. Therefore, the curriculum system for migrant workers should be constructed based on their actual demands for knowledge and skills. The curriculum should be co-developed by the community's administrative managers, full-time or part-time teachers and migrant workers themselves.

When developing curriculum for migrant workers, several aspects should be taken into consideration. First, when providing vocational skills training for migrant workers, we should make full use of local resources and try to make the training program coordinate with the local economic situation and the characteristics of the local industrial structure. Second, the training programs should put emphasis on improving migrant workers' citizenship and overall quality, and a wide range of knowledge would be covered to achieve the purpose, such as the common sense of laws and regulations, urban civilization sense, production safety knowledge, mental health knowledge and so on. According to the newest survey of labor market, the employers had increasing academic requirements when employing migrant workers. While the survey data showed that among the migrant workers, 85.23 % of them got middle school diploma or lower and only 14.77 % of them got the college bachelor degree. Therefore, the issue to promote the migrant workers' academic degree should not be embodied when designing training programs in the range of community education instead of being ignored.

Employment Service Mechanism

The current theme of community education is to enhance the migrant workers' professional quality and employment ability. To strengthen the pertinence and effectiveness of vocational training, employment information service work matching with the labor market is needed. Career guidance and service was found to be not only one of the functions of educational institutions, but also important for intervention and cooperation of government, labor management departments, community educational institutions and employers.

Community educational institutions could first of all conduct a general survey concerning the migrant workers' employment preference, educational background and specialties and other aspects. And meanwhile, they should cooperate with local labor agencies actively, stay well informed of the supply and demand information in labor market and implement market-oriented training contents and forms based on diverse requirements in different regions and fields. In addition, they should provide quality tracking and employment guidance services via cooperation with local government, such as establishing talent database for migrant workers, holding job fairs targeting migrant workers regularly, setting up service offices for migrant workers inside communities and opening job hotlines, which could help, to a certain degree, enhance the migrant workers' employment situation after participating in training programs.

Educational Stimulation Mechanism

Migrant workers' active participation is a source of power in effective development of community education. First of all, it is important to strengthen information and cultivate good social and policy environments emphasizing the importance of training migrant workers. Second, interests and benefits have been identified as the main motivators for migrant workers to participate in community education. When migrant workers felt that participation in educational programs could help improve their income, they would have greater incentives to participate in community education. It then becomes vital to reduce the cost of migrant workers training and even improve their training treatment. The recognition that educational programs would benefit the workers greatly while searching for jobs would enable more and more migrant workers to participate in community education.

Quality Monitoring Mechanism

Quality is also a foundation for education undertaking to develop in an efficient manner. At present, due to economic restrictions, investments in community education and training for migrant workers are very limited.

How can educational quality and efficient use of limited investments be ensured? One way out is to establish quality monitoring mechanism for tracking educational effects, evaluating educational efforts and eliminating some undesirable phenomena caused by inadequate supervision, such as school insufficiency, trainee drop-out, and misuse of funds. Additionally, teachers are a crucial factor in ensuring the quality of education. There are several resources for teachers in community education for migrant workers: some of them would be invited from vocational schools, adult schools, research institutions, enterprises, technology-promotion departments, and meanwhile to compensate for the shortage of teachers, community educational institutions should actively cultivate volunteers with professional knowledge in diverse fields to periodically provide educational training services. Last, education evaluation should not be ignored. The evaluation standards for migrant workers would be guided by goals to achieve high quality employment. The recurring evaluation would ensure that migrant workers benefit from course arrangements, fee standards, graduation rate, employment rate, wage status and other aspects.

Conclusion

Due to China's rapid industrialization and urbanization, it has become an inevitable trend for surplus rural labor to move to cities and towns and work in non-agricultural sectors. They are termed "migrant workers," and those who were born in 1980s or 1990s, are called new migrant workers. In the past decade, this group has made great contributions to China's modernization, while they were confronted with various problems, such as ambiguity of identity, lack of sense of belonging, low income, marginalization and exclusion from political, economic, cultural, and social security system in cities. In recent years, migrant workers related issues have attracted widespread concern, around issues of social stability, community well-being and continuous economic growth in China as well.

Differing from school education, community education provides obvious advantages in terms of migrant workers' education via a wide range of training programs. While in the implementation phase, many problems emerged, such as lack of sufficient attention, shortage of educational funds, difficulties in educational management, inappropriateness in training contents and method, weak sense of participation, and lags in educational service provision. Therefore, it is of great significance to set up an overall developmental mechanism to ensure the quality and effects of community education for migrant workers including a more efficient and effective mechanism for educational funds distribution and use, administrative and management oversight, curriculum development, educational stimulation, and quality monitoring mechanism. Obtaining these improvements will help foster better levels of community well-being for China's migrant workers.

References

Chang, Z. Z. (2009). Problems in current migrant workers training and corresponding suggestions. *The Journal of Today China Forum, 5*, 9–15.

Chen, N. L. (2005). Community education: The support to construct lifelong educational system. *The Research on Open Education, 5*, 10–16.

Chen, L. G. (2010). Research on developmental mechanism for migrant-workers targeting community education. *The Journal of China Adult Education, 6*, 23–28.

Haworth, J., & Hart, G. (2007). *Well-being, individual, community, and social perspectives*. Hampshire: Palgrave.

Merriam, S. B., & Kee, Y. (2014). Promoting community wellbeing: The case for lifelong learning for older adults. *Adult Education Quarterly, 64*(2), 128–144.

The State Council Research Group. (2006). *Survey report on China's migrant workers*. Beijing: China's Yanshi Press.

Wang, Y. H. (2006). Research on community-based education for migrant workers. *Zhuzhou Institute Journal, 1*, 7–14.

Wang, Y. Y. (2008). Community education: An effective way to solve the social exclusion of migrant workers. *The Journal of Adult Education, 4*, 7–10.

Wen, J. (2004). Peasants' integration into cities and innovation of community education. *The Journal of Continuing Education Research, 6*, 42–44.

Xi, M. H. (2007). Community's educational support strategy on rural migrant workers. *The Journal of Vocational and Technical education, 13*, 8–13.
Xu, C. X. (2007). Identity of new migrant workers and analysis of influential factors. *The Journal of Academic Discovery, 3*, 58–62.
Xue, T. S. (2011). Community education: Effective approach for new migrant workers integrating into city. *The Journal of Vocational and Technical education, 19*, 63–67.
Zeng, S. Q. (2012). Causes and strategy analysis of factors influencing new-generation migrant workers' integrating into cities. *The Journal of China Adult Education, 3*, 158–160.

Community Well-Being Through Intergenerational Cooperation

Ann-Kristin Boström, Seung Jong Lee and Youngwha Kee

Intergenerational Learning

The population of the world has grown old. It has been estimated that in 2050, there will be more people aged 60 than those aged 15 and that the mean age will have risen from 26 years of age today to 36 years of age. The causes of this are a combination of increased life expectancy and a falling birth rate. At the same time the generation gap between younger people and older people has widened, especially in industrial countries in the West, such as the USA, Great Britain and the Netherlands. The pattern in Sweden is not very different from this. Here too older people and younger people are, to a very large extent, distanced from each other (Boström 2000).

Meetings of various kinds between different generations which have been arranged to compensate for this mutual loss of interchange of experience and knowledge are known as "intergenerational programmes". Intergenerational initiatives and projects have been developed in various industrial countries in order for younger and older people to meet more frequently than is often otherwise the case.

A definition of intergenerational learning and related projects was adopted at a meeting attended by researchers from ten different countries (China, Cuba, Germany, Japan, the Netherlands, Palestine, South Africa, Sweden, Great Britain, and the USA),

A.-K. Boström
Jönköping University, Jönköping, Sweden
e-mail: ann-kristin.bostrom@hlk.hj.se

S.J. Lee
Seoul, Republic of Korea
e-mail: slee@snu.ac.kr

Y. Kee (✉)
Soongsil University, Seoul, Korea, Republic of (South Korea)
e-mail: key@ssu.ac.kr

Y. Kee et al. (eds.), *Learning and Community Approaches for Promoting Well-Being*, SpringerBriefs in Well-Being and Quality of Life Research,
DOI 10.1007/978-3-319-12439-1_4

which had been arranged by UNESCO's Institute of Education, in Hamburg, in 1999. The common definition is as follows:

> Intergenerational learning provides a conscious and continuous mutual exchange of experience and knowledge between older and younger generations.
> (Boström et al. 2000)

Intergenerational learning is a form of lifelong learning. A short description of the concept of lifelong learning is included here in order for intergenerational learning to be understood in that context.

Lifelong Learning

The concept of lifelong learning has been extensively discussed ever since the end of the 1960s. In a report prepared under the leadership of Edgar Faure, for UNESCO (1996), lifelong learning was described as being largely an issue of philosophical principles with regard to the organization of learning. The concepts of formal learning and informal learning were first introduced by Dave in 1976:

> Lifelong education is a process of accomplishing personal, social and professional development through the lifespan of individuals in order to enhance the quality of life of both individuals and their collectives. It is a comprehensive and unifying idea which includes formal, non-formal and informal learning for acquiring and enhancing enlightenment so as to attain the fullest possible development in different stages and domains of life (Dave 1976: 34)

Subsequently, Coombs and Ahmed (1974) defined lifelong learning more precisely and eventually, in 1980, the term "lifelong learning" was introduced to replaced the term "lifelong education" (Cropley 1980).

Lifelong Learning Policy in Sweden

The Swedish government has placed emphasis on lifelong learning since the end of the 1960s, when the minister responsible for education at the time, Olof Palme, introduced the concept of recurrent education at a meeting of OECD ministers of education. Policy in that time was much based on research, especially of Torsten Husén. In his articles he argues about loss of talent in the selective school systems (Husén 1960). In Educational change in Sweden he describes "…the events on the educational scene in Sweden after 1945 to be seen within a socio-economic framework" (Husén 1965, p. 181) and in an article the following year describes the most important educational problem which many countries faced at that time. Upper secondary education was the problem and the reason was twofold. One force to consider was the increasing need for highly trained manpower and the second force was the democratisation of schools. Both forces were strong political forces "…aiming at making 'the educational facilities' available to all young people irrespective of their place of residence, financial level or general social background" (Husén 1966, p. 250). In 1968 he published his article "Lifelong Learning in the Educative Society" where he discussed formal education in relation to vocational education and foresees that the future will require a longer period of formal education and education during different phases in life. As a consequence of this policy lifelong learning was implemented in the Swedish society and in the Educational system from the 1970s including popular education which had been introduced earlier during the latter part of the 20th century.

In Sweden there is no upper age-limit for commencing university studies, so that even after the official retirement age, there are opportunities available for senior citizens to commence studies at this, or some other, level. One form of popular education, are the programmes offered through what is known as "study circles," which at present are organised by 11 nationwide voluntary educational associations. Another form of popular education is that offered by what is known as folk high schools. There are about 150 throughout the country, which are run either by county councils or by trade unions, churches, temperance societies or other non-governmental organizations. The programmes they offer are largely residential and they are more comprehensive than those provided by the voluntary educational associations. The latter are usually affiliated with a political party or interest organization. Both folk high schools and voluntary educational associations receive state subsidies while remaining very free to organize the courses and to develop the contents as much as they wish.

The introduction of the policy of lifelong learning has many practical consequences. The first consequence is that lifelong learning dissolves the boundaries between policy sectors. This is because of the fact that lifelong learning concerns such sectors as education policy, labour market policy, industrial policy as well as social policy. All partners are also involved, from municipalities to county councils and the government, and from employers to social partners and employees. The second consequence is a shift in responsibility for education and learning from the public to the private and civil spheres.

Social Capital

Through changes occurring in the Western world during the 1990s, involving globalization, increased mobility, demographic transformations implying ageing populations, increased investments in IT and deregulation of markets, there has been an increased need to make use of that which people have learned outside of formal learning.

In the industrialized world, we also experience the same need for contact and the same desire to share that energy which is important when people work together in pursuit of a common goal. This is the reason for associating lifelong learning with the concept of social capital.

While, over a long period, great interest has been shown in human capital, in the industrial countries of the Western world, no such attention has been paid to the soft value to be found in society in the shape of that social capital related to human capital and increased quality of life in working life. However, the importance of higher standards in working environments and greater quality of life has been proposed in a publication from the OECD in 2001. In addition, these "soft" values have, on the other hand, long been appreciated in Asia. Here it is considered not to be possible at all to measure such matters in financial terms. In "Education of the Rising Sun—an Introduction to Education in Japan," Okamoto (1992, 2001) proposes that lifelong education ought to be of an important and spiritual nature in order for people who are working together to be able to experience greater spiritual and physical well-being in their daily lives (*kokoru*).

Social capital itself is a concept that is used both in a micro-perspective, in relationships between people (Coleman 1971, 1988, 1990) and in a macro-perspective, where the social capital in particular countries is considered. In the latter case, it concerns such issues as democracy and, as a field of research, may be typified in the work of, among others, Putnam (1995, 2000) and Inglehart (1999). The definition of social capital formulated by Coleman (1988: 98) forms the basis of most other definitions of that concept. In the view of Coleman (1988: 98), it is not a matter of one single concept; rather the concept of social capital is a compilation of various components. These have the following in common:

- They consist of some form of social structure.
- They facilitate certain aspects of social structures.
- They facilitate certain responses on the part of various actors within the structure.

He also describes how three different forms of social capital may be observed. These three entities consist of (1) the level of confidence and trust which is found in the actual social environment. (2) The second entity consists of those information channels which are found there and how information is passed between the actors. (3) The third entity described by Coleman are the shared norms and structures that are found when people are working towards the same goal, when they are doing this on the basis of common, unselfish interests.

Boström (2002, 2003a, b) has investigated the relationship between lifelong learning, intergenerational learning and social capital, by analyzing the concept and through a study of a particular project concerned with intergenerational learning, "the class granddad project." The project has shown a possibility to create social capital (Boström 2002, 2003a, b) for the participating groups of students, increased well-being and self-esteem for the older persons and thereby increasing community well-being in the local community.

The Concept of Well-Being

One study that has been used in studying well-being and health very thoroughly is called "the Malmö longitudinal study". It was started in 1938 as an investigation of intelligence and social environment. The group examined included all pupils in the third grade of Malmö's elementary and private schools—a total of 1,542 pupils—834 boys and 708 girls. The majority was born in 1928. There were follow ups 1942, 1947, 1950s, 1963–1965. A study regarding life patterns and health was made (Furu 1985) regarding the male population in the Malmö study. There was a conceptual model constructed for the analysis of the data where the historical context, the socioeconomic resources as well as the cognitive resources and their effect on home and school processes could have impact on the acquired level of education. This in turn was used to compare socioeconomic and work conditions, styles of living and psychosocial and mental conditions and their impact on neuro-endocrine processes and get medical health indicators.

The results showed that

> socioeconomically, 'the healthy' middle-aged man was not better favoured than others during his childhood. The individual's cognitive ability, however, was somewhat above the total-group average. In adulthood the healthy man is characterised by stable socioeconomic conditions, relatively speaking. His level of education is average and income is slightly above the total-group average. The group has a small proportion of divorces (5 %), low criminality, and few have received social assistance. By and large, working conditions do not differ from the total average. Most can relax after a normal working day and do not find their work uncongenial. Few have experienced continual stress during the past five years. (Furu 1985)

The Malmö longitudinal study was also used by Bang (1992) when she asked if quality of life can be measured. By doing a thorough literature review she as a start point choose 3,432 documents with the indicators: happiness, life satisfaction and quality in life. She found 143 articles and described 34 articles and how they try to measure quality in life. She continued to make research in the area and 1995 came the next book which dealt with retirement and quality of life. Network and health are the most important dimensions for the people in the Malmö longitudinal study in this qualitative study. Another perspective is taken by Schuller et al. (2004) when the benefits of learning and the impact of education on health, family life and social capital is researched and the important effects of learning to self-esteem. The model that is used is building on three kinds of capital; human capital, identity capital and social capital.

A recent trend, starting in the industrialised western countries is that for many older workers, delaying retirement does not mean continuing in the same jobs they held when they were younger. Instead many leave their old occupations behind and as they reach their fifties and sixties they start on new careers, that is more personally fulfilling but less financially rewarding than previous jobs. They may involve moves from wage-and salary jobs to self-employment or they may gradually shift into retirement much because workers would like to move from demanding, full-time work into less stressful part-time work. This trend has been extensively described by Freedman (2007) as he was looking for the work that matters for older people and he is introducing the word Boomers for the senior people of the new generations. He means that now senior people will have meaning or quality in their life, and therefore can think of changing work or workplace as they know that "*activity, recreation, and exercise are key ingredients in living healthier, happier, and longer lives.*" He also concludes that the boomers are not *celebrating their freedom from work but rather their freedom to work, in ways that hold the promise of personal fulfilment, economic benefit and social renewal.* The Boomers in this way contributes to the social capital and well-being in their community.

The Granddad Project

The Granddad project, that now has developed to a National organization is an example of how an intergenerational relationships in practice can increase social capital and by that also affect community development.

The intergenerational project which is introduced includes older men, +55 that have been working as support for young people in schools. These older men receive an introductory education at folk high schools during one semester before entering their work in the school. The course consists of both theory and practice which gives them a possibility to adapt to a new situation as most of them went to school during the last century and many things have changed.

The project developed out of the work of two men and has become a national association. There are now other projects developing out of the same model. The model has shown a possibility to create social capital (Coleman 1988; Boström 2002, 2003a) which contributes to both generational solidarity and community development.

Increasing Social Capital in the Granddad Program

The theoretical background rests on both a lifelong perspective and a lifewide perspective of learning. The lifewide perspective includes both formal and informal learning. Indicators to measure social capital were constructed and LISREL was used to analyze the results (Boström 2009, 2011). In addition, a qualitative method was used to analyze the understanding and experiences of their work on the part of the "class granddads."

The schools that were included in the study were very different. Therefore the granddads entered very different environments when they began their work. The results indicate that the work of the class granddads in itself was part of the social capital that was to be found between the class granddad and the pupils. The

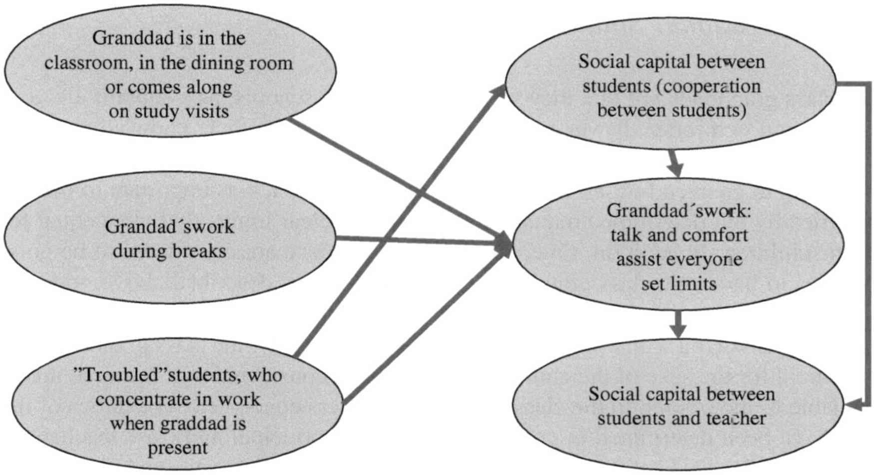

Fig. 1 Social capital in the school (Boström 2011)

responses of the pupils indicate that both boys and girls felt they had equal opportunities in the school, that they felt secure in school and that the class granddad assisted everyone. These results support the proposition that social capital increased for the pupils. The responses from the participating class granddads indicate that, even though their work was very demanding, they still experienced it as being very positive. They gained a social network together with the other adults at the school and very positive energy from the pupils. This indicates that there was also an increase in social capital and well-being for the older people (Fig. 1).

When the granddad is outside during the breaks, he is more on his own, which becomes apparent as an indicator in the data, being a strong component in itself. The work of the class granddad consists of many different parts. The pupils most often refer to the fact that he talks to the children and comforts them when this is needed (communication = a part of social capital), that he assists everyone (security = a part of social capital) and that he sets limits for what is allowed (norms and structure = a part of social capital). In other words, the work of the class granddad contains all the parts that constitute social capital, according to Coleman's definition (Coleman 1988). The work of the class granddad, both together with the teacher and outside during the breaks, also influences, in a positive way, the social capital between the pupils and the teacher. The work of the teacher is very greatly influenced by the relationships between the pupils in the group.

Intergenerational Solidarity

The class granddads felt that they were needed in the schools, as a support and as a listener and comforter. It was important for them to share their knowledge, experiences and feelings. They wanted to support the less strong but to also to stimulate the strong to go even further. Many of the granddads felt it was important to behave in a friendly but determined manner and to set very clear limits, that were equal for all the children. In addition, time, empathy and patience are considered to be good qualities to have as a class granddad. The circumstances described above, together with their own views about their work, have also meant that the class granddads have been involved with different kinds of duties. Naturally, the class granddad is in the school for the sake of the children, and has been considered as a resource that is available to the school, to the class and to the teachers concerned. The duties of the men have been determined in consultation with the principal and class teacher, on the basis of the needs of the pupils or class, and on the requirements and planning of each teacher, and, of course, taking into consideration the talents and wishes of each granddad. This is based on a selection criterion where, on the part of the male senior citizen, there has been a voluntary wish to participate and take on the role of class granddad, which is an important prerequisite in order for his work to be successful.

The teachers who worked with the granddad appreciated, that he was together with them in various activities and made it possible for the teacher to deal with the children in a better way. The teachers no longer needed to supervise the breaks to the same extent and so had more time to prepare their work for the children. They placed the most value on him having been an additional adult during breaks and who could accompany them on excursions. He was an additional adult, but there more for the children than for the teachers. He had time for the pupils and they liked him. His way of working was very flexible—nothing was impossible—he really wanted to be of assistance. When required to do so, he was also able to assist in resolving emergencies. The class granddads in each case have developed their own professional role. None is like any of the others—this reflects the flexibility in the professional role. Assistance with homework, gymnastics, woodwork, outdoor activities, each indicates that they find the activities where they can meet the children outside of the classroom where they are able to be of assistance to a large number of them. The class granddad is seen as an additional factor in providing security and this is particularly important for the "new ones" in the school or those who are graduating to the middle or senior school.

It is necessary to have a positive social environment in order for learning to take place. This may be created through providing basic security in the school environment and the opportunity to be given individual assistance when this is required. Positive social capital may provide the conditions for a learning society for all—both young and old. In a longer social perspective, the teachers and other staff felt that the class granddad built bridges between generations and improved the quality of life for the pupils by helping them feel secure in school. This is difficult to measure in a shorter perspective, but nonetheless this aspect of the effects of the work of the class granddad should not be ignored. The program has deepened the understanding of meetings between generations in those schools which have introduced a class granddad, an additional adult in their activities. It rests on earlier research on informal learning, which takes place between an older and a younger generation, a part of lifelong learning that also contributes to the well-being in the actual community.

The Granddad Program in Community Development

During the period between 1996 and 2012 the project has developed quite remarkably. From one granddad 1996 in Stockholm there are more than 1,000 granddads all over Sweden. There is one national association and there are ten regional associations. The National Association is an NGO where the board is working on voluntary basis doing the administration and coordination of the regional associations.

The regional associations are also NGOs and they sometimes also have staff when there is funding. The persons working there are the connecting hubs to choose the right people to work in schools in co-operation with the unemployment offices, for introduction of the granddads into the schools by the giving them the possibility to learn about the job in the folk high schools and for following up and organizing meetings. The regional associations work together with the regional folk high schools organize education and training during one semester for the granddads to be. There is a defined curriculum and they also have training in regional schools. After finishing the training there is an exam and if they pass they receive a certificate. The Swedish granddads are working in schools from preschool to upper secondary schools. There are both men and women working as granddads now. At the same time there have been a problem regarding the policy level as to who is responsible for their work. They are working in the schools but the schools do not achieve any funding for their work even if they produce prerequisites for learning by creating social capital in the classroom and in the school.

Complementary Actors

According to the Swedish Government (prop. 2011/2012: 1) it is important to have a well functioning and effective employment service and unemployment insurance. Therefore it is necessary to regularly follow up how the unemployed person's uses his/her efforts to find a job. Older persons who lose their jobs face a difficult labour market situation and are at risk of ending up in long-term unemployment or leaving the labour force. There are also many individuals who voluntary leave the labour force relatively early, for example, via supplementary pension schemes. The government finds it important to encourage a later labour market exit, not least to safe guard the financing of the public sector in the long run. One measure that has been taken by the Government for stimulating the labour supply among older people is the higher in-work tax credit for people who have turned 65. People aged 55–64 can also get new-starts jobs for up to 10 years.

The aim of new start jobs, is to provide work experience and references that increase the chances of getting an unsubsidised job. The long term unemployed with limited education may also be in need of further education in order to become competitive in the labour market. The Swedish granddad program combines work with learning. As a complementary agent the regional NGOs have had the task of introducing unemployed persons to work in schools as Granddads. During this period they also apply for other jobs. Getting the social network in the schools and the social well-being this creates means that employers are interested in them. Several of the granddads have got other jobs even if they also like the work in schools.

The Granddads in schools have been a good investment in the community, supported from the need and appreciated of all stakeholders that have been involved and increasing social capital and community well-being. As this work has been introduced from bottom-up it has meant hard work and flexibility regarding new prerequisites during the work that now has continued for more than 15 years. The experience and knowledge that the regional NGOs have attained during this time is important in itself as the model is new for our society.

In fact this has functioned so well that the local NGOs now also are working with young unemployed persons. They are also working in intergenerational settings by helping old people. This is specially developed in the western part of Sweden, called Halland. Here there are different projects supporting older people in three communities. The persons that involve in this are also introduced by different courses and information to be able to do a good job. They are often gathered in groups to feel that they have a network so they are not working alone. The scope of these projects are very wide but some examples could be to help older people to go outside their apartment or to help in their apartment with cleaning or other things the old people have difficulties in doing. They could also visit for small talks over a cup of coffee or for a walk in the park or to the library. Their goal is to make the existence better and give more quality to their daily life so that the old people can enjoy their life. This is also seen as something positive for the society and contributes to well-being and increased social capital. During this time they also apply for work and by having work experience and well-being from their social network they are also more interesting for employers.

There is now an increased cooperation between different agents in the area. The Government has given the Swedish Insurance Agency, in cooperation with the national board of Health and Welfare and in consultation with the Public Employment Service, a remit to further develop instruments for assessing work capacity. Hopefully this will give more impact for intergenerational work and relationships in Sweden and thereby increase the possibilities for community well-being.

References

Bang, H. (1992). *Can quality be measured? A literature review*. Stockholm: Institute of international Education, Stockholm University.

Boström, A.-K. (2000). Sweden. In A. Hutton-Yeo & T. Ohsako (Eds.), *Intergenerational programmes. Public policy and research implications: An international perspective*. Hamburg: UNESCO Institute of Education and Stoke-on-Trent: The Beth Foundation.

Boström, A.-K. (2002). Informal learning in a formal context: Problematizing the concept of social capital in a contemporary Swedish context. *International Journal of Lifelong Education, 21*, 510–524.

Boström, A.-K. (2003a). *Lifelong learning, intergenerational learning, and social capital: From theory to practice*. Stockholm: Institute of International Education, Stockholm University.

Boström, A.-K. (2003b). Intergenerational learning in Stockholm County in Sweden: A practical example of elderly men working in compulsory schools as a benefit for children. *Journal of Intergenerational Relationships, 1*(4), 7–24.

Boström, A.-K. (2009). Social capital in intergenerational meetings in compulsory schools in Sweden. *Journal of Intergenerational Relationships, 7*, 425–441.

Boström, A.-K. (2011). Lifelong learning in intergenerational settings: The development of the Swedish Granddad program from project to National association. *Journal of Intergenerational Relationships, 9*(3), 293–306.

Boström, A.-K., Hutton-Yeo, A., Ohsako, T., & Sawano, Y. (2000). A general assessment of IP initiatives in the countries involved. In A. Hutton-Yeo & T. Ohsako (Eds.), *Intergenerational programmes. Public policy and research implications: An international perspective*. Hamburg: UNESCO Institute of Education and Stoke-on-Trent: The Beth Foundation.

Coleman, J. S. (1971). *Resources for social change*. New York: Wiley.

Coleman, J. S. (1988). Social capital in the creation of human capital. *American Journal of Sociology, 94*, S95–S120.

Coleman, J. S. (1990). *Foundations of social theory*. Boston: Harvard University Press.

Coombs, P. A., & Ahmed, M. (1974). *Attacking rural poverty: How nonformal education can help*. Baltimore: John Hopkins University Press.

Cropley, A. J. (1980). *Towards a system of lifelong education*. Oxford: Pergamon Press.

Dave, R. H. (Ed.). (1976). *Foundations of lifelong education*. Oxford: Pergamon Press.

Freedman, M. (2007). *Encore, finding work that matters in the second half of life*. New York: Public Affairs.

Furu, M. (1985). *Life patterns in health. A longitudinal study of men from childhood to middle age*. Stockholm: Gleerups.

Husén, T. (1960). Loss of talent in selective school systems: The case of Sweden. *Comparative Education Review, 4*(2), 70–74.

Husén, T. (1965). Educational change in Sweden. *Comparative Education, 3*(1), 181–191.

Husén, T. (1966). *The contribution of research to the reform of secondary education*. Paedagogika Europaea.

Inglehart, R. (1999). Trust, well-being and democracy. In M. Warren (Ed.), *Democracy and trust*. Cambridge: Cambridge University Press.

Okamoto, K. (1992). *Education of the rising sun: An introduction to education in Japan*. Tokyo: Ministry of Education, Science and Culture.

Okamoto, K. (2001). *Education for the rising sun 21: An introduction to education in Japan*. Tokyo: National Federation of Social Education.

Prop. (2011/2012:1). *Budget statement*. Swedish Government: Author.

Putnam, R. D. (1995). Bowling alone: America's declining social capital. *Journal of Democracy, 6*, 65–78.

Putnam, R. D. (2000). *Bowling alone: The collapse and revival of American community*. London: Simon & Schuster.
Schuller, T., Preston, J., Hammond, C., Brasset-Grundy, A., & Brynner, J. (2004). *The impact of education on health, family life and social capital*. London: RoutledgeFalmer.
UNESCO. (1996). International commission on education for the 21st Century. *Learning: The treasure within*. Paris: Author.

Index

Y. Kee et al. (eds.), *Learning and Community Approaches for Promoting Well-Being*,
SpringerBriefs in Well-Being and Quality of Life Research,
DOI 10.1007/978-3-319-12439-1

MIX
Papier aus verantwortungsvollen Quellen
Paper from responsible sources
FSC® C105338

If you have any concerns about our products,
you can contact us on
ProductSafety@springernature.com

In case Publisher is established outside the EU,
the EU authorized representative is:
Springer Nature Customer Service Center GmbH
Europaplatz 3, 69115 Heidelberg, Germany

Printed by Libri Plureos GmbH
in Hamburg, Germany